T0269159

CAMBRIDGE LIBRARY COLLECTION

Books of enduring scholarly value

Physical Sciences

From ancient times, humans have tried to understand the workings of the world around them. The roots of modern physical science go back to the very earliest mechanical devices such as levers and rollers, the mixing of paints and dyes, and the importance of the heavenly bodies in early religious observance and navigation. The physical sciences as we know them today began to emerge as independent academic subjects during the early modern period, in the work of Newton and other 'natural philosophers', and numerous sub-disciplines developed during the centuries that followed. This part of the Cambridge Library Collection is devoted to landmark publications in this area which will be of interest to historians of science concerned with individual scientists, particular discoveries, and advances in scientific method, or with the establishment and development of scientific institutions around the world.

The Grammar of Science

First published in 1892, this important work by the mathematician Karl Pearson (1857–1936) presents a thoroughly positivist account of the nature of science. Pearson claims that 'the scientific method is the sole gateway to the whole region of knowledge', rejecting additional fields of inquiry such as metaphysics. He also emphasises that science can, and should, describe only the 'how' of phenomena and never the 'why'. A scholar of King's College, Cambridge, and later a professor at King's College and University College London, Pearson made significant contributions to the philosophy of science. Including helpful chapter summaries, this book explores in detail a number of scientific concepts, such as matter, energy, space and time. The work influenced such thinkers as Albert Einstein, who considered it to be essential reading when he created his study group, the Olympia Academy, at the age of twenty-three.

Cambridge University Press has long been a pioneer in the reissuing of out-of-print titles from its own backlist, producing digital reprints of books that are still sought after by scholars and students but could not be reprinted economically using traditional technology. The Cambridge Library Collection extends this activity to a wider range of books which are still of importance to researchers and professionals, either for the source material they contain, or as landmarks in the history of their academic discipline.

Drawing from the world-renowned collections in the Cambridge University Library and other partner libraries, and guided by the advice of experts in each subject area, Cambridge University Press is using state-of-the-art scanning machines in its own Printing House to capture the content of each book selected for inclusion. The files are processed to give a consistently clear, crisp image, and the books finished to the high quality standard for which the Press is recognised around the world. The latest print-on-demand technology ensures that the books will remain available indefinitely, and that orders for single or multiple copies can quickly be supplied.

The Cambridge Library Collection brings back to life books of enduring scholarly value (including out-of-copyright works originally issued by other publishers) across a wide range of disciplines in the humanities and social sciences and in science and technology.

The Grammar of Science

Karl Pearson

CAMBRIDGE
UNIVERSITY PRESS

University Printing House, Cambridge, CB2 8BS, United Kingdom

Cambridge University Press is part of the University of Cambridge.
It furthers the University's mission by disseminating knowledge in the pursuit of
education, learning and research at the highest international levels of excellence.

www.cambridge.org
Information on this title: www.cambridge.org/9781108077118

© in this compilation Cambridge University Press 2015

This edition first published 1892
This digitally printed version 2015

ISBN 978-1-108-07711-8 Paperback

THE CONTEMPORARY SCIENCE SERIES.

Edited by HAVELOCK ELLIS.

THE GRAMMAR OF SCIENCE.

THE
GRAMMAR OF SCIENCE.

BY

KARL PEARSON, M.A.,

Sir Thomas Gresham's Professor of Geometry.

"La critique est la vie de la science."—COUSIN.

WITH 25 FIGURES IN THE TEXT.

LONDON:
WALTER SCOTT,
24, WARWICK LANE, PATERNOSTER ROW,
1892.

To

THE MEMORY OF

Sir THOMAS GRESHAM, Knight,

WHILOM MERCHANT

OF THE

CITY OF LONDON.

PREFACE.

———◦◇◦———

THERE are periods in the growth of science when it is well to turn our attention from its imposing superstructure and to carefully examine its foundations. The present book is primarily intended as a criticism of the fundamental concepts of modern science, and as such finds its justification in the motto placed upon its title-page. At the same time the author is so fully conscious of the ease of criticism and the difficulty of reconstruction, that he has attempted not to stop short at the lighter task. No one who knows the author's views, or who reads, indeed, this book, will believe that he holds the labour of the great scientists or the mission of modern science to be of small account. If the reader finds the opinions of physicists of world-wide reputation, and the current definitions of physical concepts called into question, he must not attribute this to a purely sceptical spirit in the author. He accepts almost without reserve the great results of modern physics ; it is the language in which these results are stated that he believes needs reconsideration. This reconsideration is the more urgent because the language of physics is widely used in all branches of biological (including sociological) science. The obscurity which envelops the *principia* of science is not only due to an historical evolution marked by the authority of great names, but to the fact that science, as long as it had to carry on a difficult warfare with metaphysics and dogma, like a skilful general conceived it best to hide its own deficient organization. There can be small doubt, however, that this deficient organization will not only in time be perceived by the enemy, but that it has already had a very discouraging influence both on scientific recruits and on intelligent laymen. Anything more hopelessly illogical than the statements with regard to force and matter current in elementary text-books of science, it is difficult to imagine ; and the author, as a result of some ten years' teaching and examining, has been forced to the conclusion that these works possess little, if any, *educational* value ; they do not encourage the growth of logical clearness or form any exercise in scientific method. One result of this obscurity we probably find in the ease with which the physicist, as compared with either the pure mathematician or the historian, is entangled in the meshes of such pseudo-sciences as natural theology and spiritualism. If the constructive portion of this work appears to the reader unnecessarily dogmatic or polemical, the author would beg him to remember that it is essentially intended to arouse and stimulate the reader's own thought, rather than to inculcate doctrine : this result is often best achieved by the assertion and contradiction which excite the reader to independent inquiry.

The views expressed in this *Grammar* on the fundamental concepts of science, especially on those of force and matter, have formed part of the author's teaching since he was first called upon to think how the elements of dynamical science could be presented free from metaphysics to young students. But the endeavour to put them into popular language only dates from the author's appointment last year to Sir Thomas Gresham's professorship in geometry. The substance of this work formed the topic of two introductory courses on the *Scope and Concepts of Modern Science*. Gresham College is but the veriest shred of what its founder hoped and dreamt it would become —a great teaching university for London—but the author in writing this volume, whatever its failings, feels that so far as in him lies he is endeavouring to return to the precedent set by the earliest and most distinguished of his predecessors in the chair of geometry. To restore the chair and the college to its pristine importance is work worth doing, but it lies in other hands.

This *Grammar of Science*, imperfect as it is, would have been still more wanting but for the continual help and sympathy of several kind friends. Mr. W. H. Macaulay, of King's College, Cambridge, has given aid in many ways, ever trying to keep the author's scientific radicalism within moderate and reasonable bounds. To his friend, Mr. R. J. Parker, of Lincoln's Inn, the author is indebted for a continuation of that careful and suggestive revision which he has for the last ten years given to nearly everything the author has written. Especially, however, his thanks are due to Dr. R. J. Ryle, of Barnet, whose logical mind and wide historical reading have produced a " betterment," which gives him almost a tenant-right in these pages.

Lastly, the author has to thank his friend and former pupil, Miss Alice Lee, Demonstrator in Physics at Bedford College, London, for the preparation of the index and for several important corrections.

KARL PEARSON.

GRESHAM COLLEGE, LONDON.
January, 1892.

CONTENTS.

— ◆◆ —

CHAPTER I.

INTRODUCTORY.

		PAGE
§	1.—Science and the Present	1
§	2.—Science and Citizenship	7
§	3.—The First Claim of Science	10
§	4.—Essentials of Good Science	11
§	5.—The Scope of Science	14
§	6.—Science and Metaphysics	18
§	7.—The Ignorance of Science	23
§	8.—The Wide Domain of Science	29
§	9.—The Second Claim of Science	31
§	10.—The Third Claim of Science	35
§	11.—Science and the Imagination	36
§	12.—The Method of Science Illustrated	39
§	13.—Science and the Æsthetic Judgment	42
§	14.—The Fourth Claim of Science	44
	Summary and Literature	45

CHAPTER II.

THE FACTS OF SCIENCE.

§	1.—The Reality of Things	47
§	2.—Sense-Impressions and Consciousness	50
§	3.—The Brain as a Central Telephone Exchange	53
§	4.—The Nature of Thought	55
§	5.—Other-Consciousness as an Eject	59
§	6.—Attitude of Science towards Ejects	61
§	7.—The Scientific Validity of a Conception	64
§	8.—The Scientific Validity of an Inference	67
§	9.—The Limits to Other-Consciousness	69

PAGE

§ 10.—The Canons of Legitimate Inference .   .   .   71
§ 11.—The External Universe   73
§ 12.—Outside and Inside Myself   77
§ 13.—Sensations as the Ultimate Source of the Materials of
           Knowledge   80
§ 14.—Shadow and Reality   83
§ 15.—Individuality   86
§ 16.—The Futility of " Things-in-Themselves "   .   . 87
§ 17.—The Term Knowledge is meaningless if applied to
           Unthinkable Things   89
           Summary and Literature   90–1

CHAPTER III.

THE SCIENTIFIC LAW.

§ 1.—Foreword and *Résumé*   92
§ 2.—Of the Word Law and its Meanings   .   . 94
§ 3.—Natural Law Relative to Man   .   .   . 99
§ 4.—Man as the Maker of Natural Law .   .   . 102
§ 5.—The Two Senses of the Words " Natural Law "   . 104
§ 6.—Confusion between the Two Senses of Natural Law   106
§ 7.—The Reason behind Nature   .   .   . 109
§ 8.—True Relation of Civil and Natural Law   .   . 111
§ 9.—Physical and Metaphysical Supersensuousness   . 114
§ 10.—Progress in the Formulating of Natural Law   . 116
§ 11.—The Universality of Scientific Law   .   . 120
§ 12.—The Routine of Perceptions as possibly a Product of the
           Perceptive Faculty   122
§ 13.—The Mind as a Sorting-Machine   .   .   . 128
§ 14.—Science, Natural Theology, and Metaphysics   . 129
§ 15.—Conclusions 131
           Summary and Literature 135

CHAPTER IV.

CAUSE AND EFFECT. PROBABILITY.

§ 1.—Mechanism   136
§ 2.—Force as a Cause   140
§ 3.—Will as a Cause 143
§ 4.—Secondary Causes Involve no Enforcement .   . 144
§ 5.—Is Will a First Cause ?   147
§ 6.—Will as a Secondary Cause 148

PAGE

§ 7.—First Causes have no Existence for Science . . 151
§ 8.—Cause and Effect as the Routine of Experience . 153
§ 9.—Width of the Term Cause 156
§ 10.—The Universe of Sense-Impressions as a Universe of
    Motions 157
§ 11.—Necessity belongs to the World of Conceptions, not to
    that of Perceptions 160
§ 12.—Routine in Perception is a Necessary Condition of
    Knowledge 162
§ 13.—Probable and Provable 166
§ 14.—Probability as to Breaches in the Routine of Perceptions 170
§ 15.—The Basis of Laplace's Theory in an Experience of
    Ignorance 171
§ 16.—Nature of Laplace's Investigations . . . 176
§ 17.—The Permanency of Routine for the Future . . 177
    Summary and Literature 180

CHAPTER V.

SPACE AND TIME.

§ 1.—Space as a Mode of Perception 181
§ 2.—The Infinite Bigness of Space . . . 187
§ 3.—The Infinite Divisibility of Space . . . 190
§ 4.—The Space of Memory and Thought . . . 193
§ 5.—Conceptions and Perceptions 196
§ 6.—Sameness and Continuity 200
§ 7.—Conceptual Space, Geometrical Boundaries . . 203
§ 8.—Surfaces as Boundaries 206
§ 9.—Conceptual Discontinuity of Bodies. The Atom . 208
§ 10.—Conceptual Continuity. Ether . . . 213
§ 11.—On the General Nature of Scientific Conceptions . 214
§ 12.—Time as a Mode of Perception . . . 217
§ 13.—Conceptual Time and its Measurement . . . 222
§ 14.—Concluding Remarks on Space and Time . . 228
    Summary and Literature 229

CHAPTER VI.

THE GEOMETRY OF MOTION.

§ 1.—Motion as the Mixed Mode of Perception . . 231
§ 2.—Conceptual Analysis of a Case of Perceptual Motion.
    Point-Motion 233

PAGE

§ 3.—Rigid Bodies as Geometrical Ideals . . . 237
§ 4.—On Change of Aspect, or Rotation . . . 239
§ 5.—On Change of Form, or Strain . . . 242
§ 6.—Factors of Conceptual Motion 246
§ 7.—Point-Motion. Relative Character of Position and
    Motion 247
§ 8.—Position. The Map of the Path . . . 250
§ 9.—The Time-Chart 253
§ 10.—Steepness and Slope 257
§ 11.—Speed as a Slope. Velocity 260
§ 12.—The Velocity Diagram, or Hodograph. Acceleration 262
§ 13.—Acceleration as a Spurt and a Shunt . . . 265
§ 14.—Curvature 268
§ 15.—The Relation between Curvature and Normal Acceleration 273
§ 16.—Fundamental Propositions in the Geometry of Motion . 276
§ 17.—The Relativity of Motion. Its Synthesis from Simple
    Components 279
    Summary and Literature 284

CHAPTER VII.

MATTER.

§ 1.—"All Things Move"—but only in Conception. . . 285
§ 2.—The Three Problems 288
§ 3.—How the Physicists define Matter . . . 291
§ 4.—Does Matter occupy Space? 296
§ 5.—The "Common-sense" View of Matter—Impenetrable
    and Hard 301
§ 6.—Individuality does not denote Sameness in Substratum 303
§ 7.—Hardness not characteristic of Matter . . 308
§ 8.—Matter as Non-Matter in Motion . . . 310
§ 9.—The Ether as "Perfect Fluid" and "Perfect Jelly" 313
§ 10.—The Vortex-Ring Atom and the Ether-Squirt Atom . 316
§ 11.—A Material Loophole into the Supersensuous . 319
§ 12.—The Difficulties of a Perceptual Ether . . 323
§ 13.—Why do Bodies move? 325
    Summary and Literature 330

CHAPTER VIII.

THE LAWS OF MOTION.

§ 1.—Corpuscles and their Structure . . . 332
§ 2.—The Limits to Mechanism . . . 337

PAGE

§ 3.—The First Law of Motion 340
§ 4.—The Second Law of Motion, or the Principle of Inertia 342
§ 5.—The Third Law of Motion. Acceleration is determined
by Position 345
§ 6.—Velocity as an Epitome of Past History. Mechanism
and Materialism 351
§ 7.—The Fourth Law of Motion 354
§ 8.—The Scientific Conception of Mass . . . 357
§ 9.—The Fifth Law of Motion. The Definition of Force 359
§ 10.—Equality of Masses tested by Weighing . . . 363
§ 11.—How far does the Mechanism of the Fourth and Fifth
Laws of Motion extend ? 367
§ 12.—Density as the Basis of the Kinetic Scale . . 370
§ 13.—The Influence of Aspect on the Corpuscular Dance . 374
§ 14.—The Hypothesis of Modified Action and the Synthesis
of Motion 376
§ 15.—Criticism of the Newtonian Laws of Motion . 380
Summary and Literature 386

CHAPTER IX.

LIFE.

§ 1.—The Relation of Biology to Physics . . . 388
§ 2.—Mechanism and Life 392
§ 3.—Mechanism and Metaphysics in Theories of Heredity . 395
§ 4.—The Definition of Living and Lifeless . . 400
§ 5.—Do the Laws of Motion apply to Life? . . 404
§ 6.—Life Defined by Secondary Characteristics . . 408
§ 7.—The Origin of Life 410
§ 8.—The Perpetuity of Life, or Biogenesis . . 411
§ 9.—The Spontaneous Generation of Life, or Abiogenesis . 413
§ 10.—The Origin of Life in an " ultra-scientific" Cause . 417
§ 11.—On the Relation of the Conceptual Description to the
Phenomenal World 420
§ 12.—Natural Selection in the Inorganic World . . 422
§ 13.—Natural Selection and the History of Man . . 425
§ 14.—Primitive History describable in terms of the Principles
of Evolution 428
§ 15.—Morality and Natural Selection . . . 430
§ 16.—Individualism, Socialism, and Humanism . . 434
Summary and Literature 439

CHAPTER X.

THE CLASSIFICATION OF THE SCIENCES.

PAGE

§   1.—Summary as to the Material of Science .   .   . 441
§   2.—Bacon's "Intellectual Globe"   .   .   .   443
§   3.—Comte's "Hierarchy"   446
§   4.—Spencer's Classification   448
§   5.—Precise and Synoptic Sciences   .   .   . 452
§   6.—Abstract and Concrete Sciences.—Abstract Science . 454
§   7.—Concrete Science.—Inorganic Phenomena   .   . 459
§   8.—Concrete Science.—Organic Phenomena   .   . 465
§   9.—Applied Mathematics and Bio-Physics as Cross Links . 469
§ 10.—Conclusion   471
       Summary and Literature 475

APPENDIX.

Note   I. On the Principle of Inertia and Absolute Rotation . 477
Note II. On Newton's Third Law of Motion   .   . 480
Note III. William of Occam's Razor   .   .   . 481
Note IV. On the Vitality of Seeds 482
Note V. A. R. Wallace on Matter   .   .   . 483
Note VI. On the Sufficiency of Natural Selection to account for
         the History of Civilized Man   .   .   . 484

INDEX.

THE GRAMMAR OF SCIENCE.

CHAPTER 1.

INTRODUCTORY.—THE SCOPE AND METHOD OF SCIENCE.

§ 1.—*Science and the Present.*

WITHIN the past forty years so revolutionary a change has taken place in our appreciation of the essential facts in the growth of human society, that it has become necessary not only to rewrite history, but to profoundly modify our theory of life and gradually, but none the less certainly, to adapt our conduct to the novel theory. The insight which the investigations of Darwin, seconded by the suggestive but far less permanent work of Spencer, have given us into the development of both individual and social life, has compelled us to remodel our historical ideas and is slowly widening and consolidating our moral standards. The slowness ought not to dishearten us, for one of the strongest factors of social stability is the inertness, nay, rather active hostility, with which human societies receive all new ideas. It is the crucible in which the dross is separated from the genuine metal, and which saves the body-social from a succession of unprofitable and possibly

2

injurious experimental variations.   That the reformer should be also the martyr is, perhaps, a not over-great price to pay for the caution with which society as a whole must move ; to replace an individual man may require years, but a stable and efficient society is the outcome of centuries of development.

If we have learnt, indirectly it may be, from the writings of Darwin that the means of production, the holding of property, the forms of marriage, and the organization of the family are the essential factors which the historian has to trace in the growth of human society ; if in our history books we are ceasing to head periods with the names of monarchs and to devote whole paragraphs to their mistresses, still we are far indeed from clearly grasping the exact inter-action of the various factors of social evolution, or understanding why one becomes predominant at one or another epoch. We can indeed mark periods of great social activity and others of apparent quiescence, but it is probably only our ignorance of the exact stages of social evolution, which leads us to associate the fundamental variations in social institutions with re-formations and revolutions. We associate, it is true, the German Reformation with a replacement of collectivist by individualist standards, not only in religion but also in handicraft, art, and politics. The French Revolution in like manner is the epoch from which many are inclined to date the rebirth of those social ideas which have largely remoulded the mediæval relations of class and caste, relations little affected by the sixteenth-century Reformation. Coming nearer to our own time indeed we can measure with some degree of accuracy the social influence of the great changes in the method of production, the transition from home to capitalistic production, which transformed English life in the first half of this century, and has since made its way throughout the civilized world. But when we come to our own age, an age one of the most marked features of which is the startlingly rapid growth of the natural sciences and their far-reaching influence on the standards of both the comfort and conduct of human life, we find it impossible to compress its social history into the bald phrases by which we attempt to connote the characteristics of more distant historical epochs.

It is very difficult for us who live in the last quarter of the nineteenth century to rightly measure the relative importance of our age in the history of civilization. In the first place we can look at it only from one standpoint—that of the *past*. It needed at least an Erasmus to predict the outcome of the Reformation from all that *preceded* the Diet of Worms. Or, to adopt a metaphor, a blind man climbing a hill might have a considerable appreciation of the various degrees of steepness in the parts he had traversed, and he might even have a reasonable amount of certainty as to the slope whereon he was standing for the time being, but whether that slope led immediately to a steeper ascent, or was practically the top, it would be impossible for him to say. In the next place we are too close to our age, both in position and feeling, to appreciate without foreshortening and personal prejudice the magnitude of the changes which are undoubtedly taking place.

The contest of opinion in nearly every field of thought—the struggle of old and new standards in every sphere of activity, in religion, in commerce, in social life—touch the spiritual and physical needs of the individual far too nearly for us to be dispassionate judges of the age in which we live. That we live in an era of rapid social variation can scarcely be doubted by any one who regards attentively the marked contrasts presented by our modern society. It is an era alike of great self-assertion and of excessive altruism ; we see the highest intellectual power accompanied by the strangest recrudescence of superstition ; there is a strong socialist drift and yet not a few remarkable individualist teachers ; the extremes of religious faith and of unequivocal freethought are found jostling each other. Nor do these opposing traits exist only in close social juxtaposition. The same individual mind, unconscious of its own want of logical consistency, will often exhibit our age in microcosm.

It is little wonder that we have hitherto made small way towards a common estimate of what our time is really contributing to the history of human progress. The one man finds in our time a restlessness, a distrust of authority, a questioning of the basis of all social institutions and long-established methods—characteristics which mark for him a decadence of social unity, a collapse of the only principles which he conceives capable of guiding conduct. The other with a different temperament pictures for us a golden age in the near future, when the new knowledge shall be diffused through the people, and when the new view of human relations, which he finds everywhere taking root, shall finally have supplanted worn-out customs.

One teacher propounds what is flatly contradicted by a second. " We want more piety," cries one ; "We must have less," retorts another. " State inter-

ference in the hours of labour is absolutely needful,"
declares a third ; " It will destroy all individual initia-
tion and self-dependence," rejoins a fourth. "The
salvation of the country depends upon the technical
education of its workpeople," is the shout of one
party ; "Technical education is merely a trick by
which the employer of labour thrusts upon the nation
the expense of providing himself with better human
machines," is the prompt answer of its opponents.
"We need more private charity," say some ; "All
private charity is an anomaly, a waste of the nation's
resources and a pauperizing of its members," reply
others. " Endow scientific research and we shall know
the truth, when and where it is possible to ascertain
it ;" but the counterblast is at hand: "To endow
research is merely to encourage the research for
endowment ; the true man of science will not be held
back by poverty, and if science is of use to us, it will
pay for itself." Such are but a few samples of the
conflict of opinion which we find raging around us.
The prick of conscience and the prick of poverty
have succeeded in arousing a wonderful restlessness in
our generation—and this at a time when the advance
of positive knowledge has called in question many
of the old customs and old authorities. It is true that
there are but few remedies which have not a fair
chance to-day of being put upon their trial. Vast
sums of money are raised for every sort of charitable
scheme, for popular entertainment, for technical
instruction, and even for higher education—in short,
for religious, semi-religious, and anti-religious movements of all types. Out of this chaos ought at least
to come some good ; but how shall we set the good
against the evil which too often arises from ill-defined,

or even undefined, appropriation of those resources which the nation has spared by the hard labour of the past, or is drawing on the future's credit?

The responsibility of individuals, especially with regard to wealth, is great, so great that we see a growing tendency of the state to interfere in the administration of private charities and to regulate the great educational institutions endowed by private or semi-public benefactions in the past. But this tendency to throw back the responsibility from the individual upon the state is really only throwing it back on the social conscience of the citizens as a body—the " tribal conscience," as Professor Clifford was wont to call it. The wide extension of the franchise in both local and central representation has cast a greatly increased responsibility on the individual citizen. He is brought face to face with the most conflicting opinions and with the most diverse party cries. The state has become in our day the largest employer of labour, the greatest dispenser of charity, and, above all, the schoolmaster with the biggest school in the community. Directly or indirectly the individual citizen has to find some reply to the innumerable social and educational problems of the day. He requires some guide in the determination of his own action or in the choice of fitting representatives. He is thrust into an appalling maze of social and educational problems; and if his tribal conscience has any stuff in it, he feels that these problems ought not to be settled, so far as he has the power of settling them, by his own personal interests, by his individual prospects of profit or loss. He is called upon to form a judgment apart from his own feelings and emotions if it possibly may be—a judgment in what he conceives to be the interests of society at large. It may be a difficult thing for the large employer of labour to form a right judgment in matters of factory legislation, or for the private schoolmaster to see clearly in questions of state-aided education. None the less we should probably all agree that the tribal conscience ought for the sake of social welfare to be stronger than private interest, and that the *ideal* citizen, if he existed, would form a judgment free from personal bias.

§ 2.—*Science and Citizenship.*

How is such a judgment—so necessary in our time with its hot conflict of personal opinion and its increased responsibility for the individual citizen—how is such a judgment to be formed? In the first place it is obvious that it can only be based on a clear knowledge of facts, an appreciation of their sequence and relative significance. The facts once classified, once understood, the judgment based upon them ought to be independent of the individual mind which examines them. Is there any other sphere, outside that of ideal citizenship, in which there is habitual use of this method of classifying facts and forming judgments upon them? For if there be, it cannot fail to be suggestive as to methods of eliminating individual bias; it ought to be one of the best training grounds for citizenship. The classification of facts and the formation of absolute judgments upon the basis of this classification—judgments independent of the idiosyncrasies of the individual mind—is peculiarly the *scope and method of modern science.* The scientific man has above all things to aim at self-elimination in his judgments, to provide an argument which is as true for each individual mind as for his own. *The classification of facts, the recognition of their sequence and relative significance is the function of science*, and the habit of forming a judgment upon these facts unbiased by personal feeling is characteristic of what we shall term the scientific frame of mind. The scientific method of examining facts is not peculiar to one class of phenomena and to one class of workers; it is applicable to social as well as to physical problems, and we must carefully guard ourselves against supposing that the scientific frame of mind is a peculiarity of the professional scientist.

Now this frame of mind seems to me an essential of good citizenship, and of the several ways in which it can be acquired few surpass the careful study of some one branch of natural science. The insight into method and the habit of dispassionate investigation which follow from acquaintance with the scientific classification of even some small range of natural facts, give the mind an invaluable power of dealing with many other classes of facts as the occasion arises.[1] The patient and persistent study of some one branch of natural science is even at the present time within the reach of many. In some branches a few hours' study a week, if carried on earnestly for

[1] To decry specialization in education is to misinterpret the purpose of education. The true aim of the teacher must be to impart an appreciation of method and not a knowledge of facts. This is far more readily achieved by concentrating the student's attention on a small range of phenomena, than by leading him in rapid and superficial survey over wide fields of knowledge. Personally I have no recollection of at least 90 per cent. of the *facts* that were taught to me at school, but the notions of *method* which I derived from my instructor in Greek Grammar (the contents of which I have long forgotten), remained in my mind as the really valuable part of my school equipment for life.

two or three years, would be not only sufficient to give a thorough insight into scientific method, but would also enable the student to become a careful observer and possibly an original investigator in his chosen field, thus adding a new delight and a new enthusiasm to his life. The importance of a just appreciation of scientific method is so great, that I think the state may be reasonably called upon to place instruction in pure science within the reach of all its citizens. Indeed, we ought to look with extreme distrust on the large expenditure of public money on polytechnics and similar institutions, if the manual instruction which it is proposed to give at these places be not accompanied by efficient teaching in pure science. The scientific habit of mind is one which may be acquired by all, and the readiest means of attaining to it ought to be placed within the reach of all.

The reader must be careful to note that I am only praising the scientific habit of mind, and suggesting one of several methods by which it may be cultivated. No assertion has been made that the man of science is necessarily a good citizen, or that his judgment upon social or political questions will certainly be of weight. It by no means follows that, because a man has won a name for himself in the field of natural science, his judgments on such problems as Socialism, Home Rule, or Biblical Theology will necessarily be sound. They will be sound or not according as he has carried his scientific method into these fields. He must properly have classified and appreciated his facts, and have been guided by them, and not by personal feeling or class bias in his judgments. It is the scientific habit of mind as an

essential for good citizenship and not the scientist as
a sound politician that I wish to emphasize.

§ 3.—*The First Claim of Modern Science.*

We have gone a rather roundabout way to reach
our definition of science and scientific method. But
it has been of purpose, for in the spirit—and it is a
healthy spirit—of our age we have accustomed ourselves to question all things and to demand a reason
for their existence. The sole reason that can be
given for any social institution or form of human
activity—I mean not how they came to exist, which
is a matter of history, but why we continue to
encourage their existence—lies in this : their existence
tends to promote the welfare of human society, to
increase social happiness, or to strengthen social
stability. In the spirit of our age we are bound to
question the value of science ; to ask in what way it
increases the happiness of mankind or promotes
social efficiency. We must justify the existence of
modern science, or at least the large and growing
demands which it makes upon the national exchequer.
Apart from the increased physical comfort, apart
from the intellectual enjoyment which modern science
provides for the community—points often and loudly
insisted upon and to which I shall briefly refer later
—there is another and more fundamental justification
for the time and material spent in scientific work.
From the standpoint of morality, or from the relation
of the individual unit to other members of the same
social group, we have to judge each human activity
by its outcome in *conduct*. How, then, does science
justify itself in its influence on the conduct of men
as citizens? I assert that the encouragement of

scientific investigation and the spread of scientific knowledge by largely inculcating scientific habits of mind will lead to more efficient citizenship and so to increased social stability. Minds trained to scientific methods are less likely to be led by mere appeal to the passions, by blind emotional excitement to sanction acts which in the end may lead to social disaster. In the first and foremost place, therefore, I lay stress upon the educational side of modern science, and state my proposition in some such words as these :—

Modern Science, as training the mind to an exact and impartial analysis of facts is an education specially fitted to promote sound citizenship.

Our first conclusion, then, as to the value of science for practical life turns upon the efficient training it provides in *method*. The man who has accustomed himself to marshal facts, to examine their complex mutual relations, and predict upon the result of this examination their inevitable sequences—sequences which we term natural laws and which are as valid for every normal mind as for that of the individual investigator—such a man we may hope will carry his scientific method into the field of social problems. He will scarcely be content with mere superficial statement, with mere appeal to the imagination, to the emotions, to individual prejudices. He will demand a high standard of reasoning, a clear insight into facts and their results, and his demand cannot fail to be beneficial to the community at large.

§ 4.—*Essentials of Good Science.*

I want the reader to appreciate clearly that science justifies itself in its methods, quite apart from any serviceable knowledge it may convey. We are too apt to forget this purely educational side of science in the great value of its practical applications. We see too often the plea raised for science that it is *useful knowledge*, while grammar and philosophy are supposed to have small utilitarian or commercial value. Science, indeed, often teaches us facts of primary importance for practical life ; yet not on this account, but because it leads us to classifications and systems independent of the individual thinker, to sequences and laws admitting of no play-room for individual fancy, must we rate the training of science and its social value higher than those of grammar and philosophy. Herein lies the first, but of course not the sole, ground for the popularization of science. That form of popular science which merely recites the results of investigations, which merely communicates *useful knowledge*, is from this standpoint bad science, or no science at all. Let me recommend the reader to apply this test to every work professing to give a popular account of any branch of science. If any such work gives a description of phenomena that appeals to his ima-gination rather than to his reason, then it is bad science. The first aim of any genuine work of science, however popular, ought to be the presentation of such a classification of facts that the reader's mind is irresistibly led to acknowledge a logical sequence —a law which appeals to the reason before it captivates the imagination. Let us be quite sure that whenever we come across a conclusion in a scientific work which does not flow from the classifi-cation of facts, or which is not directly stated by the author to be an assumption, then we are dealing with bad science. Good science will always be intelligible to the logically trained mind, if that mind can read and translate the language in which science is written. The scientific method is one and the same in all branches, and that method is the method of all logically trained minds. In this respect the great classics of science are often the most intelligible of books, and if so, are far better worth reading than popularizations of them written by men with less insight into scientific method. Works like Darwin's *Origin of Species* and *Descent of Man*, Lyell's *Principles of Geology*, Helmholtz's *Sensations of Tone*, or Weismann's *Essays on Heredity*, can be profitably read and largely understood by those who are not specially trained in the several branches of science with which these works deal.[1] It may need some patience in the interpretation of scientific terms, in learning the language of science, but like most cases in which a new language has to be learnt, the comparison of passages in which the same word or term recurs, will soon lead to a just appreciation of its true meaning. In the matter of language the descriptive natural sciences such as geology or biology are more easily accessible to the layman than the exact sciences such as algebra or mechanics, where the reasoning process must often be clothed in mathematical symbols, the right interpretation of which may require months, if not years, of study. To this distinction between the descriptive and exact sciences I propose to return later, when we are dealing with the classification of the sciences.

[1] The list might be easily increased, for example by W. Harvey's *Anatomical Dissertation on the Motion of the Heart and Blood*, and by Faraday's *Experimental Researches*.

I would not have the reader suppose that the mere perusal of some standard scientific work will, in my opinion, produce a scientific habit of mind. I only suggest that it will give some insight into scientific method and some appreciation of its value. Those who can devote persistently some four or five hours a week to the conscientious study of any *one* limited branch of science will achieve in the space of a year or two much more than this. The busy layman is not bound to seek about for some branch which will give him useful facts for his profession or occupation in life. It does not indeed matter for the purpose we have now in view whether he seek to make himself proficient in geology, or biology, or geometry, or mechanics, or even history or folklore, if these be studied scientifically. What is necessary is the *thorough* knowledge of some small group of facts, the recognition of their relationship to each other, and of the formulæ or laws which express scientifically their sequences. It is in this manner that the mind becomes imbued with the scientific method and freed from individual bias in the formation of its judgments—one of the conditions, as we have seen, for ideally good citizenship. This first claim of scientific training, its education in method, is to my mind the most powerful claim it has to state support. I believe more will be achieved by placing instruction in pure science within the reach of all our citizens, than by any number of polytechnics devoting themselves to technical education, which does not rise above the level of manual instruction.

§ 5.—*The Scope of Science.*

The reader may, perhaps, feel that I am laying all

stress upon *method* at the expense of solid contents. Now this is the peculiarity of scientific method, that when once it has become a habit of mind, that mind converts *all* facts whatsoever into science. The field of science is unlimited ; its solid contents are endless, every group of natural phenomena, every phase of social life, every stage of past or present development is material for science. *The unity of all science consists alone in its method, not in its material.* The man who classifies facts of any kind whatever, who sees their mutual relation and describes their sequence, is applying the scientific method and is a man of science. The facts may belong to the past history of mankind, to the social statistics of our great cities, to the atmosphere of the most distant stars, to the digestive organs of a worm, or to the life of a scarcely visible bacillus. It is not the facts themselves which form science, but the method in which they are dealt with. The material of science is coextensive with the whole physical universe, not only that universe as it now exists, but with its past history and the past history of all life therein. When every fact, every present or past phenomenon of that universe, every phase of present or past life therein, has been examined, classified, and co-ordinated with the rest, then the mission of science will be completed. What is this but saying that the task of science can never end till man ceases to be, till history is no longer made, and development itself ceases?

It might be supposed that science has made such strides in the last two centuries, and notably in the last fifty years, that we might look forward to a day when its work would be practically accomplished. At the beginning of this century it was possible for an Alexander von Humboldt to take a survey of the entire domain of then extant science. Such a survey would be impossible for any scientist now, even if gifted with more than Humboldt's powers. Scarcely any specialist of to-day is really master of all the work which has been done in his own comparatively small field. Facts and their classification have been accumulating at such a rate, that nobody seems to have leisure to recognize the relations of sub-groups to the whole. It is as if both in Europe and America individual workers were bringing their stones to one great building and piling them on and fastening them down without regard to any general plan or to their individual neighbour's work ; only where some one has placed a great corner-stone, is it regarded, and the building then rises on this firmer foundation more rapidly than at other points, till it reaches a height at which it is stopped for want of side support. Yet this great structure, the proportions of which are beyond the ken of any individual man, possesses a symmetry and unity of its own, notwithstanding its haphazard mode of construction. This symmetry and unity lies in scientific method. The smallest group of facts, if properly classified and logically dealt with, will form a stone which has its proper place in the great building of knowledge, wholly independent of the individual workman who has shaped it. Even when two men work unwittingly at the same stone they will but modify and correct each other's angles. In the face of all this enormous progress of modern science, when in all civilized lands men are applying the scientific method to natural, historical, and mental facts, we have yet to admit that the goal of science is and must be infinitely distant.

Here, too, we may note that when from a sufficient if partial classification of facts a simple principle has been discovered which describes the relationship and sequences of the group, then this principle or law itself generally leads to the discovery of a still wider range of hitherto unregarded phenomena in the same or associated fields. Every great advance of science opens our eyes to facts which we had failed before to observe, and makes new demands on our powers of interpretation. This extension of the material of science into regions where our great-grandfathers could see nothing at all, or where they would have declared human knowledge impossible, is one of the most remarkable features of modern progress. Where they interpreted the motion of the planets of our own system, we discuss the chemical constitution of stars, many of which did not exist for them, for their telescopes could not reach them. Where they discovered the circulation of the blood, we see the physical conflict of living poisons within the blood, whose battles would have been absurdities for them. Where they found void and probably demonstrated to their own satisfaction that there was void, we conceive great systems in rapid motion capable of carrying energy through brick walls as light passes through glass. Great as the advance of scientific knowledge has been, it has not been greater than the growth of the material to be dealt with. The goal of science is clear—it is nothing short of the complete interpretation of the universe. But the goal is an ideal one—it marks the *direction* in which we move and strive, but never the point we shall actually reach.

3

§ 6.—*Science and Metaphysics.*

Now I want to draw the reader's attention to two results which flow from the above considerations, namely: that the material of science is coextensive with the whole life, physical and mental, of the universe, and furthermore that the limits to our perception of the universe are only apparent, not real. It is no exaggeration to say that the universe was not the same for our great-grandfathers as it is for us, and that in all probability it will be utterly different for our great-grandchildren. The universe is a variable quantity, which depends upon the keenness and structure of our organs of sense, and upon the fineness of our powers and instruments of observation. We shall see more clearly the important bearing of this latter remark when we come to discuss more closely in another chapter how the universe is largely the construction of each individual mind. For the present we must briefly consider the former remark, which defines the unlimited scope of science. To say that there are certain fields—for example *metaphysics*—from which science is excluded, wherein its methods have no application, is merely to say that the rules of methodical observation and the laws of logical thought do not apply to the facts, if any, which lie within such fields. These fields, if indeed such exist, must lie outside any intelligible definition which can be given of the word *knowledge.* If there are facts, and sequences to be observed among those facts, then we have all the requisites of scientific classification and knowledge. If there are no facts, or no sequences to be observed among them, then the possibility of *all* knowledge disappears. The greatest assumption of everyday life—the inference which the metaphysicians tell us is wholly beyond science—namely, that other beings have consciousness as well as ourselves, seems to have just as much or as little *scientific* validity as the statement that an earth-grown apple would fall to the ground if carried to the planet of another star. Both are beyond the range of experimental demonstration, but to assume uniformity in the characteristics of brain 'matter' under certain conditions seems as scientific as to assume uniformity in the characteristics of stellar 'matter.' Both are only working hypotheses and valuable in so far as they simplify our description of the universe. Yet the distinction between science and metaphysics is often insisted upon, and not unadvisedly, by the devotees of both. If we take any group of physical or biological facts— say, for example, electrical phenomena or the development of the ovum—we shall find that, though physicists or biologists may differ to some extent in their measurements or in their hypotheses, yet in the fundamental principles and sequences the professors of each individual science are in practical agreement among themselves. A similar if not yet so complete agreement is rapidly springing up in both mental and social science, where the facts are more difficult to classify and the bias of individual opinion is much stronger. Our more thorough classification, however, of the facts of human development, our more accurate knowledge of the early history of human societies, of primitive customs, laws, and religions, our application of the principle of natural selection to man and his communities, are converting anthropology, folklore, sociology, and psychology into true sciences. We begin to see indisputable sequences in groups of both mental and social facts. The causes which favour the growth or decay of human societies become more obvious and more the subject of scientific investigation. Mental and social facts are thus not beyond the range of scientific treatment, but their classification has not been so complete, nor for obvious reasons so unprejudiced, as those of physical or biological phenomena.

The case is quite different with metaphysics and those other supposed branches of human knowledge which claim exemption from scientific control.[1] Either they are based on an accurate classification of facts, or they are not. But if their classification of facts were accurate, the application of the scientific method ought to lead their professors to a practically identical system. Now one of the idiosyncrasies of metaphysicians lies in this : that each metaphysician has his own system, which to a large extent excludes that of his predecessors and colleagues. Hence we must conclude that metaphysics are either built on air or on quicksands—either they start from no foundation in facts at all, or the superstructure has been raised before a basis has been found in the accurate classification of facts. I want to lay special stress on this point. There is no short cut to truth, no way to gain a knowledge of the universe except through the gateway of scientific method. The hard and stony path of classify-

---

[1] It is perhaps impossible to satisfactorily define the metaphysician, but the meaning attached by the present writer to the term will become clearer in the sequel. It is here used to denote a class of writers, of whom well-known examples are : Kant, in his later uncritical period (when he discovered that the universe was created in order that man might have a sphere for moral action !) ; the post-Kantians (notably Hegel and Schopenhauer), and their numerous English disciples, who " explain " the universe without having even an elementary knowledge of physical science.

ing facts and reasoning upon them is the only way to ascertain truth. It is the reason and not the imagination which must ultimately be appealed to. The poet may give us in sublime language an account of the origin and purport of the universe, but in the end it will not satisfy our æsthetic judgment, our idea of harmony and beauty, like the few facts which the scientist may venture to tell us in the same field. The one will agree with all our experiences past and present, the other is sure, sooner or later, to contradict our observation because it is a dogma, where we are yet far from knowing the whole truth. Our æsthetic judgment demands harmony between the representation and the represented, and in this sense science is often more artistic than modern art.

The poet is a valued member of the community, for he is known to be a poet ; his value will increase as he grows to recognize the deeper insight into nature with which modern science provides him. The metaphy sician is a poet, often a very great one, but unfortunately he is not known to be a poet, because he clothes his poetry in the language of apparent reason, and hence it follows that he is liable to be a dangerous member of the community. The danger at the present time that metaphysical dogmas may check scientific research is, perhaps, not very great. The day has gone by when the Hegelian philosophy threatened to strangle infant science in Germany ;—that it begins to languish at Oxford is a proof that it is practically dead in the country of its birth. The day has gone by when philosophical or theological dogmas of any kind can throw back, even for generations, the progress of scientific investigation. There is no restriction now on research in any field, or on the publication of the

truth when it has been reached. But there is nevertheless a danger which we cannot afford to disregard,
a danger which retards the spread of scientific knowledge among the unenlightened, and which flatters
obscurantism by discrediting the scientific method.
There is a certain school of thought which finds the
laborious process by which science reaches truth too
irksome ; the temperament of this school is such that
it demands a short and easy cut to knowledge, where
knowledge can only be gained, if at all, by the long
and patient toiling of many groups of workers, perhaps
through several centuries. There are various fields at
the present day wherein mankind is ignorant, and the
honest course for us is simply to confess our ignorance.
This ignorance may arise from the want of any proper
classification of facts, or because supposed facts are
themselves inconsistent, unreal creations of man's untrained mind. But because this ignorance is frankly
admitted by science, an attempt is made to wall off
these fields as ground whereon science has no business
to trespass, where the scientific method is of no avail.
Wherever science has succeeded in ascertaining the
truth, there, according to the school we have referred to, are the " legitimate problems of science."
Wherever science is yet ignorant, there we are told
its method is inapplicable ; there some other relation
than cause and effect (than the same sequence recurring
with the like grouping of phenomena), some new, but
undefined relationship rules. In these fields we are
told problems become philosophical and can only
be treated by the method of philosophy. The philosophical method is opposed to the scientific method ;
and here I think the danger I have referred to arises.
We have defined the scientific method to consist in

the orderly classification of facts followed by the recognition of their relationship and recurring sequences. The scientific judgment is the judgment based upon this recognition and free from personal bias. If this were the philosophical method there would be no need of further discussion, but as we are told the subject-matter of philosophy is not the " legitimate problem of science," the two methods are presumably not identical. Indeed the philosophical method seems based upon an analysis which does not start with the classification of facts, but reaches its judgments by some process of internal cogitation. It is therefore dangerously liable to the influence of individual bias ; it results, as experience shows us, in an endless number of competing and contradictory systems. It is because the so-called philosophical method does not lead, like the scientific, to practical unanimity of judgments, when different individuals approach the same range of facts,[1] that science, rather than philosophy, offers the better training for modern citizenship.

§ 7.—*The Ignorance of Science.*

It must not be supposed that science for a moment denies the existence of some of the problems which have hitherto been classed as philosophical or metaphysical On the contrary, it recognizes that a great variety of physical and biological phenomena lead directly to these problems. But it asserts that the methods

[1] This statement by no means denies the existence of many moot points, unsettled problems in science ; but the genuine scientist *admits* that they are unsolved. As a rule they lie just on the frontier line between knowledge and ignorance, where the outposts of science are being pushed forward into unoccupied and difficult country.

hitherto applied to these problems have been futile, because they have been unscientific. The classifications of facts hitherto made by the system-mongers have been hopelessly inadequate or hopelessly prejudiced. Until the scientific study of psychology, both by observation and experiment, has advanced immensely beyond its present limits—and this may take generations of work—science can only answer to the great majority of 'metaphysical' problems, "I am ignorant." Meanwhile it is idle to be impatient or to indulge in system-making. The cautious and laborious classification of facts must have proceeded much further than at present, before the time will be ripe for drawing conclusions.

Science stands now with regard to the problems of life and mind in much the same position as it stood with regard to cosmical problems in the seventeenth century. Then the system-mongers were the theologians, who declared that cosmical problems were not the "legitimate problems of science." It was vain for Galileò to assert that the theologians' classification of facts was hopelessly inadequate. In solemn congregation assembled they settled that :—

"*The doctrine that the earth is neither the centre of the universe nor immovable, but moves even with a daily rotation, is absurd, and both philosophically and theologically false, and at the least an error of faith.*" [1]

It took nearly two hundred years to convince the whole theological world that cosmical problems were the legitimate problems of science and science alone,

[1] "*Terram non esse centrum Mundi, nec immobilem, sed moveri motu etiam diurno, est item propositio absurda, et falsa in Philosophia, et Theoligice considerata, ad minus erronea in fide*" (Congregation of Prelates and Cardinals, June 22, 1633).

for in 1819 the books of Galilei, Copernicus, and Keppler were still upon the index of forbidden books, and not till 1822 was a decree issued allowing books teaching the motion of the earth about the sun to be printed and published in Rome!

I have cited this memorable example of the absurdity which arises from trying to pen science into a limited field of thought, because it seems to me exceedingly suggestive of what must follow again, if any attempt, philosophical or theological, be made to define the "legitimate problems of science." Wherever there is the slightest possibility for the human mind to *know*, there is a legitimate problem of science. Outside the field of actual knowledge can only lie a region of the vaguest opinion and imagination, to which unfortunately men too often, but still with decreasing prevalence, pay higher respect than to knowledge.

We must here investigate a little more closely what the man of science means when he says: "*Here I am ignorant.*" In the first place he does not mean that the method of science is necessarily inapplicable, and accordingly that some other method is to be sought for. In the next place, if the ignorance really arises from the inadequacy of the scientific method, then we may be quite sure that no other method whatsoever will reach the truth. The ignorance of science means the enforced ignorance of mankind. I should be sorry myself to assert that there is any field of either mental or physical perceptions which science may not in the long course of centuries enlighten. Who can give us the assurance that the fields already occupied by science are alone those in which knowledge is possible? Who, in the words of Galilei, is willing to set limits to the human intellect? It is true that this view is not held by several leading scientists, both in this country and Germany. They are not content with saying, " We *are* ignorant," but they add, with regard to certain classes of facts, " Mankind must *always* be ignorant." Thus in England Professor Huxley has invented the term *Agnostic*, not so much for those who are ignorant as for those who limit the possibility of knowledge in certain fields. In Germany Professor E. du Bois-Reymond has raised the cry: *"Ignorabimus"* —"We shall be ignorant," and both his brother and he have undertaken the difficult task of demonstrating that with regard to certain problems human knowledge is impossible.[1] We must, however, note that in these cases we are not concerned with the limitation of the scientific method, but with the denial of the possibility that any method whatever can lead to knowledge. Now I venture to think that there is great danger in this cry: "We *shall* be ignorant." To cry "We are ignorant," is safe and healthy, but the attempt to demonstrate an endless futurity of ignorance appears a modesty which approaches despair. Conscious of the past great achievements and the present restless activity of science, may we not do better to accept as our watchword that of Galilei : '' Who is willing to set limits to the human intellect ? "—interpreting it by what evolution has taught us of the continual growth of man's intellectual powers.

Scientific ignorance may, as I have remarked (p. 22), either arise from an insufficient classification of facts, or be due to the unreality of the facts with which science has been called upon to deal. Let us take for example a number of fields of thought which

---

[1] See especially Paul du Bois-Reymond : *Ueber die Grundlagen der Erkenntniss in den exacten Wissenschaften.* Tübingen, 1890.

were very prominent in mediæval times, such as alchemy, astrology, witchcraft. In the fifteenth century nobody doubted the "facts" of astrology and witchcraft. Men were ignorant as to how the stars exerted their influence for good or ill; they did not know the exact mechanical process by which all the milk in a village was turned blue by a witch. But for them it was nevertheless a fact that the stars did influence human lives, and a fact that the witch had the power of turning the milk blue. Have we solved the problems of astrology and witchcraft to-day?

Do we now know how the stars influence human lives, or how witches turn milk blue? Not in the least. We have learnt to look upon the facts themselves as unreal, as vain imaginings of the untrained human mind; we have learnt that they could not be described scientifically because they involved notions which were in themselves contradictory and absurd. With alchemy the case was somewhat different. Here a false classification of real facts was combined with inconsistent sequences—that is, sequences not deduced by a rational method. So soon as science entered the field of alchemy with a true classification and a true method, alchemy was converted into chemistry and became an important branch of human knowledge. Now it will, I think, be found that the fields of inquiry, where science has not yet penetrated and where the scientist still confesses ignorance, are very like the alchemy, astrology, and witchcraft of the Middle Ages. Either they involve facts which are in themselves unreal—conceptions which are self-contradictory and absurd, and therefore incapable of analysis by the scientific or any other method,—or, on the other hand, our ignorance

arises from an inadequate classification and a neglect of scientific method.

This is the actual state of the case with those
mental and spiritual phenomena which are said to
lie outside the proper scope of science, or which
appear to be disregarded by scientific men. No
better example can be taken than the range of phenomena which are entitled *Spiritualism*. Here science is
asked to analyze a series of facts which are to a great
extent unreal, which arise from the vain imaginings
of untrained minds and from atavistic tendencies
to superstition. So far as the facts are of this character, no account can be given of them, because, like
the witch's supernatural capacity, their unreality will
be found at bottom to make them self-contradictory.
Combined, however, with the unreal series of facts
are probably others, connected with hypnotic conditions, which are real and only incomprehensible because there is as yet scarcely any intelligent classification or true application of scientific method. The
former class of facts will, like astrology, never be
reduced to law, but will one day be recognized as
absurd ; the other, like alchemy, may grow step by step
into an important branch of science. Whenever, therefore, we are tempted to desert the scientific method
of seeking truth, whenever the silence of science
suggests that some other gateway must be sought to
knowledge, let us inquire first whether the elements
of the problem, of whose solution we are ignorant,
may not after all, like the facts of witchcraft, arise
from a superstition, and be self-contradictory and
incomprehensible because they are unreal.

If on inquiry we ascertain that the facts cannot
possibly be of this class, we must then remember that

it may require long ages of increasing toil and investigation before the classification of the facts can be so complete that science can express a definite judgment on their relationship. Let us suppose that the Emperor Karl V. had said to the learned of his day: "I want a method by which I can send a message in a few seconds to that new world, which my mariners take weeks in reaching. Put your heads together and solve the problem." Would they not undoubtedly have replied that the problem was impossible? To propose it would have seemed as ridiculous to them as the suggestion that science should straightway solve many problems of life and mind seems to the learned of to-day. It required centuries spent in the discovery and classification of new facts before the Atlantic cable became a possibility. It may require the like or even a longer time to unriddle those psychical and biological enigmas to which I have referred ; but he who declares that they can never be solved by the scientific method is to my mind as rash as the man of the early sixteenth century would have been had he declared it utterly impossible that the problem of talking across the Atlantic Ocean should ever be solved.

§ 8.—*The Wide Domain of Science.*

If I have put the case of science at all correctly, the reader will have recognized that modern science does much more than demand that it shall be left in undisturbed possession of what the theologian and metaphysician please to term its "legitimate field." It claims that the whole range of phenomena, mental as well as physical—the entire universe—is its field. It asserts that the scientific method is the sole gateway to the whole region of knowledge. The word science is here used in no narrow sense, but applies to all reasoning about facts which proceeds, from their accurate classification, to the appreciation of their relationship and sequence. The touchstone of science is the universal validity of its results for all normally constituted and duly instructed minds Because the glitter of the great metaphysical systems becomes dross when tried by this touchstone, we are compelled to classify them as interesting works of the imagination, and not as solid contributions to human knowledge.

Although science claims the whole universe as its field, it must not be supposed that it has reached, or ever can reach, complete knowledge in every department. Far from this, it confesses that its ignorance is more widely extended than its knowledge. In this very confession of ignorance, however, it finds a safeguard for future progress. Science cannot give its consent to man's development being some day checked again by the barriers which dogma and myth would wish to erect round territory that science has not yet effectually occupied. It cannot allow theologian or philosopher, those Portuguese of the intellect, to establish a right to the foreshore of ignorance, and so to hinder the settlement in due time of vast and yet unknown continents of thought. In the like barriers erected in the past science finds some of the greatest difficulties in the way of intellectual progress and social advance at the present. It is the want of impersonal judgment, of scientific method, and of accurate insight into facts, due largely to a non-scientific training, which renders clear thinking so rare, and random and irresponsible judgments so common, in the mass of our citizens to-day. Yet these citizens, owing to the growth of democracy, have graver problems to settle than probably any which have confronted their forefathers since the days of the Revolution.

§ 9.—*The Second Claim of Science.*

Hitherto the sole ground on which we have considered the appeal of modern science to the citizen is the *indirect* influence it has upon conduct owing to the more efficient mental training which it provides. But we have further to recognize that science can on occasion adduce facts having far more *direct* bearing on social problems than any theory of the state propounded by the philosophers from the days of Plato to those of Hegel. I cannot bring home to the reader the possibility of this, better than by citing some of the conclusions to which the theory of heredity elaborated by the German biologist Weismann introduces us. Weismann's theory lies on the borderland of scientific knowledge; his results are still open to discussion, his conclusions to modification.[1] But to indicate the manner in which science can directly influence conduct, we may assume for the time being Weismann's main conclusions to be correct. One of the chief features of his theory is the non-inheritance

[1] His theory of the "continuity of the germ plasm" is in many respects open to question, but his conclusion as to acquired characteristics being uninherited stands on firmer ground. See Weismann : *Essays on Heredity and Kindred Biological Problems*, Oxford, 1889. A good criticism will be found in C. Ll. Morgan's *Animal Life and Intelligence*, chap. v. A summary in W. P. Ball's *Are the Effects of Use and Disuse Inherited?* The reader should also consult P. Geddes and J. A. Thomson, *The Evolution of Sex*, and a long discussion in *Nature*, vols. xl. and xli. (*sub indice*, Weismann, *Heredity*).

by the offspring of characteristics acquired by the parents in the course of life. Thus good or bad habits acquired by the father or mother in their lifetime are not inherited by their children. The effects of special training or of education on the parents have no direct influence on the child before birth. The parents are merely trustees who hand down their commingled stocks to their offspring. From a bad stock can come only bad offspring, and if a member of such a stock is, owing to special training and education, an exception to his family, his offspring will still be born with the old taint.[1] Now this conclusion of Weismann's—if it be valid, and all we can say at present is that the arguments in favour of it are remarkably strong—radically affects our judgment on the moral conduct of the individual, and on the duties of the state and society towards their degenerate members. No degenerate and feeble stock will ever be converted into healthy and sound stock by the accumulated effects of education, good laws, and sanitary surroundings. Such means may render the individual members of the stock passable if not strong members of society, but the same process will have to be gone through again and again with their offspring, and this in ever-widening circles, if the stock, owing to the conditions in which society has placed it, is able to increase in numbers. The removal of that process of natural selection which in the struggle for existence crushed out feeble and degenerate stocks, may be a real danger to

---

[1] Class, poverty, localization do much to approximately isolate stock, to aggregate the unfit even in modern civilization. The mingling of good and bad stock due to dispersion leads solely to *panmixia*, it degenerates the good as much as it improves the bad.

society, if society relies solely on changed environment for converting its inherited bad into an inheritable good. If society is to shape its own future—if we are to replace the stern processes of natural law, which have raised us to our present high standard of civilization, by milder methods of eliminating the unfit—then we must be peculiarly cautious that in following our strong social instincts we do not at the same time weaken society by rendering the propagation of bad stock more and more easy.

If this theory of Weismann's be correct—if the bad man can by the influence of education and surroundings be made good, but the bad stock can never be converted into good stock—then we see how grave a responsibility is cast at the present day upon every citizen, who directly or indirectly has to consider problems relating to the state endowment of education, the revision and administration of the Poor Law, and, above all, the conduct of public and private charities annually disposing of immense resources. In all problems of this kind the blind social instinct and the individual bias at present form extremely strong factors of our judgment. Yet these very problems are just those which, affecting the whole future of our society, its stability and its efficiency, require us, as good citizens, above all to understand and obey the laws of healthy social development.

The example we have considered will not be futile, nor its lessons worthless, should Weismann's views after all be inaccurate. It is clear that in social problems of the kind I have referred to, the laws of heredity, whatever they may be, must profoundly influence our judgment. The conduct of parent to child, and of society to its anti-social members, can never be placed

4

on a sound and permanent basis without regard to what science has to tell us on the fundamental problems of inheritance. The "philosophical" method can never lead to a real theory of morals. Strange as it may seem, the laboratory experiments of a biologist may have greater weight than all the theories of the state from Plato to Hegel! The scientific classification of facts, biological or historical, the observation of their correlation and sequence, the resulting absolute, as opposed to the individual judgment—these are the sole means by which we can reach truth in such a vital social question as that of heredity. In these considerations alone there appears to be sufficient justification for the national endowment of science, and for the universal training of our citizens in scientific methods of thought. Each one of us is now called upon to give a judgment upon an immense variety of problems, crucial for our social existence. If that judgment confirms measures and conduct tending to the increased welfare of society, then it may be termed a moral, or, better, a social judgment. It follows, then, that to ensure a judgment's being moral, method and knowledge are essential to its formation. It cannot be too often insisted upon that the formation of a moral judgment—that is, one which the individual is reasonably certain will tend to social welfare—does not depend solely on the readiness to sacrifice individual gain or comfort, to act unselfishly : it depends in the first place on knowledge and method. The first demand of the state upon the individual is not for self-sacrifice, but for self-development. The man who gives a thousand pounds to a vast and vague scheme of charity, may or may not be acting socially ; his self-sacrifice, if it be such, proves nothing ; but the man who gives a vote, either directly or even indirectly, in the choice of a representative, after forming a judgment based upon *knowledge* is undoubtedly acting socially, and is fulfilling a higher standard of citizenship.

§ 10.—*The Third Claim of Science.*

Thus far I have been examining more particularly the action of science with regard to social problems. I have endeavoured to point out that it cannot legitimately be excluded from any field of investigation after truth, and that, further, not only is its *method* essential to good citizenship, but that its *results* bear closely on the practical treatment of many social difficulties. In this I have endeavoured to justify the state endowment and teaching of pure science as apart from its technical applications. If in this justification I have laid most stress on the advantages of scientific method — on the training which science gives us in the appreciation of evidence, in the classification of facts, and in the elimination of personal bias, in all that may be termed exactness of mind—we must still remember that ultimately the *direct* influence of pure science on practical life is enormous. The observations of Newton on the relation between the motions of a falling stone and the moon, of Galvani on the convulsive movements of frogs' legs in contact with iron and copper, of Darwin on the adaptation of woodpeckers, of tree-frogs, and of seeds to their surroundings, of Kirchhoff on certain lines which occur in the spectrum of sunlight, of other investigators on the life-history of bacteria—these and kindred observations have not only revolutionized our conception of the universe, but they have revolutionized, or are revolutionizing, our practical life, our means of transit, our social conduct, our treatment of disease. What at the instant of its discovery appears to be only a sequence of purely theoretical interest, becomes the basis of discoveries which in the end profoundly modify the conditions of human life. It is impossible to say of any result of pure science, that it will not some day be the starting-point of wide-reaching technical applications. The frog's legs of Galvani and the Atlantic cable seem wide enough apart, but the former was the starting-point of the series of investigations which ended in the latter. In the recent discovery of Hertz that the action of electro-magnetism is propagated in waves like light—in his confirmation of Maxwell's theory that light is only a special phase of electro-magnetic action—we have a result which, if of striking interest to pure science, seems yet to have no immediate practical application. But that man would indeed be a bold dogmatist who would venture to assert that the results which may ultimately flow from this discovery of Hertz's will not, in a generation or two, do more to revolutionize life than the frog's legs of Galvani had done when they had led to the perfection of the electric telegraph.

§ 11.—*Science and the Imagination.*

There is another aspect from which it is right that we should regard pure science—one that makes no appeal to its utility in practical life, but touches a side of our nature which the reader may have thought that I have entirely neglected. There is an element in our being which is not satisfied by the formal processes of reasoning ; it is the imaginative or æsthetic side, the side to which the poets and philosophers appeal, and one which science cannot, to be scientific, disregard. We have seen that the imagination must not replace the reason in the deduction of relation and law from classified facts. But, none the less, disciplined imagination has been at the bottom of all great scientific discoveries. All great scientists have, in a certain sense, been great artists ; the man with no imagination may collect facts, but he cannot make great discoveries. If I were compelled to name the Englishmen who during our generation have had the widest imaginations and exercised them most beneficially, I think I should put the novelists and poets on one side and say Michael Faraday and Charles Darwin. Now it is very needful to understand the exact part imagination plays in pure science. We can, perhaps, best achieve this result by considering the following proposition : Pure science has a further strong claim upon us on account of the exercise it gives to the imaginative faculties and the gratification it provides for the æsthetic judgment. The exact meaning of the terms " scientific fact " and " scientific law " will be considered in later chapters, but for the present let us suppose an elaborate classification of such facts has been made, and their relationships and sequences carefully traced. What is the next stage in the process of scientific investigation ? Undoubtedly it is the use of the imagination. The discovery of some single statement, some brief *formula* from which the whole group of facts is seen to flow, is the work not of the mere cataloguer, but of the man endowed with creative imagination. The single statement, the brief formula, the words of which replace in our minds a wide range of relationships between isolated phenomena, is what we term a scientific *law*. Such a law, relieving our memory from the burden of individual sequences, enables us, with the minimum of intellectual fatigue, to grasp a vast complexity of natural or social phenomena.   The discovery of law is therefore the peculiar function of the creative imagination.   But this imagination has to be a *disciplined* one.   It has in the first place to appreciate the whole range of facts, which require to be resumed in a single statement ; and then when the law is reached—often by what seems solely the inspired imagination of genius—it must be tested and criticised by its discoverer in every conceivable way, till he is certain that the imagination has not played him false, and that his law is in real agreement with the whole group of phenomena which it resumes.   Herein lies the keynote to the scientific use of the imagination.   Hundreds of men have allowed their imagination to solve the universe, but the men who have contributed to our real understanding of natural phenomena have been those who were unstinting in their application of criticism to the product of their imaginations.   It is such criticism which is the essence of the scientific use of the imagination, which is, indeed, the very life-blood of science.[1]

No less an authority than Faraday writes :—

"The world little knows how many of the thoughts and theories which have passed through the mind of a scientific investigator have been crushed in silence and secrecy by his own severe criticism and adverse examination ; that in the most successful instances not a tenth of the suggestions, the hopes, the wishes, the preliminary conclusions have been realized."

---

[1] " *La critique est la vie de la science,*" says Victor Cousin.

§ 12.—*The Method of Science Illustrated.*

The reader must not think that I am painting any ideal or purely theoretical method of scientific discovery. He will find the process described above accurately depicted by Darwin himself in the account he gives us of his discovery of the law of natural selection. After his return to England in 1837, he tells us,[1] it appeared to him that :—

" By collecting all facts which bore in any way on the variation of animals and plants under domestication and nature, some light might perhaps be thrown on the whole subject. My first note-book was opened in July, 1837. I worked on true Baconian principles,[2] and, without any theory, collected facts on a wholesale scale, more especially with respect to domesticated productions, by printed enquiries, by conversation with skilful breeders and gardeners, and by extensive reading. When I see the list of books of all kinds which I read and abstracted, including whole series

---

[1] *The Life and Letters of Charles Darwin*, vol. i. p. 83.

[2] It is from men like Laplace and Darwin, who have devoted their lives to natural science, rather than from workers in the pure field of conception, like Mill and Stanley Jevons, that we must seek for a true estimate of the Baconian method. Beside Darwin's words we may place those of Laplace on Bacon :—

"Il a donné pour la recherche de la vérité, le précepte et non l'exemple. Mais en insistant avec toute la force de la raison et de l'éloquence, sur la nécessité d'abandonner les subtilités insignifiantes de l'école, pour se livrer aux observations et aux expériences, et en indiquant la vraie méthode de s'élever aux causes générales des phénomènes, ce grand philosophe a contribué aux progrès immenses que l'esprit humain a faits dans le beau siècle où il a terminé sa carrière" (*Théorie analytique des Probabilités*, Œuvres T. vii. p. clvi.). The carpenter who uses a tool is a better judge of its efficiency than the smith who forges it. For a good sketch of the estimation in which Bacon was held by his *scientific* contemporaries see the introduction to Prof. Fowler's edition of the *Novum Organum*.

of Journals and Transactions, I am surprised at my own industry. I soon perceived that selection was the keystone of man's success in making useful races of animals and plants. But how selection could be applied to organisms living in a state of nature remained for some time a mystery to me."

Here we have Darwin's scientific classification of facts, what he himself terms his " systematic inquiry." Upon the basis of this systematic inquiry comes the search for a law. This is the work of the imagination ; the inspiration in Darwin's case being apparently due to a perusal of Malthus' *Essay on Population*. But Darwin's imagination was of the disciplined scientific sort. Like Turgot, he knew that if the first thing is to invent a system, then the second is to be disgusted with it. Accordingly there followed the period of self-criticism, which lasted four or five years, and it was no less than *nineteen* years before he gave the world his discovery in its final form. Speaking of his inspiration that natural selection was the key to the mystery of the origin of species, he says :—

" Here, then, I had at last got a theory by which to work ; but I was so anxious to avoid prejudice, that I determined not for some time to write even the briefest sketch of it. In June, 1842 [*i.e.*, four years after the inspiration], I first allowed myself the satisfaction of writing a very brief abstract of my theory in pencil in 35 pages ; and this was enlarged during the summer of 1844 into one of 230 pages, which I had fairly copied out and still possess."

Finally an abstract from Darwin's manuscript was published with Wallace's Essay in 1858, and the *Origin of Species* appeared in 1859.

In like manner Newton's imagination was only

paralleled by that power of self-criticism which led him to lay aside a demonstration touching the gravitation of the moon for nearly eighteen years, until he had supplied a missing link in his reasoning. But our details of Newton's life and discoveries are too meagre for us to see his method as closely as we can Darwin's, and the account I have given of the latter is amply sufficient to show the actual application of scientific method, and the real part played in science by the disciplined use of the imagination.[1]

[1] That the classification of facts is often largely guided by the imagination as well as the reason must be fully admitted. At the same time an accurate classification, either due to the scientist himself or to previous workers, must exist in the scientist's mind before he can proceed to the discovery of law. Here, as elsewhere, the reader will find that I differ very widely from Stanley Jevons' views as developed in his *Principles of Science.* I cannot but feel that Chapter xxvi. of that work would have been recast had the author been acquainted with Darwin's method of procedure. The account given by Jevons of the Newtonian method seems to me to lay insufficient stress upon the fact that Newton had a wide acquaintance with physics *before* he proceeded to use his imagination and test his theories by experiment—that is, to a period of self-criticism. The reason that pseudo-scientists cumber the reviewer's table with idle theories, often showing great imaginative power and ingenuity, is not solely want of self-criticism. Their theories, as a rule, are not such as the scientist himself would ever propound and criticise. Their impossibility is obvious, because their propounders have neither formed for themselves, nor been acquainted with others' classifications of the groups of facts which their theories are intended to summarise. Newton and Faraday *started* with full knowledge of the classifications of physical facts which had been formed in their own days, and proceeded to further conjoint theorizing and classifying. Bacon, of whom Stanley Jevons is, I think, unreasonably contemptuous, lived at a time when but little had been done by way of classification, and he was wanting in the scientific imagination of a Newton or a Faraday. Hence the barrenness of his method in his own hands. The early history of the Royal Society's meetings shows how essentially the period of collection and classification of facts preceded that of valuable theory.

With Stanley Jevons' last chapter on *The Limits of Scientific Method* the present writer can only express his complete disagreement ; many

§ 13.—*Science and the Æsthetic Judgment.*

We are justified, I think, in concluding that science does not cripple the imagination, but rather tends to exercise and discipline its functions. We have still, however, to consider another phase of the relationship of the imaginative faculty to pure science. When we see a great work of the creative imagination, a striking picture or a powerful drama, what is the essence of the fascination it exercises over us? Why does our æsthetic judgment pronounce it a true work of art? Is it not because we find concentrated into a brief statement, into a simple formula or a few symbols, a wide range of human emotions and feelings? Is it not because the poet or the artist has expressed for us in his representation the true relationship between a variety of emotions, which we, in a long course of experience, have been consciously or unconsciously classifying? Does not the beauty of the artist's work lie for us in the accuracy with which his symbols resume innumerable facts of our past emotional experience? The æsthetic judgment pronounces for or against the interpretation of the creative imagination according as that interpretation embodies or contradicts the phenomena of life, which we ourselves have observed.[1] It is only satisfied when the artist's formula contradicts none of the emotional phenomena which it is intended to resume. If this account of the æsthetic judgment be at all a true one, the reader will have re-

of its arguments appear to him unscientific, if it were not better to term them anti-scientific.

[1] How important a part length and variety of emotional experience play in the determination of the æsthetic judgment is easily noted by investigating the favourite authors and pictures of a few friends of diverse ages and conditions.

marked how exactly parallel it is to the scientific judgment.[1] But there is really more than mere parallelism between the two. The laws of science are, as we have seen, products of the creative imagination. They are the mental interpretations—the formulæ under which we resume wide ranges of phenomena, the results of observation on the part of ourselves or of our fellowmen. The scientific interpretation of phenomena, the scientific account of the universe, is therefore the only one which can permanently satisfy the æsthetic judgment, for it is the only one which can never be contradicted by our observation and experience. It is necessary to strongly emphasise this side of science, for we are frequently told that the growth of science is destroying the beauty and poetry of life. It is undoubtedly rendering many of the old interpretations of life meaningless, because it demonstrates that they are false to the facts which they profess to describe. It does not follow from this, however, that the æsthetic and scientific judgments are opposed; the fact is, that with the growth of our scientific knowledge the basis of the æsthetic judgment is changing and must change. There is more real beauty in what science has to tell us of the chemistry of a distant star, or in the life-history of a protozoon, than in any cosmogony produced by the creative imagination of a pre-scientific age. By " more real beauty " we are to understand that the æsthetic judgment will find more satisfaction, more permanent delight in the former than in the latter. It is this continual gratification of the æsthetic judgment which is one of the chief delights of the pursuit of pure science.

[1] The curious reader may be referred to Wordsworth's " General View of Poetry " in his preface to the *Lyrical Ballads*, 1815.

§ 14.—*The Fourth Claim of Science.*

There is an insatiable desire in the human breast to resume in some short formula, some brief statement, the facts of human experience. It leads the savage to "account" for all natural phenomena by deifying the wind and the stream and the tree. It leads civilized man, on the other hand, to express his emotional experience in works of art, and his physical and mental experience in the formulæ or so-called laws of science. Both works of art and laws of science are the product of the creative imagination, both afford material for the gratification of the æsthetic judgment. It may seem at first sight strange to the reader that the laws of science should thus be associated with the creative imagination in man rather than with the physical world outside him. But as we shall see in the course of the following chapters the laws of science are products of the human mind rather than factors of the external world. Science endeavours to provide a mental *résumé* of the universe, and its last great claim to our support is the capacity it has for satisfying our cravings for a brief description of the history of the world. Such a brief description, a formula resuming all things, science has not yet found and may probably never find, but of this we may feel sure, that its method of seeking for one is the sole possible method, and that the truth it has reached is the only form of truth which can permanently satisfy the æsthetic judgment. For the present, then, it is better to be content with the fraction of a right solution, than to beguile ourselves with the whole of a wrong solution. The former is at least a step towards the truth, and shows us the direction in which other steps may be taken.

The latter cannot be in entire accordance with our past or future experience, and will therefore ultimately fail to satisfy the æsthetic judgment. Step by step that judgment, restless under the growth of positive knowledge, has discarded creed after creed, and philosophic system after philosophic system. Surely we might now be content to learn from the pages of history that only little by little, slowly line upon line, man, by the aid of organized observation and careful reasoning, can hope to reach knowledge of the truth, that science, in the broadest sense of the word, is the sole gateway to a knowledge which can harmonize with our past as well as with our possible future experience. As Clifford puts it : " Scientific thought is not an accompaniment or condition of human progress, but human progress itself."

SUMMARY.

1. The scope of science is to ascertain truth in every possible branch of knowledge. There is no sphere of inquiry which lies outside the legitimate field of science. To draw a distinction between the scientific and philosophical methods is obscurantism.

2. The scientific method is marked by the following features:—(a) Careful and accurate classification of facts and observation of their correlation and sequence ; (b) The discovery of scientific laws by aid of the creative imagination ; (c) Self-criticism and the final touchstone of equal validity for all normally constituted minds.

3. The claims of science to our support depend on:—(a) The efficient mental training it provides for the citizen ; (b) The light it brings to bear on many important social problems ; (c) The increased comfort it adds to practical life ; (d) The permanent gratification it yields to the æsthetic judgment.

LITERATURE.

BACON, Francis.—Novum Organum, London, 1620. A good edition by T. Fowler. Clarendon Press, 1878.

BOIS-REYMOND, E. du.—Ueber die Grenzen des Naturerkennens. Veit & Co., Leipzig, 1876.

BOIS-REYMOND, P. du.—Ueber die Grundlagen der Erkenntniss in den exacten Wissenschaften. H. Laupp, Tübingen, 1890.

CLIFFORD, W. K.—Lectures and Essays. Macmillan, 1879. (" Aims and Instruments of Scientific Thought," " The Ethics of Belief," and " Virchow on the Teaching of Science.")

HAECKEL, E.—Freie Wissenschaft und freie Lehre. E. Schweizerbart, Stuttgart, 1878.

HALDANE, J. S.—" Life and Mechanism," Mind, ix. pp. 27–47 ; also Nature, vol. xxvii., 1883, p. 561, vol. xxiv., 1886, p. 73 ; and also Haldane, R. B., Proceedings of the Aristotolean Society, 1891, vol. i. no. 4, part i. pp. 22–27.

HELMHOLTZ, H.—On the Relation of the Natural Sciences to the Totality of the Sciences, translated by C. H. Schaible. London, 1869.

This occurs also in the Popular Lectures, translated by Atkinson and others, First Series, p. 1. Longmans, 1881.

HERSCHEL, Sir John. — A Preliminary Dissertation on Natural Philosophy. London, 1830.

JEVONS, W. Stanley.—The Principles of Science : A Treatise on Logic and Scientific Method, 2nd ed. Macmillan, 1877.

PEARSON, K.—The Ethic of Freethought : A Selection of Essays and Lectures (" The Enthusiasm of the Market-place and of the Study "). Fisher Unwin, 1888.

VIRCHOW, R.—Die Freiheit der Wissenschaft im modernen Staat (Versammlung deutscher Naturforscher). München, 1877.

CHAPTER II.

The Facts of Science.

§1.—*The Reality of Things.*

In our first chapter we have frequently spoken of the classification of *facts* as the basis of the scientific method; we have also had occasion to use the words *real* and *unreal, universe* and *phenomenon.* It is proper, therefore, that before proceeding further we should endeavour to clear up our ideas as to what these terms signify. We must strive to define a little more closely in what the material of science consists. We have seen that the legitimate field of science embraces all the mental and physical facts of the universe. But what are these facts in themselves and what is for us the criterion of their reality?

Let us start our investigation with some " external object," and as apparent simplicity will be satisfied by taking a familiar requisite of the author's calling, namely, a blackboard, let us take it.[1] We find an outer rectangular frame of brownish-yellow colour, which on closer inspection we presume to be wood, surrounding an inner fairly smooth surface painted black. We can measure a certain height, thickness, and breadth, we notice a certain degree of hardness, weight, resistance

---

[1] The blackboard as an " object-lesson " is such a favourite instance with the writer, that the reader will perhaps pardon him the use of it here. *Seine Mundart klebt jedem an.*

to breaking, and, if we examine further, a certain temperature, for the board feels to us cold or warm. Now although the blackboard at first sight appears a very simple object, we see that it at once leads us up to a very complex group of properties. In common talk we attribute all these properties to the blackboard, but when we begin to think over the matter carefully we shall find that it is by no means so simple as it seems to be. To begin with, I receive certain impressions of size and shape and colour by means of my organs of sight, and these enable me to pronounce with very considerable certainty that the object is a blackboard made of wood and coated with paint, even before I have touched or measured it. I *infer* that I shall find it hard and heavy, that I could if I pleased saw it up, and that I should find it to possess various other properties which I have learnt to *associate* with wood and paint. These inferences and associations are something which I *add* to the sight-impressions; and which I myself contribute from my past experience and put into the object—blackboard. I might have reached my conception of the blackboard by impressions of touch and not by those of sight. Blindfolded I might have judged of its size and shape, of its hardness and surface texture, and then have inferred its probable use and appearance, and associated with it all blackboard characteristics. In both cases it must be noted that a *sine quâ non* of the existence of an *actual* blackboard is some immediate sense-impression to start with. The sense-impressions which determine the reality of the external object may be very few indeed, the object may be largely constructed by inferences and associations, but *some* sense-impressions there must be if I am to term the object real, and not a product merely of my imagination. The existence of a certain number of sense-impressions leads me to infer the possibility of my receiving others, and this possibility I can, if I please, put to the test.

I have heard of the Capitol at Washington, and although I have never been to America, I am convinced of the reality of America and the Capitol— that is, I believe certain sense-impressions would be experienced by me if I put myself in the proper circumstances. In this case I have had indirect sense-impressions, contact with Americans, and with ships and chattels coming from America, which lead me to believe in the " reality " of America and of what my eyes or ears have told me of its contents. In constructing the Capitol it is clear that past experience of a variety of kinds is largely drawn upon. But it must be noted that this past experience is itself based upon sense-impressions of one kind or another. These sense-impressions have been as it were stored in the memory. A sense-impression, if sufficiently strong, leaves in our brain some more or less permanent trace of itself, which is rendered manifest in the form of association whenever an immediate sense-impression of a like kind recurs. The stored effects of past sense-impressions form to a great extent what we are accustomed to speak of as an " external object." On this account such an object must be recognized as largely constructed by ourselves ; we add to a greater or less number of immediate sense-impressions an associated group of stored sense-impresses. The proportion of the two contributions will depend largely on the keenness of our organs of sense and on the length and variety of our experience.

5

Owing to the large amount we ourselves contribute to most external objects, Professor Lloyd Morgan, in the able discussion of this matter in his *Animal Life and Intelligence* (p. 312) proposes to use the term *construct* for the external object. What for our present purpose, then, it is very needful to bear in mind is this : an external object is in general a construct—that is, a combination of immediate with past or stored sense-impressions. The reality of a thing depends upon the possibility of its occurring as a group of immediate sense-impressions.[1]

§ 2.—*Sense-Impressions and Consciousness.*

This conception of reality as based upon sense-impressions requires careful consideration and some reservations and modifications. Let us examine a little more closely what we are to understand by the word sense-impression. In turning round quickly in my chair, I knock my knee against a sharp edge of the table. Without any thought of what I am doing my hand moves down and rubs the bruised part, or the knee may cause me so much discomfort that I get up, think of what I shall do, and settle to apply some arnica. Now the two actions on my part appear of totally different character—at least on first

---

[1] The division between the real and unreal, and again between the real and ideal, is less distinct than many may think. For example, the planet Neptune passed from the ideal to the real, but the atom is still ideal. The ideal passes into the real when its perceptual equivalent is found, but the unreal can never become real. Thus the concepts of the metaphysicians, Kant's *thing in itself* or Clifford's *mind stuff* are in my sense of the words unreal (not ideal), they cannot become immediate sense-impressions, but the physical hypotheses as to the nature of matter are ideal (not unreal) for they do not lie absolutely outside the field of possible sense-impressions.

examination. In both cases physiologists tells us that as a primary stage a message is carried from the affected part by what is termed a *sensory nerve* to the brain. The manner in which this nerve conveys its message is without doubt physical, although its exact *modus operandi* is still unknown. At the brain what we term the sense-impression is formed, and there most probably some physical change takes place which remains with a greater or less degree of persistence in the case of those stored sense-impresses which we term memories. Everything up to the receipt of the sense-impression by the brain is what we are accustomed to term physical or mechanical, it is a legitimate inference to suppose that what from the psychical aspect we term memory, has also a physical side, that the brain takes for every memory a permanent physical impress, whether by change in the molecular constitution or in the elementary motions of the brain-substance, and that such physical impress is our stored sense-impress.[1] These physical impresses play an important part in the manner in which future sense-impressions of a like character are received. If these immediate sense-impressions be of sufficient strength, or amplitude as we might perhaps venture to say, they will call into some sort of activity a number of physical impresses due to past sense-impressions allied, or, to use a more suggestive word, *attuned* to the immediate sense-impression. The immediate sense-impression is conditioned by the physical impresses of the past,

---

[1] The closest physical analogies to the "permanent impresses" termed memory are the *set* and *after-strain* of the elastician. To assert that they are more than analogies would be to usurp the function of the physiologist.

and the general result is what has been termed a
"construct."

Besides the *sensory* nerves which convey the messages to the brain, there are other nerves which proceed from the brain and control the muscles termed
motor nerves. Through these motor nerves a message
is sent to my arm bidding it rub my bruised knee.
This message may be sent immediately or after my
fingers have been dipped in arnica. In the latter case
a very complex process has been gone through. I have
realized that the sense-impression corresponds to a
bruised knee, that arnica is good for a bruise, that a
bottle of arnica is to be found in a certain cupboard,
and so forth. Clearly the sense-impression has been
conditioned by a number of past impresses before
the motor nerve of the arm is called into play to rub
the knee. The process is described as thinking, and
as a variety of past experiences may come into play,
the ultimate message to the motor nerves appears to
us voluntary, and we call it an act of *will*, however
much it is really conditioned by the stored sense-impresses of the past. On the other hand, when, without apparently exciting any past sense-impresses,
the message from the sensory nerve no sooner reaches
the brain than a command is sent along the motor
nerve for the hand to rub the knee, I am said to act
involuntarily, from instinct or habit. The whole process may be so rapid, I may be so absorbed in my work,
that I never realized the message from the sensory
nerve at all. I do not even say to myself, " I have
knocked my knee and rubbed it." Only a spectator,
perhaps, has been conscious of the whole process of
knee-knocking and rubbing. Now this is in many
respects an important result. I can receive a sense-

impression without recognizing it, or a sense-impression does not involve consciousness. In this case there is no group of stored sense-impresses, no chain of what we term thoughts intervening between the immediate sense-impression and the message to the motor nerve. Thus what we term consciousness is largely, if not wholly, due to the stock of stored sense-impresses, and to the manner in which these condition the messages given to the motor nerves when a sensory nerve has conveyed a message to the brain. The measure of consciousness will thus largely depend on (1) the extent and variety of past sense-impressions, and (2) the degree to which the brain can permanently preserve the impress of these sense-impressions, or what might be termed the complexity and plasticity of the brain.

§ 3.—*The Brain as a Central Telephone Exchange.*

The view of brain activity here discussed may perhaps be elucidated by comparing the brain to the central office of a telephone exchange, from which wires radiate to the subscribers A, B, C, D, E, F, &c., who are senders, and to W, X, Y, Z, &c., who are receivers of messages. A, having notified to the company that he never intends to correspond with anybody but W, his wire is joined to W's, and the clerk remains unconscious of the arrival of the message from A and its dispatch to W, although it passes through his office.[1] There is indeed no call-bell. This

---

[1] If these wires were connected *outside* the office, we should have an analogy to certain possibilities of reflex action, which arise from sensory and motor nerves being linked before reaching the brain—*e.g.*, a frog's leg will be moved so as to rub an irritated point on its back even after the removal of the brain.

corresponds to an instinctive exertion following unconsciously on a sense-impression. Next the clerk finds
by experience that B invariably desires to correspond
with X, and consequently whenever he hears B's call-
bell he links him mechanically to X, without stopping
for a moment his perusal of *Tit-Bits*. This corresponds to a habitual exertion following unconsciously
on a sense-impression. Lastly, C, D, E, and F may
set their bells ringing for a variety of purposes ; the
clerk has in each case to answer their demands, but
this may require him to listen to the special com-
munications of these subscribers, to examine his lists,
his post-office directory, or any other source of information stored in his office. Finally, he shunts their
wires so as to bring them in circuit with those of Y and
Z, which seem to best suit the nature of the demands.
This corresponds to an exertion following consciously
on the receipt of a sense-impression. In all cases
the activity of the exchange arises from the receipt of
a message from one of a possibly great, but still finite
number of senders, A, B, C, D, &c. ; the originality of
the clerk is confined to immediately following their
behests or to satisfying their demands to the best of
his ability by the information stored in his office.
The analogy of course must not be pressed too far—
in particular senders and receivers must be considered
distinct, for sensory and motor nerves do not appear
to interchange functions. But the conception of the
brain as a central exchange certainly casts considerable
light not only on the action of sensory and motor
nerves, but also on thought and consciousness. Without sense-impressions there would be nothing to store ;
without the faculty of receiving permanent impress,
without memory, there would be no possibility of

thought ; and without this thought, this hesitation between sense-impression and exertion, there would be no consciousness. When an exertion follows immediately on a sense-impression we speak of the exertion as involuntary, our action as subject to the mechanical control of the "external object" to which we attribute the sense-impression. On the other hand, when the exertion is conditioned by stored sense-impresses we term our action voluntary. We speak of it as determined from "within ourselves," and assert the "freedom of our will." In the former case the exertion is conditioned solely by the immediate sense-impression ; in the latter it is conditioned by a complex of impressions partly immediate and partly stored. The past training, the past history and experience which mould character and determine the will, are really based on sense-impressions received at one time or another, and hence we may say that exertion, whether immediate or deferred, is the product directly or indirectly of sense-impressions.

§ 4.—*The Nature of Thought.*

There are still one or two points to be noted here. In the first place the immediate sense-impression is to be looked upon as the spark which kindles thought, which brings into play the stored impresses of past sense-impressions. But the complexity of the human brain is such, its stored sense-impresses are linked together in so many and diverse ways—partly by continual thinking, partly by immediate sense-impressions occurring in proximity and so linking together apparently discordant groups of past impresses—that we are not always able to recognize the relation be-

tween an immediate sense-impression and the resulting train of thought.   Nor, on the other hand, can we
always trace back a train of thought to the immediate
sense-impression from which it started.   Yet we may
take it for certain that elements of thought are ultimately the permanent impress of past sense-impressions, and that thought itself is started by immediate
sense-impressions.[1]

This statement must not be in any way supposed
to narrow the material of thought to those combinations of "external objects" which we associate with
immediate sense-impressions.   Thought once excited,
the mind passes with wonderful activity from one
stored impress to another, it classifies these impresses, analyzes or simplifies their characteristics,
and forms general notions of properties and modes.
It proceeds from the direct—what might perhaps be
termed the physical—association of memory, to the
indirect or mental association ; it passes from *perceiving* to *conceiving*.   The mental association, or recognition of relation between the stored impresses of past
sense-impressions has probably, if we could follow it,
as definite a physical side as the physical association
of immediate sense-impressions and past impresses.
But the physical side of the impress is only a
reasonable inference from the physical nature of the
immediate sense-impression, and we must therefore
content ourselves at present by considering it highly
probable that every process of thought has a physical

---

[1] The exact train of thought which follows an immediate sense-impression depends largely on the physical condition of the brain at the
time of its receipt, and is further largely conditioned by the mode in
which stored sense-impresses have been excited in the past, *i.e.*, the
memory exercised.

aspect, even if we are very far as yet from being able to trace it out.

This process of mental association we can only recognize as certainly occurring in our individual selves. The reason why we infer it in others we shall consider later. The amount of it, however, in our individual selves must largely depend on the variety and extent of our stored impresses, and further on the individual capacity for thinking, or on the form and development of the physical organ wherein the process of thinking takes place, on the brain. The brain in the individual man is probably considerably influenced by heredity, by health, by exercise, and by other factors, but speaking generally the physical instruments of thought in two normal human beings are machines of the same type, varying indeed in efficiency, but not in kind or function. For the same two normal human beings the organs of sense are also machines of the same type and thus within limits only capable of conveying the same sense-impressions to the brain. Herein consists the similarity of the universe for all normal human beings. The same type of physical organ receives the same sense-impressions and forms the same "constructs." Two normal perceptive faculties construct practically the same universe. Were this not true the results of thinking in one mind would have no validity for a second mind. The universal validity of science depends upon the similarity of the perceptive and reasoning faculties in normal civilized men.

The above discussion of the nature of thought is not of course to be looked upon as final, or as offering any real explanation of the psychical side of thought. It is merely intended to suggest the manner in which we may consider thought to be conditioned by its physical aspects. What the actual relations between the psychical and physical sides of thought are, we do not know, and, as in all such cases, it is best to directly confess our ignorance. It is no use, indeed only dangerous, in the present state of our knowledge with regard to psychology and the physics of the brain, to fill the void of ignorance by hypotheses which can neither be proven nor refuted. Thus if we say that thought and motion are the same thing seen from different sides, we make no real progress in our analysis for we can form no conception whatever as to what the nature in itself of this thing may be. Indeed, if we go further and compare thought and motion to the concave and convex sides of the same surface, we may do positive harm rather than good ; for convexity and concavity are not when accurately defined by the mathematician different qualities, but only degrees of the same quantity, curvature, passing the one into the other through zero-curvature or flatness. On the other hand the distinction between the psychical and physical aspects of brain activity seems to be essentially one of quality, not of degree. It is better to content ourselves in the present state of our knowledge by remarking that in all probability sense-impressions lead to certain physical (including under this term possible chemical) activities of the brain, and that these activities are recognized by each individual *for himself only* under the form of thought. Each individual recognizes his own consciousness, perceives that the interval between sensation and exertion is occupied by a certain psychical process. We recognize consciousness in our individual selves, we *assume* it to exist in others,

§ 5.—*Other—Consciousness as an Eject.*

The assumption just referred to is by no means of the same nature as that which we make every moment in the formation of what we have termed constructs from a limited group of immediate sense-impressions. I see the shape, size, and colour of the blackboard, and I *assume* that I shall find it hard and heavy. But here the assumed properties are capable of being put to the direct test of immediate sense-impression. I can touch and lift the blackboard and complete my analysis of its properties. Even the Capitol in Washington, of which I have had no direct sense-impression, is capable of being put to the same sort of direct test. Another man's consciousness, however, can never, it is said, be directly perceived by sense-impression, I can only *infer* its existence from the apparent similarity of our nervous systems, from observing the same hesitation in his case as in my own between sense-impression and exertion, and from the similarity between his activities and my own. The inference is really not so great as the metaphysicians would wish us to believe. It is an inference ultimately based on the physical fact of the interval between sense-impression and exertion ; and though we cannot as yet physically demonstrate another person's consciousness, neither can we demonstrate physically that earth-grown apples would fall at the surface of the planet of a fixed star or that atoms really are a stage in the resolution of matter. It may be suggested that if our organs of sense were finer, or our means of locomotion more complete, we might be able to see atoms or to carry earth-grown apples to a fixed star—in other words, to test physically, or by immediate sense-impression these inferences. But :—

" When I come to the conclusion that *you* are conscious, and that there are objects in your consciousness similar to those in mine, I am not inferring any actual or possible feelings of my own, but *your* feelings, which are not, and cannot by any possibility become, objects in my consciousness." [1]

To this it may be replied, that, were our physiological knowledge and surgical manipulation sufficiently complete, it is conceivable that it would be possible for me to be conscious of your feelings, to recognize your consciousness as a direct sense-impression ; let us say, for example, by connecting the *cortex* of your brain with that of mine through a suitable commissure of nerve-substance. The possibility of this physical verification of other-consciousness does not seem more remote than that of a journey to a fixed star. Indeed, there are some who think that without this hypothetical nerve connection the processes popularly termed, " anticipating another person's wishes," "reading his thoughts," &c., have in them the elements of a sense-impression of other-consciousness, and are not entirely indirect inferences from practical experience.

Clifford has given the name *eject* to existences which, like other-consciousness, are only inferred, and the name is a convenient one. At the same time it seems to me doubtful whether the distinction between *object* (what might possibly come to my consciousness as a direct sense-impression) and *eject* is so marked as he would have us to believe. The complicated physical motions of another person's brain, it is admitted, might possibly be objective realities to me ; but on the other hand might not the hypothetical brain

[1] W. K. Clifford, " On the Nature of Things in Themselves," *Lectures and Essays*, vol. ii. p. 72.

commissure render me just as certain of the workings of another person's consciousness as I am of my
own? In this respect, therefore, it does not seem
necessary to assert that consciousness lies outside the
field of science, or must perforce escape the methods
of physical experiment and research. We may be far
enough removed from knowledge at the present time,
but I see no logical hindrance to our asserting that in
the dim future we might possibly obtain objective
acquaintance with what at present appears merely as
an eject. We can do this indeed without any
dogmatic assumption that psychical effects can all be
reduced to physical motion. Psychical effects are
without doubt excited by physical action, and our
only assumption is the not unreasonable one, that a
suitable physical link might transfer an appreciation
of psychical activity from one psychical centre to
another.

§ 6.—*Attitude of Science towards Ejects.*

Indeed in some respects other-consciousness appears
less beyond our reach than many inferred existences.
Some physicists infer the existence of atoms, although
they have had no experience of any individual atom,
because the hypothesis of their existence enables them
to briefly resume a number of sense-impressions. We
infer the existence of other-consciousness for a precisely
similar reason; but in this case we have the advantage
of knowing at least one individual consciousness,
namely, our own. We see in ourselves how it links
sense-impression and deferred exertion. While the
atom, like other-consciousness, might possibly some
day, attain to objective reality, there are certain conceptions dealt with by science, for which, as we shall

see in the sequel, this is impossible. For example, our geometrical ideas of curves and surfaces are of this character. None the less, although they might with greater logic be termed *ejects* than, perhaps, other-consciousness, there are few who would deny that they have their ultimate origin in sense-impressions, from which they have been extracted or isolated by the process of mental generalization, to which we have previously referred (p 56). A still more marked case of conceptions, which we are incapable of verifying directly by any form of immediate sense-impression, is that of historical facts. We believe that King John really signed *Magna Charta*, and that there was a period when snow-fields and glaciers covered the greater part of England, yet these conceptions can never have come to our consciousness as direct sense-impressions, nor can they be verified in like manner. They are conclusions we have reached by a long chain of inferences, starting in direct sense-impressions and ending in that which, unlike atom and other-consciousness, can by no possibility be verified directly by immediate sense-impression. When, therefore, we state that all the contents of our mind are ultimately based on sense-impressions, we must be careful to recognize that the mind has by classification and isolation proceeded to conceptions which are widely removed from sense-impressions capable of immediate verification. The contents of the mind at any instant are very far from being identical with the range of actual or possible sense-impressions at that instant. We are perpetually drawing inferences from our immediate and stored sense-impresses as to things which lie beyond immediate verification by sense ;—that is, we infer the existence of things which do not belong to the objective world, or which at any rate cannot be directly verified by immediate sense-impression as belonging to it at the present moment. Strange as it may seem, science is largely based upon inferences of this kind ; its hypotheses lie to a great extent beyond the region of the immediately sensible, and it chiefly deals with conceptions drawn from sense-impressions, and not with sense-impressions themselves.

This point needs to be specially emphasized, for we are often told that the scientific method applies only to the external world of phenomena, and that the legitimate field of science lies solely among immediate sense-impressions. The object of the present work is to insist on a directly contrary proposition, namely, that science is in reality a classification and analysis of the contents of the mind ; and the scientific method consists in drawing just comparisons and inferences from stored sense-impresses and the conceptions based upon them. Not till the immediate sense-impression has reached the level of a conception, or at least a perception, does it become material for science. In truth, the field of science is much more consciousness than an external world. In thus vindicating for science its mission as interpreter of conceptions rather than as investigator of a " natural law " ruling an " external world of material," I must remind the reader that science still considers the whole contents of the mind to be ultimately based on sense-impressions. Without sense-impressions there would be no consciousness, no conceptions for science to deal with. In the next place we must be careful to note that not every conception, still less every inference, has scientific validity.

§ 7.—*The Scientific Validity of a Conception.*

In order that a conception may have scientific validity, it must be self-consistent, and deducible from the perceptions of the normal human being. For instance, a centaur is not a self-consistent conception ; as soon as our knowledge of human and equine anatomy became sufficiently developed, the centaur became an unthinkable thing—a self-negating idea. As the man-horse is seen to be a compound of sense-impressions, which are irreconcilable anatomically, so the man-god, whose cruder type is Hercules, is also seen to be a chimera, a self-contradictory conception, as soon as we have clearly defined the physical and mental characteristics of man. But even if an individual mind has reached a conception, which at any rate for that mind is perfectly self-consistent, it does not follow that such a conception must have scientific validity, except as far as science may be concerned with the analysis of that individual mind. When a person conceives that one colour—green—suffices to describe the flowers and leaves of a rose-tree in my garden, I know that his conception may not, after all, be self-inconsistent, it may be in perfect harmony with his sense-impressions. I merely assert that his perceptive faculty is *abnormal*, and hold him to be colour-blind. I may study the individual abnormality scientifically, but his conception has no scientific validity, for it is not deducible from the perceptions of the normal human being. Here indeed we have to proceed very cautiously if we are to determine what self-consistent conceptions have scientific validity. Above all, we must note that a conception does not cease to be valid because it has not been deduced by the majority of normal human beings from their perceptions. The conception that a new individual originates in the union of a male and female cell may never have actually been deduced by a majority of normal human beings from their perceptions. But if any normal human being be trained in the proper methods of observation, and be placed in the right circumstances for investigating, he will draw from his perceptions this conception and not its negation. It is in this sense, therefore, that we are to understand the assertion that a conception to have scientific validity must be *deducible* from the perceptions of the normal human being.

The preceding paragraph shows us how important it is that the observations and experiments of science should be repeated as often and by as many observers as possible, in order to ensure that we are dealing with what has validity for all normal human beings, and not with the results of an abnormal perceptive faculty. It is not only, however, in experiments or observations which can be repeated easily, but still more in those which it is very difficult or impossible to repeat that a great weight of responsibility lies upon the recorder and the public which is called upon to accept his results. An event may have occurred in the presence of a limited number of observers. That the event itself cannot recur, and is totally out of accord with our customary experience, are not sufficient grounds for disregarding it scientifically. Yet what an onus is laid on the individual observers to test whether their perceptive faculties were normal on the occasion, and whether their conceptions of what took place were justified by their perceptions! Still greater onus is laid on men at large to criticize and probe the evidence given

6

by such observers, to question whether they were men trained to observe, and calm and collected at the time of the reported event. Were they not, perhaps, in an exalted state of mind, biased by pre-conceptions or hindered by the physical surroundings from clear perception? In short, were or were not their perceptive faculties in a normal condition and under normal circumstances? It can scarcely be questioned that when the truth or falsehood of an event or observation may have important bearings on conduct, over-doubt is more socially valuable than over-credulity.[1] In an age like our own, which is essentially an age of scientific inquiry, the prevalence of doubt and criticism ought not to be regarded with despair or as a sign of decadence. It is one of the safeguards of progress ;—*la critique est la vie de la science*, we must again repeat. One of the most fatal (and not so impossible) futures for science, would be the institution of a scientific hierarchy which would

---

[1] A good example of another class of experiment, that which it is difficult or unadvisable to repeat frequently, may be drawn from Brown-Séquard's researches on the inheritance by guinea-pigs of diseases acquired by their parents during life. These researches were conducted on a large scale and with great expenditure of time and animal life. (Brown-Séquard kept upwards of five hundred guinea-pigs at once.) Yet we must confess that if these experiments were conducted with every precaution that self-criticism might suggest, the "degrading effect" of inflicting disease and pain on this large amount of animal life would have been more than compensated by the light which the experiments might have cast on the socially important problem of the inheritance of acquired characteristics. Unfortunately Brown-Séquard's conceptions and inferences do not appear to many biologists valid, and there lies upon this investigator the onus of proving that (1) all possible precautions for the accuracy of the results were actually taken, and (2), being taken, that the experiments were such as could reasonably have been supposed capable of solving the problems proposed.

brand as heretical all doubt as to its conclusions, all criticism of its results.

§ 8.—*The Scientific Validity of an Inference.*

Much of what we have just said with regard to the scientific validity of conceptions holds with regard to the scientific validity of inferences, for conceptions pass imperceptibly into inferences. The scope of the present work will only permit us to discuss briefly the limits of legitimate inference and induction. For a fuller discussion the reader must be referred to treatises on logic, in particular to the chapters on inference and induction in Stanley Jevons' *Principles of Science* (chapters iv.–vii., x.–xii., especially). In the first place the inference which is scientifically valid is that which could be drawn by every logically trained normal mind, if it were in possession of the conceptions upon which the inference has been based. Stress must here be laid on the distinction between "*could* be drawn" and "actually *would* be drawn." There are many minds which have clearly defined conceptions, but refuse either from inertia or emotional bias to draw the inferences from them which can be drawn. A scientific inference—witness Darwin's natural selection,—however logical, often takes years to overcome the inertia of the scientific world itself, and longer still may be the period before it forms an essential factor of the thought of the majority of normal-minded human beings. Yet, while logically trained minds which are able to draw inferences frequently neglect to do so, the illogically trained, on the other hand, unfortunately devote a large part of their ill-regulated energies to the production of cobwebs of inference; and this with such rapidity that the logical broom fails to keep pace with their activity. The mediæval superstitions are scarce discredited, before they reappear as theosophy or spiritualism.

The assumption which lies at the bottom of most popular fallacious inference might pass without reference, for it is obviously absurd, were it not, alas! so widely current. The assumption is simply this : that the strongest argument in favour of the truth of a statement is the absence or impossibility of a demonstration of its falsehood. Let us note some of its products:—All the constituents of material bodies are to be found in the atmosphere ; it is impossible to assert that these constituents could not be brought together.[1] *Ergo*, the Mahatmas of Thibet, can take upon themselves material forms in St. John's Wood.—Science cannot demonstrate that the uniform action of material causes precludes the hypothesis of a benevolent Creator. *Ergo*, the impulses and hopes of men receive confirmation from science.—Consciousness is found associated with matter ; we cannot demonstrate that consciousness is not found with *all* forms of matter. *Ergo,* all matter is conscious, or matter and mind are never found except in conjunction, and we may legitimately speak of the " consciousness of society " and the "consciousness of the universe." These are but a few samples of the current method of fallacious inference—usually, be it remarked, screened beneath an unlimited flow of words, and not thus exhibited in its naked absurdity. When we recognize how widely inferences of this

---

[1] "That is a noteworthy fact which I have not fully appreciated before," remarks the untrained mind, and is already more than half-converted to theosophy.

character affect conduct in life, and yet grasp how unstable must be the basis of such conduct, how liable to be shaken to the foundations by the first stout logical breeze, then we understand how honest doubt is far healthier for the community, is more social, than unthinking inference, light-hearted and over-ready belief. Doubt is at least the first stage towards scientific inquiry ; and it is better by far to have reached that stage than to have made no intellectual progress whatever.

§ 9.—*The Limits to Other-Consciousness.*

We cannot better illustrate the limits of legitimate inference than by considering the example we have last cited, and asking how far we may infer the existence of consciousness and of thought. We have seen (p. 52) that consciousness is associated with the process which *may* intervene in the brain between the receipt of a sense-impression from a sensory nerve and the dispatch of a stimulus to action through a motor nerve. Consciousness is thus associated with machinery of a certain character, which we term the brain and nerves. Further, it depends upon the lapse of an interval between sense-impression and exertion, this interval being filled, as it were, with the mutual resonance and cling-clang of stored sense-impresses and the conceptions drawn from them. Where no like machinery, no like interval can be observed, there we have no right to infer any consciousness. In our fellow-men we observe this same machinery and the like interval, and we infer consciousness, it may be as an eject, but as an eject which, as we have seen (p. 60), might not inconceivably, however improbably, become some day an object. In the lower forms of life we observe machinery approximately like our own, and a shorter and shorter interval between sense-impression and exertion ; we may reasonably infer consciousness, if in reduced intensity. We cannot, indeed, put our finger on a definite type of life and say here consciousness ends, but it is completely illogical to infer its existence where we can find no interval between sense-impression and exertion, or where we can find no nervous system. Because we cannot point to the exact form of material life at which consciousness ceases, we have no more right to infer that consciousness is associated with all life, still less with all forms of matter, than we have to infer that there must always be wine mixed with water, because so little wine can be mixed with water that we are unable to detect its presence. Will, too, as we have seen, is closely connected with consciousness ; it is the feeling in our individual selves when exertion flows from the stored sense-impresses " within us," and not from the immediate sense-impression which we term " without us." We are justified, therefore, in inferring the feeling of will as well as consciousness in nervous systems more or less akin to our own; we may throw them out from ourselves, *eject* them into certain forms of material life. But those who eject them into matter, where no nervous system can be found, or even into existences which they postulate as immaterial, are not only exceeding enormously the bounds of scientific inference, but forming conceptions which, like that of the centaur, are inconsistent in themselves. From will and consciousness associated with material machinery, we can infer nothing whatever as to will and consciousness without that machinery. We are passing by the trick of a common-name to things of which we can postulate absolutely nothing, and of which we are only unable to deny the existence when we give to that term a meaning wholly opposed to the customary one.[1]

§ 10.—*The Canons of Legitimate Inference.*

We cannot here discuss more fully the limits of belief and legitimate inference. We shall, however, to some extent return to the subject when considering *Causation* and *Probability* in Chapter IV. But it may not be without service to state certain canons of legitimate inference with a few explanatory remarks, leaving the reader, if he so desire, to pursue the subject further in Stanley Jevons' *Principles of Science*, or in Clifford's essay on *The Ethics of Belief.* We ought first to notice that the use of the word *belief* in our language is changing : formerly it denoted something taken as definite and certain on the basis of some external authority ; now it has grown rather to denote credit given to a statement on a more or less sufficient balancing of probabilities.[2]

The change in usage marks the gradual transition of the basis of conviction from uncriticizing faith to

[1] Consciousness without a nervous system is like a horse without a belly—a chimera, of which in customary language we deny the " existence." We cannot demonstrate that a horse without a belly may not exist " outside " the physical universe, only it would not be a horse and would exist " nowhere." The existence of something of which we can postulate nothing at nowhere can never be inferred from conceptions based on sense-impressions. Such a horse would be like Meister Eckehart's deity who was a non-god, a non-spirit, a non-person, a non-idea, and of whom, he says, any assertion must be more false than true.

[2] Compare the older use in Biblical passages, such as " Jacob's heart

weighed probability.   The canons we have referred to
are the following :—

1. Where it is impossible to apply man's reason,
that is to criticize and investigate at all, there it is not
only unprofitable, but antisocial to believe.

Belief is thus to be looked upon as an adjunct to
knowledge: as a guide to action where decision is
needful, but the probability is not so overwhelming as
to amount to knowledge.   To believe in a sphere
where we cannot reason is antisocial, for it is a matter
of common experience that such belief prejudices
action in spheres where we can reason.

2. We may infer what we cannot verify by direct
sense-impression only when the inference is from
known things to unknown things of the like nature
in similar surroundings.

Thus we may not infer an "infinite" consciousness
outside the physical surroundings of finite consciousness ; we may not infer man in the moon, however
like in nature to ourselves, because the physical surroundings in the moon are not such as we find man in
here, &c., &c.

3. We may infer the truth of tradition when its
contents are of like character and continuous with
men's present experience, and when there is reasonable ground for supposing its source to lie in persons
knowing the facts and reporting what they knew.

The tradition that Wellington and Blücher won the
battle of Waterloo fulfils the necessary conditions,

---

fainted for he believed them not," and "Except ye see signs and
wonders ye will not believe," or in Locke's definition of belief as
adherence to a proposition of which one is *persuaded* but does not know
to be true, with such modern usage as : "I believe that you will find a
cab on the stand, and that the train starts at half-past eight."

while the miracle of Karl the Great and the adder fulfils neither condition.

4. While it is reasonable in the minor actions of life, where rapidity of decision is important, to infer on slight evidence and believe on small balances of probability, it is opposed to the true interests of society to take as a permanent standard of conduct a belief based on inadequate testimony.

This canon suggests that the acceptance, as habitual guides to conduct, of beliefs based on insufficient evidence, must lead to the want of a proper sense of the individual's responsibility for the important decisions of life. I have no right to believe at seven o'clock that a cab will be on the stand at eight o'clock, if my catching the train at half-past is of vital importance to others.

§ 11.—*The External Universe.*

Before we draw from our present discussion any conclusions as to the facts of science we must return once more to the immediate sense-impression and examine its nature a little more closely. We are accustomed to talk of the "external world," of the "reality" outside us. We speak of individual objects having an existence independent of our own. Stored sense-impressions, our thoughts and memories, although most probably they have beside their psychical element a close correspondence with some physical change or impress in the brain, are yet spoken of as *inside* ourselves. On the other hand, although if a sensory nerve be divided anywhere short of the brain we lose the corresponding sense-impression, we yet speak of many sense-impressions such as form

and texture as existing outside ourselves. How close
can we then actually get to this supposed world outside ourselves? Just as near as but no nearer than
the brain terminals of the sensory nerves. We are
like the clerk in the central telephone exchange who
cannot get nearer to his customers than his end of
the telephone wires. We are, indeed, worse off than
the clerk, for to carry out the analogy properly we
must suppose him *never to have been outside the
telephone exchange, never to have seen a customer or
any one like a customer—in short, never, except through
the telephone wire, to have come in contact with the outside universe.* Of that "real" universe outside himself
he would be able to form no direct impression; the
real universe for him would be the messages which
flowed from the ends of the telephone wires in his
office. About those messages and the ideas raised
in his mind by them he might reason and draw his
inferences; and his conclusions would be correct—for
what? For the world of telephonic messages, for the
type of messages which go through the telephone.
Something definite and valuable he might know with
regard to the spheres of action and of thought of his
telephonic subscribers, but outside those spheres he
could have no experience. Pent up in his office he
could never have seen or touched even a telephonic
subscriber *in himself.* Very much in the position of
such a telephonic clerk is the conscious *ego* of each
one of us seated at the brain terminals of the sensory
nerves. Not a step nearer than those terminals can
the *ego* get to the "outer world," and what in and
for themselves are the subscribers to its nerve exchange it has no means of ascertaining. Messages in
the form of sense-impressions come flowing in from

that "outside world" and these we analyze, classify, store up, and reason about. But of the nature of "things-in-themselves" of what may exist at the other end of our system of telephone wires we know nothing.

But the reader, perhaps, remarks, "I not only see an object, but I can *touch* it. I can trace the nerve from the tip of my finger to the brain. I am not like the telephone clerk, I can follow my network of wires to their terminals and find what is at the other end of them." Can you, reader? Think for a moment whether your *ego* has for one moment got away from his brain-exchange. The sense-impression that you call touch was just as much as sight felt only at the brain end of a sensory nerve. What has told you also of the nerve from the tip of your finger to your brain? Why sense-impressions also, messages conveyed along optic or tactile sensory nerves. In truth, all you have been doing is to employ one subscriber to your telephone exchange to tell you about the wire that goes to a second, but you are just as far as ever from tracing out for yourself the telephone wires to the individual subscriber and ascertaining what his nature is in and for himself. The immediate sense-impression is just as far removed from what you term the "outside world" as the stored impress. If our telephone clerk had recorded by aid of a phonograph certain of the messages from the outside world on past occasions, then if any telephonic message on its receipt set several phonographs repeating past messages, we have an image analogous to what goes on in the brain. Both telephone and phonograph are equally removed from what the clerk might call the "real outside world," but they enable him through their sounds to construct a universe; he projects those sounds, which are really inside his office, outside his office and speaks of them as the external universe. This outside world is constructed by him from the contents of the inside sounds, which differ as widely from things-in-themselves, as language, the symbol, must always differ from the thing it symbolizes. For our telephone clerk sounds would be the real world, and yet we can see how conditioned and limited it would be by the range of his particular telephone subscribers and by the contents of their messages.

So it is with our brain ; the sounds from telephone and phonograph correspond to immediate and stored sense-impressions. These sense-impressions we project as it were outwards and term the real world outside ourselves. But the things-in-themselves which the sense-impressions symbolize, the "reality," as the metaphysicians wish to call it, at the other end of the nerve remains unknown and is unknowable. Reality of the external world lies for science and for us in form and colour and touch—sense-impressions as widely divergent from the thing "at the other end of the nerve" as the sound of the telephone from the sub-scriber at the other end of the wire. We are cribbed and confined in this world of sense-impressions like the exchange clerk in his world of sounds, and not a step beyond can we get. As his world is conditioned and limited by his particular network of wires, so ours is conditioned by our nervous system, by our organs of sense. Their peculiarities determine what is the nature of the outside world which we construct. It is the similarity in the organs of sense and in the perceptive faculty of all normal human beings which makes the outside world the same, or *practically* the same for them all.[1] To return to the old analogy, it is as if two telephone exchanges had very nearly identical groups of subscribers. In this case a wire between the two exchanges would soon convince the imprisoned clerks that they had something in common and peculiar to themselves. That conviction corresponds in our comparison to the recognition of other-consciousness.

§ 12.—*Outside and Inside Myself.*

We are now in a position to see clearly what is meant by " reality " and the " external world." Any group of immediate sense-impressions we project outside ourselves and hold to be part of the external world. As such we call it a *phenomenon*, and in practical life term it *real*. Together with the immediate sense-impression we often include stored sense-impresses, which experience has taught us to associate with the immediate sense-impression. Thus we assume the blackboard to be *hard*, although we may only have seen its shape and colour. What we term the real world is thus partly based on immediate sense-impressions, partly on stored sense-impresses ; it is what has been called a *construct*. For an individual the distinction between the real world and his thought of it is the presence of some immediate sense-impression. Thus the distinction of what is " outside " and what is " inside " myself at any instant depends entirely on the amount of immediate sense-impression. This has been very cleverly represented by the well-known German scientist, Professor Ernst Mach,

[1] Not *exactly* the same, for the range of the organs of sense and the powers of perception vary somewhat with different individual men, and probably enormously, if we take other life into account.

in the accompanying sketch. The professor is lying
on his sofa, and having closed his right eye, the picture
represents what is presented to his left eye :—

"In a frame formed by the ridge of my eyebrow, by

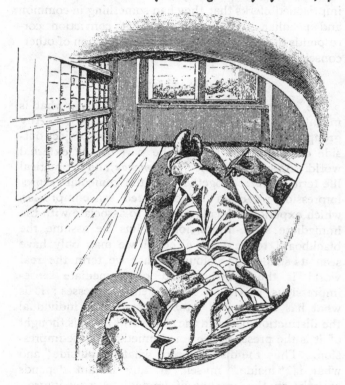

FIG. 1.

my nose, and my moustache, appears a part of my
body, so far as it is visible, and also the things and
space about it. . . . If I observe an element, A, within
my field of vision, and investigate its connection with
another element, B, within the same field, I go out of

the domain of physics into that of physiology or psychology, if B, to use the apposite expression that a friend of mine employed upon seeing this drawing, passes through my skin." [1]

From our standpoint, neglecting for simplicity the immediate contributions of any other senses than that of sight, the picture represents that part of the Professor's sense-impressions which for the instant forms his "outside world"; the rest was "inside"—existed for him only as stored sense-impresses.

There is no better exercise for the mind than the endeavour to reduce the perception we have of "external things" down to the simple sense-impressions by which we feel them. The arbitrary distinction between outside and inside ourselves is then clearly seen to be one merely of everyday practical convenience. Take a needle; we say it is thin, bright, pointed, and so forth. What are these properties but a group of sense-impressions relating to form and colour associated with conceptions drawn from past sense-impressions? Their immediate source is the activity of certain optic nerves. These sense-impressions form for us the *reality* of the needle. Nevertheless, they and the resulting construct are projected outside ourselves, and *supposed* to reside in an external thing, "the needle." Now by mischance we run the needle into our finger; another nerve is excited and an unpleasant sense-impression, one which we term painful, arises. This, on the other hand, we term "in ourselves," and do not project into the needle. Yet the colour and form which constitute for us the needle are just as much sense-impressions within us as the

---

[1] " The Analysis of the Sensations—Anti-metaphysical," *The Monist*, vol. i. p. 59.

pain produced by its prick. The distinction between
ourselves and the outside world is thus only an arbitrary, if a practically convenient, division between one
type of sense-impression and another. The group of
sense-impressions forming what I term *myself* is only
a small subdivision of the vast world of sense-impressions. My arm is paralyzed, I still term it part of
me ; it mortifies, I am not quite so certain whether it
is to be called part of me or not ; the surgeon cuts
it off, it now ceases to be a part of that group of
sense-impressions which I term " myself." Obviously
the distinction between " outside " and " inside," between one individuality and a second, is only a
practical one. How many of the group of sense-
impressions we term a tree are light and atmosphere
effects ? What might be termed the limits of the
group of sense-impressions which we term an individual cannot be scientifically drawn. But to this
point we shall return later.

§ 13.—*Sensations as the Ultimate Source of the Materials*
of Knowledge.

When we find that the mind is entirely limited to
the one source, sense-impression, for its contents, that
it can classify and analyze, associate and construct
but always with this same material, either in its immediate or stored form, then it is not difficult to
understand what, and what only, can be the facts of
science, the subject-matter of knowledge. Science,
we say at once, deals with conceptions drawn from
sense-impressions, and its legitimate field is the whole
content of the human mind. Those who assert that
science deals with the world of external phenomena
are only stating a half-truth. Science only appeals to

the world of phenomena—to immediate sense-impressions—with the view of testing and verifying the accuracy of its conceptions and inferences, the ultimate basis of which lies as we have seen in such immediate sense-impressions. Science deals with the mental, the "inside" world, and the aim of its processes of classification and inference is precisely that of instinctive or mechanical association, namely, to enable the exertion, best calculated to preserve the race and the individual, to follow on the sense-impression with the least expenditure of time and of intellectual energy. Science is in this respect an economy of thought— a delicate tuning in the interests of the mind of the organs which receive sense-impressions and those which expedite activity.

Turn the problem round and ponder over it as we will, beyond the sense-impression, beyond the brain terminals of the sensory nerves we cannot get. Of what is beyond them, of " things-in-themselves," as the metaphysicians term them, we can know but one characteristic, and this we can only describe as a capacity for producing sense-impressions, for sending messages along the sensory nerves to the brain. This is the sole scientific statement which can be made with regard to what lies beyond sense-impressions. But even in this statement we must be careful to analyze our meaning. The methods of classification and inference, which hold for sense-impressions and for the conceptions based upon them, cannot be projected outside our minds, away from the sphere in which we know them to hold, into a sphere which we have recognized as unknown and unknowable. The laws, if we can speak of laws, of this sphere must be as unknown as

7

its contents, and therefore to talk of its contents as
producing sense-impressions is an unwarranted inference, for we are asserting *cause and effect*—a law of
phenomena or sense-impressions—to hold in a region
beyond our experience.[1] We *know* ourselves, and we
know around us an impenetrable wall of sense-impressions. There is no necessity, nay, not even logic, in
the statement that behind sense-impressions there are
" things-in-themselves " *producing* sense-impressions.
Of this supersensuous sphere we may philosophize and
dogmatize unprofitably, but we can never know use-
fully. It is indeed an unjustifiable extension of the
term knowledge to apply it to something which cannot be part of the mind's contents. What is behind
or beyond sense-impressions may or may not be of
the same character as sense-impressions, we cannot
say. We feel the *surface* of a body to be soft, but its
core may be either hard or soft, we cannot say ; we
can only legitimately call it a soft-surfaced body. So
it is with sense-impressions and what may be behind
them; we can only say sense-impression-stuff, or, as
we shall term it, with a somewhat divergent meaning
from the customary, *sensation*. By sensation we shall
accordingly understand that of which the only knowable side is sense-impression. Our object in using
the word *sensation* instead of sense-impression will be
to express our ignorance, our absolute agnosticism,
as to whether sense-impressions are " produced " by
unknowable " things - in - themselves," or whether
behind them may not be something of their own
nature.[2] The outer world is for science a world of

---

[1] This will appear clearer when we have discussed the scientific
meaning of *cause and effect*.   See Chapter IV.

[2] Herein lies the arid field of metaphysical discussion.   Behind

sensations, and sensation is known to us only as sense-impression.

§ 14.—*Shadow and Reality.*

The reader who comes to these problems for the first time may feel inclined to assert that if this world of sense-impressions is the world of scientific knowledge, then science is dealing with a world of shadows and not of real substances. And yet, if such a reader will think over what happens when he knocks his elbow against the table, I think he will agree that it is the sense-impressions of hardness, and perhaps of pain, which are for him the realities, while the table, as a " source of these sense-impressions," is the shadow. Should he impatiently retort: " I see the table—four-legged, brass-handled, with black oak top shining under the elbow-grease of a past generation —there is the reality," let him stop for a moment to inquire whether his reality is not a construct from immediate and stored sense-impressions, of exactly the same character as the previous sense-impression of hardness. He will soon convince himself that the *real* table lies for him in the permanent association of a certain group of sense-impressions, and that the shadow table is what might be left were this group abstracted.

sense-impressions, and as their source, the materialists place *Matter ;* Berkeley placed *God ;* Kant, and after him Schopenhauer, placed *Will ;* and Clifford placed *Mind-stuff.* Professor E. Mach in the paper referred to on p. 79 has reduced the outer world to its known surface, sense-impression, which he terms sensation—leaving no possible unknowable *plus* which we intend to signify by our use of the word sensation. Such a theory cannot lead to scientific error, but it does not seem a justifiable inference from sense-impression. The variety of inferences cited above shows the quagmire which has to be avoided, especially when the inferences are drawn with a view of influencing judgment in the world of sense.

Let us return for a moment to our old friend the blackboard, represented for us by a complex of properties (p. 48). In the first place we have size and shape, then colour and temperature, and, lastly, other properties like hardness, strength, weight, &c. Clearly the blackboard consists for us in the permanent association of these properties, in a construct from our sense-impressions. Take away the size and shape, leaving all the other properties, and the group has ceased to be the blackboard, whatever else it may be. Suppose the colour to go and again the blackboard has ceased to be. Finally, if the hardness and weight were to vanish, we might *see* the ghost of a blackboard, but we should soon convince ourselves that it was not the "reality" we had termed blackboard. Now, as the reader may be thinking that this blackboard has had too long an existence, at least in our pages, let us employ a carpenter to pull it to pieces and construct out of it a four-legged table. To cloak the obvious deficiences of such a table we will cause it to be coated with a thick layer of Aspinall's enamel. We have now a four-legged red table. It is no longer a blackboard, and any person not knowing its origin would think us quite mad if we termed it a blackboard. We should probably, however, make ourselves intelligible to him by stating that the "same material" as was once in a blackboard is now in the red table. For practical purposes this is very proper and convenient, but will it help us to an accurate conception of individuality, if we say the blackboard and the table are the *same* thing? New paint and probably nails have been added; the carpenter may have supplied some additional wood; nay, more, if we begin to use our table a leg may come off and a new one be put on ; after a time a fresh top would be an advantage, thus even the "material" of the table may cease to be same as that of the blackboard. Or again, since our table is probably a bad one, we will break it up and burn it, and so the blackboard will be converted into various gases and some ashes. What has now become of it ? Size and shape, temperature and colour, hardness and strength have all gone. It is true that the chemist asserts that, if we could completely collect the gases and ashes, one sense-impression at least, that of weight, would remain the same in these and the original blackboard. But can we define sameness to consist in the permanence of some one sub-group of sense-impressions, notwithstanding the divergence of the majority? That permanence may be a link in the succession of our sense-impressions, but it can hardly be taken as a basis for defining individuality. *If* the gases and ashes could be collected ! They have, indeed, been scattered to the winds and in course of time may be absorbed by other vegetable life, ultimately, perhaps, to reappear as other blackboards, or even in legs of mutton. What has become of the "thing-in-itself" behind the group of sense-impressions we termed the original blackboard ? Surely there is less permanence in it than in our sense-impressions of the blackboard —far less than in that purely mental conception of sameness of weight. Is it not clear that the reality of the blackboard consisted for us in the permanent grouping together of certain sense-impressions, and that that reality has disappeared for ever, except as a group of stored sense-impresses ?

§ 15.—*Individuality*.

Let us look again at this matter from a slightly different standpoint.   Let us consider a close friend, and then suppose his height, his figure, the familiar features of his face changed ; let his entire round of physical characteristics be profoundly modified, or vanish altogether.   Next let us imagine his gifts, his prejudices, the little weaknesses which really endear him to us, his views on literature, politics, and social problems, all his conceptions of human life changed or removed entirely.   In short, all the sense-impressions which constitute our friend gone.   Clearly the friend would have ceased for us to be, his individuality would have disappeared.   The "reality" of the friend consists for us not in some shadowy "thing-in-itself," but in the persistency of the majority of the group of sense-impressions by which we identify him.   We are accustomed to speak, for practical purposes, of the boy and the man as the same individual, but the body and mind have changed so enormously that the man would probably feel the boy a perfect stranger if he were brought into his presence.   We experience an uncomfortable sense of strangeness in looking at portraits of ourselves taken twenty or thirty years ago.   The properties of youth and man are, indeed, so widely different, that though for practical purposes we call them the same person, we suspect that they would cut each other if they chanced to meet in the street.   Clearly an individual is not characterized by any sameness in the thing-in-itself, but by the permanency in certain groupings of sense-impressions; this is the basis of our identification.

§ 16.—*The Futility of "Things-in-themselves."*

If at different times we meet with two groups of sense-impressions which differ very little from each other, we term them the same object or individual, and in practical life the test of identity is sameness in sense-impressions. The individuality of an object consists for us in the sameness of the great majority of our sense-impressions at two instants of time. In the case of growth, or rapid change in a group of sense-impressions, these instants must be taken closer and closer together as the rapidity increases. A stored impress of this sameness is then formed in the mind of the observer, and this constitutes in the case of the "external world" the recognition of individuality, in the case of the "internal world" the feeling of the continuity of the *ego*.

The considerations of this section upon what we are to understand by an individual thing are more important than they may appear to the reader at first sight. Are we forced to assume a shadowy "thing-in-itself" behind a group of sense-impressions in order to account for the permanency of objects, their existence as individuals ? We have seen by the examples cited that the thing-in-itself would have to be supposed as transient as the sense-impressions, the permanency of which it is introduced to explain.[1] We are not, however, thrown back on any metaphysical inquiry as to things-in-themselves, in order to define for practical and scientific purposes the sameness of objects.

[1] Unless, indeed, we follow the crude materialism of Büchner, who takes the special sense-impressions which we term material to be the basis of all other sense-impressions, or to be the thing-in-itself. The individuality of the object is then thrown back on the sameness of the *unknown* elements of matter : see Chapter VII,

Looking out of my window I see in a *certain* corner of my garden an ash-tree, with boughs of a *certain* form and shape, the sun is playing upon it and a *certain* light and shade is visible, the wind is turning over the leaves of the western branches. All this forms a complex group of sense-impressions. I close my eyes, and on opening them I have again a complex group of sense-impressions, but slightly differing from the last, for the sun has left some leaves and fallen on others, and the wind is still ; but there is a sameness in the great majority of the sense-impressions of the two groups, and accordingly I term them one and the same individual tree—the ash-tree in my garden. If any one tells me that the sameness is due to some "thing-in-itself" which introduces the permanency into the group of sense-impressions, I can as little accept or deny his assertion as he forsooth can demonstrate anything about this shadowy thing-in-itself. He may call it *Matter*, or *God*, or *Will*, or *Mind-stuff*, but to do so serves no useful purpose, for it lies beyond the field of conception based on sense-impressions, beyond the sphere of logical inference or human knowledge. It is idle to postulate shadowy unknowables behind that real world of sense-impression in which we live. So far as they affect us and our conduct they are sense-impressions ; what they may be beyond is fantasy, not fact ; if indeed it be wise to assume a *beyond*, to postulate that the surface of sense-impressions which shuts us in, must of necessity shut something beyond out. Such unknowables do not assist us in grasping why groups of sense-impressions remain more or less permanently linked together. Our experience is that they are so linked, and their association is at the present, and may ever remain, as mysterious as is now the process by which stored impressions are involuntarily linked together in the brain. Why is the thought "garden" in my mind invariably followed by the thought "cats"? The psychical basis of the association is not what I mean. I recognize it in the repeated experience of the havoc which the feline race has wrought in *my own* garden. But what is the *physical* nexus between the two conceptions as impresses in my brain? No one can say ; and yet this problem should be easier to answer than that of the nexus between the immediate sense-impressions we term objects. When physiological psychology has answered the former problem, then it will perhaps cease to be foolish for us to discuss the latter. Meanwhile let us confess our ignorance and work where a harvest may even at present be gathered.

§ 17.—*The Term Knowledge is Meaningless if applied to Unthinkable Things.*

We are now, I think, in a position to clearly grasp what we mean by the facts of science ; we see that its field is ultimately based upon sensations. The familiar side of sensations, sense-impressions, excite the mind to the formation of constructs and conceptions, and these again, by association and generalization, furnish us with the whole range of material to which the scientific method applies. Shall we say that there are limits to the scientific method—that our power of knowledge is imprisoned within the narrow bounds of sense-impression? The question is an absurd one until it has been demonstrated that a definition can be found for knowledge, which shall include what does not lie in the plane of men's thought. Our only experience of thought is associated with the brain of man ; no inference can possibly be legitimate which carries thought any further than nervous systems akin to his. But human thought has its ultimate source in sense-impressions, beyond which it cannot reach. We can therefore only show that our knowledge is of necessity limited by demonstrating that there are problems within the sphere of man's thought, the only sphere where thought can be legitimately said to exist, which can never be solved. Such a demonstration I, for one, have never met with, and I believe that it can never be given. We must one and all confess that within the sphere of thinkable things our knowledge is still the veriest thread. We may even go so far as to assert that unto complete knowledge we shall never attain in finite time ; but this admission differs widely from the assertion that knowledge is possible as to things outside thought, but yet, however possible, must be unattainable. Such an assertion must seem hopelessly absurd unless we use knowledge as a term for some relationship which exists between things outside thought. But even this strained use of the term, apart from its confusion, leads us no further than the statement that an un-meaning x exists among an unthinkable y and z.

SUMMARY.

1. Immediate sense-impressions form permanent impresses in the brain which psychically correspond to memory. The union of im-mediate sense-impressions with associated stored impresses leads to the formation of "constructs," which we project "outside ourselves," and term phenomena. The real world lies for us in such constructs and not in shadowy things-in-themselves. "Outside" and "inside" one-self are alike ultimately based on sense-impressions; but from these

sense-impressions by association, mechanical and mental, we form conceptions and draw inferences. These are the facts of science, and its field is essentially the contents of the mind.

2. When an interval elapses between sense-impression and exertion filled by cerebral activity marking the revival and combination of stored impresses we are said to think or to be conscious. Other-consciousness is an inference, which, not yet having been verified by immediate sense-impression, we term an *eject* ; it is conceivable, however, that it could become an object. Consciousness has no meaning beyond nervous systems akin to our own; it is illogical to assert that all matter is conscious, still more that consciousness can exist outside matter.

3. The term knowledge is meaningless when extended beyond the sphere in which we may legitimately infer consciousness, or when applied to things outside the plane of thought, *i.e.*, to metaphysical terms dignified by the name of conceptions although they do not ultimately flow from sense-impressions.

LITERATURE.

These notices being only intended to indicate easily readable matter for lay students, it would be idle to provide here a list of philosophical classics. I therefore refer with some hesitation to Kant's Kritik der reinen Vernunft (Eng. Trans. by Max Müller). At the same time I know no elementary treatise on Kant's view of "things-in-themselves." As moderate in length and easily intelligible I cite:—

BERKELEY, G.—An Essay towards a New Theory of Vision, 1709 ; A Treatise Concerning the Principles of Human Knowledge, 1710 ; and Three Dialogues Between Hylas and Philonous, 1713. (All to be found in vol. i. of Wright's edition of the Works of G. B., 1843.)

CLIFFORD, W. K.—Lectures and Essays ("Body and Mind" and "On the Nature of Things-in-themselves"). Further : Seeing and Thinking. Macmillan's "Nature" Series, 2nd ed., 1880.

HUXLEY, T. H.—Hume. Macmillan, 1879.

MACH, E.—Beiträge zur Analyse der Empfindungen, 1886. Further : "The Analysis of the Sensations—Anti-metaphysical," The Monist, vol. i. pp. 48–68 ; "Sensations and the Elements of Reality," Ibid. pp. 393–400.

MORGAN, C. LL.—Animal Life and Intelligence, chaps. viii. and ix. Arnold, 1891.

PEARSON, K.—The Ethic of Freethought ("Matter and Soul").

CHAPTER III.

THE SCIENTIFIC LAW.

§ 1.—*Résumé and Foreword.*

THE discussions of our first two chapters have turned upon the nature of the method and material of modern science. The material of science corresponds, we have seen, to all the constructs and concepts of the mind. Certain of these constructs associated with immediate sense-impressions we project outwards and speak of as physical facts or phenomena ; others, which are obtained by the mental processes of isolation and co-ordination from stored sense-impresses, we are accustomed to speak of as mental facts. In the case of both these classes of facts, the scientific method is the sole path by which we can attain to knowledge. The very word knowledge, indeed, only applies to the product of the scientific method in this field. Other methods, here or elsewhere, may lead to fantasy, as that of the poet or of the metaphysician, to belief or to superstition, but never to knowledge. As to the scientific method, we saw in our first chapter that it consists in the careful and often laborious classification [1] of facts, in the com-

---

[1] The reader must be careful to recollect that *classification* is not identical with collection. It denotes the systematic association of kindred facts, the collection, not of all, but of relevant and crucial facts.

parison of their relationships and sequences, and finally in the discovery by aid of the disciplined imagination of a brief statement or *formula*, which in a few words resumes the whole range of facts. Such a formula, we have seen, is termed a *scientific law*. The object served by the discovery of such laws is the economy of thought ; the suitable association of conceptions drawn from stored and correlated sense-impresses, permits the fitting exertion to follow with the minimum of thought upon the receipt of an immediate sense-impression. The knowledge of scientific law enables us to replace or supplement mechanical association, or instinct, by mental association, or thought. It is the *forethought*, by aid of which man in a far higher degree than other animals is able to make the fitting exertion on the receipt of a novel group of sense-impressions.

We are accustomed to speak of scientific law, or at any rate of one form of it termed "natural law," as something universally valid ; we hold it to be as true for all men as for its original propounder. Nay, there are not wanting those who assert that natural law has a validity quite independent of the human minds which formulate, demonstrate, or accept it. We can easily observe that there is really something *sui generis* about the validity of natural law. The philosopher, who propounds a new system, or the prophet who proclaims a new religion, may be absolutely convinced of the truth of his statement ; but it is the result of experience from time immemorial that he cannot *demonstrate* that truth so that conviction is produced in the mind of every rational being. A philosophic or a religious formula—for example, the idealism of Berkeley, the scepticism of Hume, or the self-renun-

ciation of the mediæval mystics,—however sure its
teachers may be that it is capable of rational demonstration, really appeals to the individual temperament, and is accepted or rejected according to the
emotional sympathies of the individual. On the
other hand a formula, like that which Newton propounded for the motion of the planetary system, will
be accepted by every rational mind which has once
understood its terms and clearly analyzed the facts
which it resumes.[1] This is sufficient to indicate that
there must be some wide difference between philosophic and scientific systems, between theological and
scientific formulæ. I shall endeavour in this chapter
to ascertain wherein this difference lies, to discover
what is the meaning of the word law when used
scientifically, and in what sense we can say that
scientific law has universal validity.

§ 2.—Of the Word Law and its Meanings.

The term *law* probably recalls to the reader, in the
first place, the rules of conduct proclaimed by the
state and enforced under more or less heavy penalties
against certain classes of its citizens. Austin, the
most luminous English writer on jurisprudence,[2] who
has devoted a very large portion of his well-known
work to a discussion of the meaning of the word *law*,
remarks :—

"A law, in the most general and comprehensive
acceptation in which the term, in its literal meaning,

---

[1] *One* system of planetary gravitation is accepted throughout the civilized world, but more than a dozen distinct theological systems and
almost as many philosophical schools hardly suffice even for our own
country.

[2] *Lectures on Jurisprudence,* 4th ed.   London, 1879.

is employed, may be said to be a rule laid down for the guidance of an intelligent being by an intelligent being having power over him."

He further goes on to observe that where there is such a rule there is a command, and where there is a *command* a corresponding *duty*. From this standpoint Austin proceeds to discuss the various types of law, such as civil, moral, and divine law. It will be at once seen that with Austin's definition of law there is no place left for law in the scientific sense. He himself recognizes this, for he writes :—

"Besides the various sorts of rules which are included in the literal acceptation of the term law, and those which are by a close and striking analogy, though improperly, termed laws, there are numerous applications of the term law, which rest upon a slender analogy and are merely metaphorical or figurative. Such is the case when we talk of *laws* observed by the lower animals ; of *laws* regulating the growth or decay of vegetables ; of *laws* determining the movements of inanimate bodies or masses. For where *intelligence* is not, or where it is too bounded to take the name of *reason*, and therefore is too bounded to conceive the purpose of a law, there is not the *will* which law can work on, or which duty can incite or restrain. Yet through the misapplications of a *name*, flagrant as the metaphor is, has the field of jurisprudence and morals been deluged with muddy speculation " (p. 90).

Now Austin was absolutely in the right to emphasize the immense distinction between the use of the term *law* in science and its use in jurisprudence. There can be no doubt that the use of the same name for two totally different conceptions has led to a great deal of confusion. But on the one hand, if the flagrant misapplication of the scientific meaning of the word law to the fields of jurisprudence and morals has deluged them with "muddy speculation," there is equal certainty on the other hand that the misapplication of the legal and moral sense of the term has been equally disadvantageous to clear thinking in the field of science. Austin probably had in his mind when he wrote the above passage, works like Hegel's *Philosophy of Law*, in which we find the conception of the permanent and absolute character of scientific law applied to build up a system of absolute civil and moral law which somehow realizes itself in human institutions. To the mind which has once thoroughly grasped the principle of evolution in its special factor of natural selection, the civil and moral laws of any given society at a particular time must appear as ultimate results of the struggle for existence between that society and its neighbours. The civil and moral codes of a community at any time are those which are on the average best adapted to its current needs, and best calculated to preserve its stability. They are very plastic, and change in every age with the growth and variation of social conditions. What is lawful is what is not prohibited by the laws of a particular society at a particular time; what is moral is what tends to the welfare of a particular society at a particular time. We are all well acquainted with the continual change of civil law; in fact we keep up an important body, termed Parliament, whose chief function it is to modify and adapt our laws, so that they shall be best fitted at each period to assist the community in its struggle for

existence. Of the changes in moral law we are, perhaps, less conscious, but they are none the less real.
There are very few acts which have not been moral at
some period in the growth of one or other society, and
there are in fact many questions with regard to which
our moral judgment is totally different from that of
our grandfathers. It is the relativity, or variability with
age and community, of civil and moral law, which led
Austin, I think, to speak somewhat strongly of the
speculation which confuses such law with law in the
absolute sense of science. A law in the legal or moral
sense holds only for individuals and individual communities, and is capable of modification or repeal. A
law of science will be seen in the sequel to hold for
all normal human beings so long as their perceptive
and reasoning faculties remain without material modification. The confusion of these two ideas is productive of that "muddy speculation" which finds analogies
between natural laws and those of the spiritual or
moral world.

Now if we find that two quite distinct ideas unfortunately bear the same name we ought, in order to
avoid confusion, to re-name one of them, or failing this
we ought on all occasions to be quite sure in which of
the two senses we are using the name. Accordingly,
in my first chapter, in order to keep clear of the double
sense of the word law, I endeavoured to replace it,
when spoken of scientifically, by some such phrase as
the "brief statement or formula which resumes the
relationship between a group of facts." Indeed it
would be well, were it possible, to take the term
formula, as already used by theologians and mathematicians, and use it in place of scientific or natural
law. But the latter term has taken such root in our

8

language that it would be hard indeed to replace it now. Besides, if the word law is to be used in one sense only, we may ask why it is the scientist rather than the jurist who is to surrender his right to the word? The jurists say that historically they have the older claim to the word—that civil law existed long anterior to scientific law. This is perfectly true in a certain sense,[1] because the earliest attempts to codify laws for the conduct of men living in communities preceded any conscious recognition of scientific law. Now this leads us directly to a very important distinction, which, if it be neglected, is the source of much confusion. Does law exist before it receives expression and recognition? According to Austin, law in the juridical sense certainly does not, for such a law involves a " command," and a " corresponding duty "—that is, expression and recognition. What are we to say, then, with regard to scientific law—does it really exist before man has given expression to it ? Has the word any meaning when unassociated with the mind of man ? I hold that we must definitely answer " no " to both these questions, and I believe that the reader who has carefully followed my second chapter will see at once the grounds for this statement. A scientific law is related to the perceptions and conceptions formed by the perceptive and reasoning faculties in man ; it is meaningless except in association with these ; it is the *résumé* or *brief expression* of the relationships and sequences of certain groups of these perceptions and conceptions, and exists only when formulated by man.

[1] For our final conclusions as to the historical right to the word, see p. 114.

§ 3.—*Natural Law relative to Man.*

Let us take that branch of scientific law which deals with the so-called "outside world"—natural law. We have seen that this outside world is a *construct.* It consists of objects constructed partly from immediate sense-impressions, and partly from stored impresses. For this reason the "outside world" is essentially conditioned by the perceptive and retentive faculties in man. Even the metaphysicians, who postulate "things-in-themselves," admit that sense-impressions in nowise *resemble* them, and that *man's* sense-impressions so far from being the entire product of "things-in-themselves," are probably but the smallest portion of their "capacity for producing" sense-impression. Hence to talk about natural law as existing in "things-in-themselves" and apart from man's mind is again to assert an unmeaning x among an unthinkable y and z (p. 90). If nature for man is conditioned by his perceptive and retentive faculties, then natural law is conditioned by them also. It has no relation to something above and beyond man, but solely to the special products of his perceptive faculty. We have no right to infer its existence for things without a perceptive faculty, or even for perceptive faculties not closely akin to man's. I believe that a great deal of the obscurity involved in popular ideas about "Nature" would have been avoided had this been borne in mind.

A good instance of the *relativity* of natural law is to be found in the so-called *Second Law of Thermodynamics.* This law resumes a wide range of human experience, that is, of sequences observed in *our* sense-impressions, and embraces a great number of conclu-

sions not only bearing on practical life, but upon that
dissipation of energy which is even supposed to fore-
shadow the end of all life. The appreciation of the
relativity of natural law is so important that the
reader will, I trust, pardon me for citing the entire
passage in which Clerk-Maxwell discusses this
instance [1] :—

" One of the best-established facts in thermo-dynamics is that it is impossible in a system enclosed in an
envelope which permits neither change of volume nor
passage of heat, and in which both the temperature
and pressure are everywhere the same, to produce
any inequality of temperature or of pressure without
the expenditure of work. This is the second law of
thermo-dynamics, and it is undoubtedly true so long
as we can deal with bodies only in mass, and have no
power of perceiving or handling the separate molecules of which they are made up. But if we conceive
a being whose faculties are so sharpened that he can
follow every molecule in its course, such a being,
whose attributes are still as essentially finite as our
own, would be able to do what is at present impossible to us. For we have seen that the molecules in a
vessel of air at uniform temperature are moving with
velocities by no means uniform, though the mean
velocity of any great number of them, arbitrarily
selected, is almost exactly uniform. Now let us suppose that such a vessel is divided into two portions, A
and B, by a division in which there is a small hole,
and that a being,[2] who can see the individual mole-

[1] *Theory of Heat*, 3rd ed. p. 308. Longmans, 1872.
[2] This " being " has become known to fame as " Clerk-Maxwell's
demon," but it must be noted that Clerk-Maxwell supposes the being's
attributes " essentially finite as our own "—a peculiarity not usually
associated with demons.

cules, opens and closes this hole, so as to allow only the swifter molecules to pass from A to B, and only the slower ones to pass from B to A. He will thus, without expenditure of work, raise the temperature of B and lower that of A, in contradiction to the second law of thermo-dynamics."

To render this passage clear to the lay reader, we have only to add that in this kinetic theory the temperature of a gas depends upon the mean speed of its molecules. Now the second law of thermo-dynamics resumes with undoubted correctness a wide range of human experience, and is, to that extent, as much a law of nature as that of gravitation. But the kinetic theory of gases, whether it be hypothetical or not, enables us to conceive a demon having a perceptive faculty differing rather in degree than quality from our own, for whom the second law of thermo-dynamics would not necessarily be a law of nature. Such a conception enables us to grasp how relative what we term nature is to the faculty which perceives it. Scientific law does not, any more than sense-impression, lie in a universe outside and unconditioned by ourselves. Clerk-Maxwell's demon would perceive nature as something totally different from our nature, and to a less extent this is in great probability true for the animal world, and even for man in different stages of growth and civilization. The worlds of the child and of the savage differ widely from that of normal civilized man. One half of the perceptions which the latter links together in a law of nature may be wanting to the former. Our law of the tides could have no meaning for a blind worm on the shore, for whom the moon had no exis-

¹ This point is well brought out by Prof. Lloyd Morgan in his *Animal*

tence.[1] By the contents and the manner of perception
the law of nature is essentially conditioned for each
perceptive faculty. To speak, therefore, of the universal validity of a law of nature has only meaning
in so far as we refer to a certain type of perceptive
faculty, namely, that of a normal human being.

§ 4.—*Man as the Maker of Natural Law.*

The other problem with which we are concerned is
the existence or non-existence of a scientific law
before it has been postulated. Here the reader will
feel inclined to remark : " Admitted that ' Nature ' is
conditioned by man's perceptive faculty, surely the
sequences of man's perceptions follow the same law
whether man has formulated that law in words or
not ? The law of gravitation ruled the motion of
the planets ages before Newton was born." Yes and
no, reader; the answer must depend on how we define
our terms. The sequences involved in man's perception of the motion of the heavenly bodies were
doubtless much the same to Ptolemy and Newton ;
to primitive man and to ourselves the motion of the
sun is a common perception, but a sequence of sense-
impressions is not in itself a law. That planets

Life and Intelligence. After pointing out the widtly different character
of the sense organs in man and insects he continues :—

"Remember their compound eyes with mosaic vision, coarser by far
than our retinal vision, and their ocelli of problematical value, and the
complete absence of muscular adjustments in either one or the other.
Can we conceive that, with organs so different, anything like a similar
perceptual world can be elaborated in their insect mind? I for one
cannot. Admitting therefore that their perceptions may be fairly sur-
mised to be analogous, that their world is the result of construction, I
do not see how we can for one moment suppose that the perceptual
world they construct can in any accurate sense be said to resemble ours"
(pp. 298-9, 356-7, 361).

move, that a chick takes its origin from the egg, may
be sequences of sense-impressions, they may be facts
to be dealt with scientifically, but they are not laws
in themselves, at least not in any useful interpretation
of the word. The changes of the whole planetary
system might be perceived, and even those perceptions translated into words with a fulness surpassing
that of our most accurate modern observer, and yet
neither the sequence of perceptions in itself nor the
description involve the existence of any law. The
sequence of perceptions has to be compared with
other sequences, classification and generalization have
to follow; conceptions and ideas, pure products of the
mind, must be formed, before a description can be given
of a range of sequences which, by its conciseness and
comprehensiveness, is worthy of the name of scientific
law.

Let it be noted that in this it is not only the
process of reaching scientific law which is mental, but
that the law itself when reached involves an association of natural facts or phenomena with mental
conceptions, lying quite outside the particular field
of those phenomena. Without the mental conceptions the law could not be, and it only comes into
existence when these mental conceptions are first
associated with the phenomena. The law of gravitation is not so much the discovery by Newton of a
rule guiding the motion of the planets as his invention
of a method of briefly describing the sequences of sense-
impressions, which we term planetary motion. He did
this in terms of a purely mental conception, namely,
mutual acceleration.[1] Newton first brought the idea

[1] The reader will find mutual acceleration fully defined and discussed
in Chapter VIII.

of mutual acceleration of a certain type into association with a certain range of phenomena, and was thus
enabled to state a formula, which, by what we may
term mental shorthand, resumes a vast number of
observed sequences.   The statement of this formula
was not so much the discovery as the *creation* of the
law of gravitation.   A natural law is thus seen to be
a *résumé* in mental shorthand, which replaces for us
a lengthy description of the sequences among our
sense-impressions.   Law in the scientific sense is
thus essentially a product of the human mind and
has no meaning apart from man.   It owes its existence
to the creative power of his intellect.   There is more
meaning in the statement that man gives laws to
Nature than in its converse that Nature gives laws to
man.

§ 5.—*The Two Senses of the Words " Natural Law."*

We have now traced at least one point of analogy
between juridical and scientific law which I think
escaped Austin, namely, both are the product of
human intelligence.   But we have at the same time
seen the wide distinction between the two.   The
civil law involves a command and a duty ; the
scientific law is a description, not a prescription.
The civil law is valid only for a *special* community
at a *special* time ; the scientific law is valid for *all*
normal human beings, and is unchangeable so long
as their perceptive faculties remain at their present
stage of development.[1]   For Austin, however, and

---

[1] The perceptive faculty is probably, even on the average, varying
slightly, however insensibly.   Still, the perceptive faculty is now
among men permanent in type, as compared with the changes it must
have undergone during man's evolution from a lowly form of life.

for many other philosophers too, the law of nature was not the mental formula, but the repeated sequence of perceptions. This repeated sequence of perceptions they projected out of themselves, and considered as part of an external world unconditioned by and independent of man. In this sense of the word, a sense unfortunately far too common to-day, natural law could exist before it was recognized by man. In this sense natural law has a much older ancestry than civil law, of which it appears to be the parent. For tracing historically the growth of civil law, we find its origin in unwritten custom. The customs which the struggle for existence have gradually developed in a tribe become in course of time its earliest laws. Now, the farther we go back in the development of man, through more and more complete barbarism to a simply animal condition, the more nearly we find customs merging in instinctive habits. But the instinctive habit of a gregarious animal is very much akin to what Austin would have termed a natural law. The laws relating to property and marriage in the civilized states of to-day can be traced back with more or less continuity to the instinctive habits of gregarious animals. The historical origin, therefore, of civil law is to be sought in natural law in its older sense. Indeed this fact was recognized by the early Roman jurists, who refer to a *lex naturæ* as existing alongside the civil law. This law of nature they considered animals as well as men to have a knowledge of, and they made special reference to it in relation to marriage and the birth of children. Now it is clear that, however flagrant in Austin's opinion the metaphor may be when we speak of the *laws* observed by animals, still the use of the word law in this sense is a very old one even among jurists themselves.

§ 6.—*Confusion between the two Senses of Natural Law.*

But the Roman lawyers merely took the idea of natural law from the Greek philosophers, and it is specially to the Stoics that we owe a conception of law which is of value as illustrating the kind of obscurity which still attaches to the word natural law in many minds. The Stoics defined nature as the universe of things, and they declared this universe to be guided by reason. But reason, because it is a directive power, forbidding and enjoining, they called law. Now, the law of nature they considered to take in some manner its rise in nature itself—there was no source of law to nature outside nature—and they accordingly defined this law of nature as a force inherent in the universe. They further asserted that since reason cannot be twofold, and since man has reason as well as the universe, the reason in man and the universe must be the same, and therefore the law of nature must be the law by which men's actions ought to be guided.

The string of dogma and unwarranted inference marking this argument—which, however, has only reached us at second-hand [1]—is characteristic enough. Yet the argument is noteworthy, for we find in it the three meanings of the term law with which we have been dealing hopelessly confused. The Stoics pass from the scientific law to the *lex naturæ*,—the mere sequence of phenomena,—and then to the civil or moral law without in the least observing the magni-

---

[1] Marcus Aurelius, iv. 4, and Cicero, *De legibus* i. 6–7. *Cf.* T. C. Sandars, *The Institutes of Justinian*, p. xxii. Longmans, 1878.

tude of their spring; and what these early philosophers accomplished in this way has been surpassed by the devotees of philosophy and natural theology in later ages. One example will, perhaps, suffice for our present investigation. Richard Hooker, a divine of the sixteenth century, who achieved a remarkable reputation for himself by stating paradoxes based on a confusion between natural and moral law, thus defines *law* in general :—

"That which doth assign unto each thing the kind, that which doth moderate the force and power, that which doth appoint the form and measure of working, the same we term a Law" (*Ecclesiastical Polity*, bk. I. ii.).

Hooker further considers that all things, including nature, have some operations "not violent or casual." This leads him to assert that such operations have "some fore-conceived end." Hence he holds that nature is guided by law, and that this law is a product of reason. Unlike the Stoics, Hooker placed this reason in a worker, God, outside and not inherent in Nature, otherwise his doctrine and the conclusions he draws from it closely resemble theirs. He was, however, aware of the elastic character of his definition of law, for he writes :—

"They, who thus are accustomed to speak, apply the name Law unto that only rule of working which a superior authority imposeth ; whereas we, somewhat more enlarging the sense thereof, term any kind of rule or canon whereby actions are framed, a law" (bk. I. iii.).

The views of Hooker and the Stoics thus briefly sketched deserve careful consideration by the reader, as they suggest the type of fallacy into which we fall

by ill-defined use of the term natural law.[1] In the
first place these philosophers start from the conception of natural law as the mere concatenation of
phenomena, the succession or routine of sense-impressions. In the next place as materialists they project
these sense-impressions into a real outside world,
unconditioned by and independent of man's perceptive faculty. Then they infer reason behind the
concatenation of phenomena. Now, reason is known
to us only in association with consciousness, and we
find consciousness only with the accompaniment of
a certain type of nervous organism. Thus to infer
reason in what has been previously postulated as
outside and independent of this type of nervous
organism is unjustifiable ; it may be dogma, but it is
not logic. It makes little difference whether, with
the Stoic, we assert that reason is inherent in nature,
or, like Hooker, place the lawgiver outside nature as
at once its creator and director. Both assertions lie
completely outside the field of knowledge, and, as we
have said of the like statements before, they logically
refer to an unmeaning x existing among an unthinkable y and z (*i.e.*, " realities " unconditioned by man's
perceptive faculty).

[1] The study of fallacy in concrete examples ought to play a greater
part in our educational curriculum. Certain works have a permanent
value in this respect. I can conceive no better exercises for a student
of logic or jurisprudence than an analysis of the paralogisms in Book I
of Hooker's *Ecclesiastical Polity ;* for a student of physics, than a discovery of the fallacies in Mr. Grant Allen's *Force and Energy ;* or for
both than a critical study of Drummond's *Natural Law in the Spiritual
World ;* while a more difficult study in pseudo-science will be found in
the first part of J. G. Vogt's *Das Wesen der Elektrizität und des
Magnetismus.* The power of criticism and the logical insight thus
attainable are in many respects as advantageous as the appreciation
of method which results from the perusal of genuine science.

§ 7.—*The Reason behind Nature.*

But how, it may be asked, has the conception that reason exists behind phenomena become so widespead ? Why have so many philosophers and theologians, nay, even scientists,[1] used the "argument from design"? The duty of science does not end with showing an argument to be fallacious ; it has to investigate the origin of the fallacy and show the nature of the process by which it has arisen.   In the present case I do not think we have far to seek. Briefly stated, the " argument from design " consists in the production of evidence from the laws of nature, tending to exhibit those laws as the product of a rational being or of reason in one or another form. Now, although in the law of nature defined as a mere concatenation of phenomena, as a sequence of sense-impressions, there is, so far as I can perceive, no evidence of reason in any intelligible sense of the word, yet in the law of science, and in that branch of it which in this work we have termed natural law, there is every evidence of reason.   So soon as man begins to form conceptions from his sense-impressions, to combine, to isolate, and to generalize, then he begins to project his *own* reason into phenomena, to replace in his mind the sense-impresses of past concatenations of phenomena by those brief *résumés* or formulæ which describe the sequences of sense-impressions in mental shorthand.   He begins to confuse the scientific law, the product of his own reason, with the mere concatenation of phenomena, the natural law in the sense of Hooker and the Stoics.   As he projects his

---

[1] *E.g.*, Sir G. G. Stokes, in his otherwise most suggestive and masterly *Burnett Lectures on Light.*

sense-impressions outside himself, and forgets that they are essentially conditioned by his own perceptive faculty, so he unconsciously severs himself from the products of his own reason, projects them into phenomena, only to refind them again and wonder what reason put them there. Here, in the double sense of the word natural law, lies the origin of much obscure speculation.

The reason we find in natural phenomena is surely put there by the only reason of which we have any experience, namely, the human reason. The mind of man in the process of classifying phenomena and formulating natural law introduces the element of reason into nature, and the logic man finds in the universe is but the reflection of his own reasoning faculty. A dog, if able to recognize the instinct which guides his actions, might very naturally suppose instinct and not reason to be the basis of natural phenomena, reflecting his own source of action into all he observed around him. Indeed, it seems to me more logical to find instinct than reason behind the setting and rising of the sun, for instinct at least does not presuppose consciousness. Perhaps if our dog were a Stoic dog the instinct would seem to him inherent in the universe itself, while had he been reared at the parsonage he would certainly fancy his kennel the product of an instinct extramundane. But both dog and man, in thus arguing beyond the sphere of legitimate inference, are also breaking a fundamental canon of the scientific method. This canon is practically due to Newton, and forbids us to seek superfluous causes for natural phenomena.[1] We ought

---

[1] *Causas rerum naturalium non plures admitti debere, quam quæ & vera sint & earum Phænomenis explicandis sufficiunt. Natura enim simplex*

not to look for new causes to explain any group of phenomena until we have shown that no known cause is capable of explaining it. In our next chapter we shall see more clearly what is to be understood by the words " cause " and " explanation," but for the present Newton's canon suffices to show us that the Stoics were unscientific in seeking for unknown or unknowable "reasons" inherent in nature, until they had demonstrated that the only rational faculty known to them—namely, that of man—was insufficient to account for the rational element they professed to observe in nature. What is reason ? Where may we infer its existence ? Can we proceed from this admissible reason to the rational element in natural law ? —these are the questions the Stoics ought logically to have asked themselves. Our wonder ought not to be excited by the idea that so vast a range of phenomena are ruled (*sic !*) by so simple a law as that of gravitation, but we ought to express our astonishment that the human mind is able to express by so brief a description such wide sequences of sense-impressions. This capacity of itself suggests some harmony, some relation between the perceptive and reasoning faculties in man—a matter to which I shall return later.

§ 8.—*True Relation of Civil and Natural Law.*

Proceeding from Austin's definition of law, we have found it necessary to distinguish between two different ideas frequently confused under the term "natural law," namely, the mere concatenation of

---

est & rerum causis superfluis non luxuriat. Principia. (Editio Princeps, 1687, p. 402.) This "simplicity of nature" is, of course, a dogma, but the *regula philosophandi* which forbids us to revel in superfluous causes is fundamental to our view of science as an economy of thought.

phenomena and the mental formula which gives brief expression to their sequences. Before we devote our undivided attention to the latter as the scientific conception of natural law, it may be of interest to clear up one or two remaining points with regard to civil and scientific law. While Austin, thinking rather of natural law in the old sense, states that any relation between the two is merely metaphorical, both the Stoics and Hooker conceive that the reason, or the lawgiver to be recognized behind phenomena, ought to guide man's moral conduct. Now, if these philosophers were looking upon natural law as the product of the human reason there would be little to require further comment; but, as we have seen, this is far from the case. The Stoics tell us that reason cannot be twofold, that it must be the same reason in both man and the universe, and that therefore the civil law of man is identical with natural law.[1] The inference is of course unjustifiable, for the *same* reason may be at work in two quite distinct fields. It is important to notice, however, that in one sense civil and moral laws are natural products ; they are products of particular phases of human growth. This growth is itself capable of treatment by the scientific method, and the sequence of its stages can be expressed by scientific formulæ, or,—looking at civil and moral law as objective phenomena,—by natural laws. Thus civil law is a natural product, and not identical with natural law—any more than the particular configuration of the planetary system

[1] Up to the " sameness of the reason " there is little exception to be taken to the argument, but few of us would agree with the dictum of that ancient and upright judge, Sir John Powell, that "nothing is law that is not reason."

at this moment is identical with the law of gravitation. We are now, I think, in a position to draw a clear distinction between civil (or moral) law and natural law. Civil law takes its origin in natural law in the old sense (p. 105), while its growth and variation can, in broad outline at least, be described in the brief formulæ of science, or in natural laws in the scientific sense. Civil and moral laws are the natural product of societies, and of classes within society, struggling in the early days for self-preservation, and in these later days for a maximum of individual comfort.

A civil law, according to Austin, is a rule laid down for the guidance of an intelligent being by an intelligent being having power over him. Such a rule varies with every age and every society. On the other hand, a natural law is not laid down by one intelligent being for another ; it involves no command or corresponding duty, and it is valid for all normal human beings. It has taken centuries for men to arrive at a full appreciation of this distinction, and it would be well could the distinction be now emphasized by the specialization of the word *law* in one or other of its senses. We sadly need separate terms for the routine of sense-impressions, for the brief description or formula of science and for the canon of social conduct, or, in other words, for the perceptive order, the descriptive order, and the prescriptive order. Historically we cannot say that any of these orders has the higher claim to the title *law*, for the Roman ideas of law must at least be traced back to their Greek parentage. Here, in the Greek word νόμος, law, the confusion centres, and at the same time the historical origin of the confusion becomes apparent.

9

This word shows us that civil law originated in custom, and yet Plato derives it from "distribution of mind."[1] Anything from the harmony of nature to the strains of a song was for the Greek *law*. In the conception of order or sequence, therefore, we see the historical origin of law in all its senses, and thus no claim to priority on the part of either jurist or scientist can be historically proven. No individual writer can hope with success to remould such old-established usage as is associated with the word law, and all he can strive to do is to keep clearly distinct in the mind of his readers the sense in which the word on each occasion is used.[2]

§ 9.—*Physical and Metaphysical Supersensuousness.*

Having now analyzed our ideas of law, and reached a definition of law in its scientific sense, it may be well, even at the cost of repetition, to discuss at greater length our conclusions and their application to our theory of life. From the material provided by the senses, either directly or in the form of stored sense-impresses, we draw conceptions. About these conceptions we reason, endeavouring to ascertain their relationships and to express their sequences in those brief statements or formulæ which we have termed scientific laws. In this process we often analyze the material of sense-impressions into elements which are not in themselves capable of form-

[1] *The Laws*, iv. 714, and see also iii. 700, and vii. 800.
[2] For the remainder of this work I shall, for convenience, however speak of natural law in the old sense, or, as a mere routine of perceptions, as law in the *nomic* sense. Law in the nomic sense is thus no product of the reason, but a pure *order* of perceptions, while Bramhall's coinage *anomy* may be conveniently used for a breach in the routine of perceptions.

ing distinct sense-impressions ; we reach conceptions which are not capable of direct verification by the senses ; that is to say, we can never, or at least we cannot at present, assert that these elements have objective reality (see our p. 50). Thus physicists reduce the groups of sense-impressions which we term material substances to the elements *molecule* and *atom*, and discuss the motion of these elements, which have never been, and perhaps never can become, direct sense-impressions. No physicist ever saw or felt an individual atom. Atom and molecule are intellectual conceptions by aid of which physicists classify phenomena, and formulate the relationships between their sequences. From a certain standpoint, therefore, these conceptions of the physicist are *supersensuous*, that is, they do not at present represent direct sense-impressions ; but the reader must be careful not to confuse this kind of supersensuousness with that of the metaphysician. The physicist looks upon the atom in one or other of two different ways : either the atom is real, that is, capable of being a direct sense-impression, or else it is ideal, that is, a purely mental conception by aid of which we are enabled to formulate natural laws.[1] It is either a product of the perceptive faculty, or of the reflective or reasoning faculty in man. It may pass from the latter to the former, from the ideal stage to the real ; but till it does so, it remains merely a conceptual basis for classifying sense-impressions, it is not an actuality. On the other hand, the metaphysician asserts an existence for the supersensuous which is unconditioned by the perceptive or reflective faculties in man. His supersensuous is at once incapable of being a sense-

[1] That is, it is part of the physicist's mental shorthand.

impression, and yet has a real existence apart from the imagination of men. It is needless to say that such an existence involves an unproven and undemonstrable dogma. Nevertheless, the magnitude of the gulf between the supersensuous of the physicist and that of the metaphysician is frequently neglected, and we are told that it is as logical to discuss "things-in-themselves" as molecules and atoms !

§ 10.—*Progress in the Formulating of Natural Law.*

By the formation of conceptions, which may or may not have perceptual equivalents in the sphere of sense-impression, the scientist is able to classify and compare phenomena. From their classification he passes to formulæ or scientific laws describing their sequences and relationships. The wider the range of phenomena embraced, and the simpler the statement of the law, the more nearly we consider that he has reached a "fundamental law of nature." The progress of science lies in the continual discovery of more and more comprehensive formulæ, by aid of which we can classify the relationships and sequences of more and more extensive groups of phenomena. The earlier formulæ are not necessarily wrong, they are merely replaced by others which in briefer language describe more facts.

We cannot do better than examine this process very briefly in a special case, namely, the motion of the planetary system. An easily observed part of this motion was the daily passage of the sun, its rising in the East and setting in the West. A primitive description of the motion consisted in the statement that the *same* sun which set in the West passed, hidden by northern mountains, along the surface of the *flat* earth and rose again in the East. The description was clearly very insufficient, but it was a first attempt at a scientific formula. An obvious improvement was soon made by limiting the surface of the earth and supposing the sun to go below the solid earth. The motion of the sun taken in conjunction with the motion of the stars led early astronomers to conclude that the earth was fixed in mid-space, and sun and stars were daily carried round it. The description thus improved was still far from complete ; the sun was observed to vary its position with regard to the fixed stars. Gradually and laboriously facts were accumulated, and in time those early astronomers concluded that the sun went round yearly in the same circle, this circle itself being carried round with the starry heavens once in a day. This formula embraced a wider field of phenomena than the earlier ones, and probably was as exact a description as men's perceptions of earth and sun allowed when it was invented. Hipparchus improved it by placing the earth not exactly in the centre of the sun's circle, and thus more accurately described certain apparent irregularities in the sun's motion. A still more complete description was adopted by Ptolemy (A.D. 140) nearly three hundred years after Hipparchus, who, fixing the spherical earth, considered sun and moon to move in circles yearly round the earth, and the other planets in circles, whose centres again described circles round the earth. The whole of this system revolved daily round the earth with the stars. This, the famous Ptolemaic system, remained for many centuries the current formula, and even to this day the eccentrics of Hipparchus and

epicycles of Ptolemy are not without service as elements of the more modern description. It would be
wrong, I think, to say that the Ptolemaic system was
an erroneous *explanation*, it was simply an insufficient attempt to *describe* in brief and accurate
language a too limited range of phenomena. Then
at the end of the Middle Ages came Copernicus who
got rid of the cumbersome sphere carrying the fixed
stars by simply considering the earth to rotate
round its axis and of the epicycles, if not of the
eccentrics, by treating the sun, not the earth, as
the central point of the system. Here was an immense advance in brevity and accuracy of description ;
but still more facts remained to be included, more
difficulties to be analyzed and overcome. This work
was largely done by Keppler, who conceived the earth
and planets to move in certain curves termed ellipses,
of which the sun occupied a non-central point termed
the focus. The formula of Keppler is one of the
greatest achievements of the scientific method ; it was
the work of a disciplined imagination analyzing a
laborious and minute classification of facts.[1] A more
wide-embracing statement than that of Keppler was
not only possible, however, but required ; and this was
provided by Newton in a single formula which embraces
not only the motion of the planets, but that of their
moons and of bodies at their surfaces. This formula is
the well-known law of gravitation, but it is just as
much a *description* of what takes place in planetary
motion as Keppler's laws are a description—it is
simply a briefer, more accurate, and more wide-em-

---

[1] The elaborate observations of Tycho Brahé. Keppler not only
stated the form of the planetary path, but the mode of its description
in his famous three laws,

bracing statement. The one can just as fitly as the other be termed a natural law.

The law of gravitation is a brief description of *how* every particle of matter in the universe is altering its motion with reference to every other particle. It does not tell us *why* particles thus move ; it does not tell us *why* the earth describes a certain curve round the sun. It simply resumes, in a few brief words, the relationships observed between a vast range of phenomena. It economizes thought by stating in mental shorthand that routine of our perceptions which forms for us the universe of gravitating matter.

We have in the law of gravitation an excellent example of a scientific law. We see in its evolution the continual struggles of the human mind to reach a more and more comprehensive and exact formula, and at last Newton reaches one so simple and so wide-embracing that many have thought nothing further can be achieved in this direction. " Here," says Paul du Bois-Reymond, " is the limit to our possible knowledge." If the reader once grasps the characteristics of this law of Newton's he will understand the nature of all scientific law. Men study a range of facts—in the case of nature the material contents of their perceptive faculty—they classify and analyze, they discover relationships and sequences, and then they describe in the simplest possible terms the widest possible range of phenomena. How idle is it, then, to speak of the law of gravitation, or indeed of any scientific law, as *ruling* nature. Such laws simply *describe*, they never *explain* the routine of our perceptions, the sense-impressions we project into an " outside world."

The scientific law, while thus the product of a rational analysis of facts, is always liable to be replaced by a wider generalization. Such replacement of one formula by another is indeed the regular course of scientific progress. The only final test we have of the truth of any law, of the sufficiency of its description, the only proof that our intellect has been keen enough to reach a formula extending to the whole range of facts it professes to resume, is the actual comparison of the results of the formula with the facts themselves—that is, historical observation or physical experiment. This test is all that marks the division between scientific hypothesis and scientific law, and the scientific law itself must, with every increase of our perceptive powers, return to the position of hypothesis and be anew put to the test of experience. Yet what philosophic system, what fantasy of the metaphysical mind in the region of the supersensuous has stood like Newton's formula of gravitation without the least change, the least variation in its statement, for more than two hundred years ? Assuredly none ; they have all shifted their ground with every advance of man's positive knowledge. They have not stood the test of experience ; they are phantasms, not truth ; for, as Sir John Herschel has said :—

" The grand, and indeed only, character of truth is its capability of enduring the test of universal experience, and coming unchanged out of every possible form of fair dicussion."

§ 11.—*The Universality of Scientific Law.*

The universality, the absolute character, which we attribute to scientific law is really relative to the human mind. It is conditioned :—

1. By the perceptive faculty. The outside world,

the world of phenomena, must be practically the same for all normal human beings.

2. By the reflective faculty. The processes of association and logical inference, and the inner world of stored impresses and conceptions must be practically the same for all normal human beings.

Now, when we classify a number of things together and give them the same name, we can only mean to signify that they closely resemble each other in structure and action. Hence when we speak of *human* beings we are referring to a class which in the normal civilized condition have perceptive and reflective faculties nearly akin. It is therefore not surprising that normal human beings perceive the same world of phenomena, and reflect upon it in much the same manner. The "universality" of natural law, the "absolute validity" of the scientific method, depends on the resemblance between the perceptive and reflective faculties of one human mind and those of a second. Human minds are, within limits, all receiving and sifting-machines of one type. They accept only particular classes of sense-impressions—being like automatic sweetmeat-boxes which if well constructed refuse to act for any coin but a penny—and having received their material they arrange and analyze it, provided they are in working order, in practically the same manner. If they do not arrange and analyze it in this manner we say, that the mind is disordered, the reason wanting, the person mad. The sense-impressions of a madman may be as much reality for him as our sense-impressions are for us, but his mind does not sift them in the normal human fashion, and for him, therefore, our laws of nature are without meaning.

§ 12.—*The Routine of Perceptions is possibly a Product of the Perceptive Faculty.*

The idea of the human mind as a sorting-machine is not without suggestion with regard to another important matter, namely, the routine nature of our sense-impressions. How far does this routine of sense-impressions depend upon the perceptive faculty? How far does it lie outside that faculty in the unknown and unknowable beyond of sensation (p. 82)? The question is one to which at present no definite answer can be given, and perhaps one to which no answer can ever be found. If, with the materialists, we make matter the thing-in-itself, we throw the routine back on something behind sense-impressions, and, therefore, unknowable. Precisely the same happens if, with Berkeley, we attribute the routine to the immediate action of a deity. Materialist and idealist are here at one in casting the routine of sense-impression into the unknowable. But the business of the scientist is to know, and therefore he will not lightly assent to throwing anything into the unknowable so long as known "causes" have not been shown to be insufficient. The scientific tendency would therefore be to consider the routine of our perceptions as due in some way to the structure of our perceptive faculty before we appeal to any supersensuous aid. Far, indeed, as science at present stands from any definite solution of the problem, there are yet one or two points which it may not be unprofitable to consider.

In the first place, have we any evidence that the perceptive faculty is a *selective* machine? We have already seen that it is possible at times for us to be unconscious of sensations which on other occasions

we may keenly appreciate (p. 53). We have seen
that the outside world constructed by an insect in all
probability differs widely from our own (p. 101). To
assume, therefore, sensations which form no part of
our consciousness, perhaps no part of any consciousness, is not an illogical inference, for we proceed only
from the known to what is like the known (p. 72),
to an eject which might have been, or may one day be,
an object.[1] No better way of realizing the different
selective powers of diverse perceptive faculties can
be found than a walk with a dog. The man looks
out upon a broad landscape, and the signs of life and
activity he sees in the far distance may have deep
meaning for him. The dog surveys the same landscape indifferently, but his whole attention is devoted
to matters in his more immediate neighbourhood, of
which the man is only indirectly conscious through
the activity of the dog. Many things may be going
on in the distance, which, if at hand, would have
considerable interest for the dog: some way off the man
perceives the rabbits in the field skirting the copse,
further off still a flock of sheep on the high-road, and
behind them the shepherd with his collie—all these
remain unobserved by the dog, or if observed, un-
reasoned on. Clearly the sense-impressions corresponding to the distant landscape are far less complex
and intense in the dog than in the man. The perceptive faculty in the dog selects certain sense-
impressions, and these form for it reality ; that of
the man selects another and probably far more

[1] " A feeling can exist by itself without forming part of a consciousness," writes Clifford in a paper, the main conclusion of which seems
to me, however, quite unproven. (" On the Nature of Things-in-themselves," *Lectures and Essays*, vol. i. p. 84).

complex range, which form in turn reality for him.
Both may be again compared to automatic sweetmeat-boxes, which only work on the insertion oi
coins of definite and different value. Objective
reality does not consist of the same sense-impressions
for man and dog.

If we pass downwards from man to the lowest
forms of life, we shall find the range of sensations
perceived becoming less and less complex till they
cease altogether as perceptions with the cessation of
consciousness. Hence, if we accept the theory of the
evolution of man from the lowliest types of life, we
see a wide field of variation in the matter of the
perceptive faculty open to him. Man will evolve a
power of perceiving those sensations, the perception
of which will on the whole help him in the struggle
for existence.[1]

Now, step by step with the perceptive faculty the
reflective or reasoning faculty is developed ; the
power of sifting and arranging perceptions, the power
of rapidly passing from sense-impression to fitting
exertion (p. 55), is seen to be a factor of paramount
importance to man in the battle of life. Without
our being able at present to clearly understand the
relation between the perceptive and reflective faculties
in man, the nature of their co-ordination, it is still
reasonable to suppose a close relation between the
two ; the one largely selects those perceptions which
the other is capable of analyzing and resuming in
brief formulæ or laws. Within sufficiently wide limits
the intensity of the perceptive faculty appears in all

[1] Light and vision, sound and hearing, extension and touch, are
known not to be identical in range. See Sir William Thomson's
Popular Lectures and Addresses, vol. i. pp. 278–90.

forms of life proportional to the reasoning faculty. [1]
A world of sense-impressions in no way amenable to
man's reason would be very prejudicial to man's preser-
vation. In such plight a man, like an idiot or insane
person, would be incapable of analysis, or would analyze
wrongly; the fitting exertion would not follow on the
sense-impression, and this man would have small
chance of surviving among men whose perceptive and
reasoning faculties were attuned. Possibly some sorts
of idiocy and madness are a kind of atavism, a return
to variations of the human mind in which perceptive
and reflective faculties are not co-ordinated—variations
which on the whole have been eliminated in the
struggle for existence. If this interpretation be at all
a correct one—if, namely, the perceptive faculty can
be so moulded in the process of evolution as to accept
some and reject other sense-impressions; if, further,
the perceptive and reflective faculties have been developed in co-ordination, so that the former accepts
what, in wide limits, can be analyzed by the latter—
then we have advanced some way towards understanding why the routine of perceptions can be
expressed in brief formulæ by the human reason.
The relation between natural law in the nomic
(p. 44 *footnote*) and the scientific senses becomes more
intimate, when we thus attribute the routine of the
perceptions to the machinery of the perceptive faculty.

It will not, however, do to press this interpretation

---

[1] That woman has greater perceptive, man greater reflective power,
is one of those futilities which has been used as an excuse for hind-
rances to woman's development of both faculties. Exceptions of
course there are, but the general rule seems to be that the deeper the
intellectual power in both sexes, the wider is the range of perceptions,
the more delicately sensitive is the nervous system.

too far; or at least we must be careful to remember
that, while the perceptive faculty has developed the
power of solely perceiving sense-impressions *capable*
of being dealt with by the reflective faculty, it
does not follow that they have already been dealt
with by the latter faculty. Otherwise we shall be
abruptly confuted by the fact that there are many
sense-impressions which we perceive and yet have
not classified and reduced to simple formulæ.
There are many phenomena of which we can at
present only confess our ignorance. Compare,
for example, what we know of the tides and
the weather. Had Odysseus and his men been
stranded high and dry by a spring tide on the
Thrinacian Isle they would probably have offered a
hecatomb to Poseidon praying him to send another
spring tide on the morrow. A modern mariner, more
wise and less pious than Odysseus, would have consumed the kine of Helios in peace for a fortnight,
and then have taken his departure with comparative
ease. On the other hand the modern mariner, like
Odysseus of old, might still pray for calm weather,
thus projecting his inability to formulate a scientific
law into want of routine and possible anomy (p. 114)
in the sequence of his perceptions. If we believe in
the capacity of the reflective faculty for ultimately
reducing to a brief formula or law all types of
phenomena, if we believe in the co-ordination of
perception and reflection, then the weather will not
probably appear a very strong argument against our
hypothesis. It must at least be confessed that the
discovery of a hundred or a five hundred years' period
in the weather would sadly discomfort those who
delight in assuming that some group of perceptions

at least must be beyond the analysis of the reflective faculty. Yet such a discovery would not now be more remarkable than that of the Chaldean *Saros* or eclipse period,[1] must have been to those who looked upon eclipses as an arbitrary interference with their perceptions, and prayed vigorously for a restoration of the light of sun or moon. The coeval development of the perceptive and reflective faculties associated with a power of selecting sensations in the former is possibly an important, but it may not be the sole, factor in the marvellous power which the reason possesses of describing wide ranges of phenomena by simple laws. There is another point which undoubtedly deserves notice. Our sense-impressions are indeed complex in their grouping, but they come to us by very few and comparatively simple channels, namely, through the organs of sense. The simplicity of the scientific law may therefore be partly conditioned by the simplicity of the modes in which sense-impressions are received.

The arguments of this section are, of course, very far from conclusive. They are only meant to suggest the possibility that the perceptive faculty in itself determines largely or entirely the routine of our perceptions. If this be true it will seem less of a marvel that the co-ordinated reflective faculty should be able to describe the "outside universe" by comparatively simple formulæ. On the whole this seems a more scientific hypothesis than those which make the routine depend on supersensuous entities, and which then—to account for the power of the human reason

---

[1] The Chaldeans had discovered that eclipses of the sun and moon recur in a cycle of eighteen years and eleven days, and were thus able to predict the dates of their occurrence.

to analyze nature—endow those entities with reason akin to man's, thus postulating thought and consciousness apart from that material machinery which alone justifies our inferring its existence. The hypothesis we have discussed, unproven as it may be, postulates reason no further than we may logically infer it, and at the same time attempts to account for the power of analyzing the routine of the perceptions, which is undoubtedly possessed by the human reflective faculty.

§ 13.—*The Mind as a Sorting-Machine.*

It is not hard to imagine by extension of existing machinery a great stone-sorting machine of such a character that, when a confused heap of stones was thrown in pell-mell at one end, some sizes would be rejected, while the remainder would come out at the other end of the machine sifted and sorted according to their sizes. Thus a person who solely regarded the final results of the machine might consider that only stones of certain sizes had any existence, and that such stones were always arranged according to their sizes. In some such way as this, perhaps, we may look upon that great sorting-machine—the human perceptive faculty. Sensations of all kinds and magnitudes may flow into it, some to be rejected at once, others to be sorted all orderly, and arranged in place and time. It may be the perceptive faculty itself, which, without our being directly conscious of it, contributes the ordered sequence in time and space to our sense-impressions. The routine of perception may be due to the recipient, and not characteristic of the material. If anything like this be the case, then (granted a co-ordination of perceptive and reasoning faculties), it will be less surprising that, when the human mind comes to analyze phenomena in time and space, it should find itself capable of briefly describing the past, and of predicting the future sequences of all manner of sense-impressions. From this standpoint the nomic natural law is an unconscious product of the machinery of the perceptive faculty, while natural law in the scientific sense is the conscious product of the reflective faculty, analyzing the process of perception, the working of the sorting-machine. The whole of *ordered* nature is thus seen as the product of one mind—the only mind with which we are acquainted—and the fact that the routine of perceptions can be expressed in brief formulæ ceases to be so mysterious as when we postulate a twofold reason, one type characteristic of "things-in-themselves," beyond our sense-impressions, and another associated with the material machinery of nervous organization.

§ 14.—*Science, Natural Theology, and Metaphysics.*

The reader, I trust, will treat these suggestions as suggestions and no more. What we are sure of is a certain routine of perceptions and a capacity in the mind to resume them in the mental shorthand of scientific law. What we have no right to infer is that order, mind, or reason—all human characteristics or human conceptions falling on this side of sense-impressions—exist on the other side of sense-impressions, in the unknown plus of sensations or in things-in-themselves. Whatever there may be on that outside, we cannot logically infer it to be like anything whatever on this side. Scientifically we must remain agnostic. If, however, it is possible to conceive the order, the routine of perceptions as being due to anything on this side of sense-impression, we shall have withdrawn from the beyond the last anthropomorphical element, and left it that chaos behind sense-impression, whereof to use the word knowledge would be the height of absurdity.

To positive theology, to *revelation*, science has no rejoinder. It works in a totally different plane. Only when belief enters the sphere of possible knowledge, the plane of reality, must science sternly remonstrate ; only when belief replaces knowledge as a basis of conduct is science driven to criticize not the reality, but the morality of belief. Quite different, however, is the relation of science to natural theology and metaphysics, when they assert that reason can help us to some knowledge of the supersensuous. Here science is perfectly definite and clear ; natural theology and metaphysics are pseudo-science. The mind is absolutely confined within its nerve-exchange; beyond the walls of sense-impression it can logically infer nothing. Order and reason, beauty and benevolence, are characteristics and conceptions which we find solely associated with the mind of man, with this side of sense-impressions. Into the chaos of sensations we cannot scientifically project them; we have no ground whatever for asserting that any human conception will suffice to describe what may exist there, for it lies outside the barrier of sense-impressions from which all human conceptions are ultimately drawn. Briefly chaos is all that science can logically assert of the supersensuous—the sphere outside knowledge, outside classification by mental concepts. If the Brahmins believe that the world arose from the instinct of an infinite spider, for so it has been *revealed* to them, we may wonder what the conceptions *instinct* and *spider* may be in their minds, and remark that their belief is without meaning for us. But if they assert that the phenomenal world gives in itself evidence of being spun from the bowels of this monster, then we pass from the plane of belief to that of reason and science, and promptly demolish their fantasy.

§ 15.—*Conclusions.*

It may seem to the reader that we have been discussing at unjustifiable length the nature of scientific law. Yet therein we have reached a point of primary importance, a point over which the battles of system and creed have been long and bitter. Here the materialists have thrown down the gauntlet to the natural theologians, and the latter in their turn have endeavoured to deck dogma with the mantle of science. The world of phenomena for the materialists was an outside world unconditioned by man's perceptive faculty, a world of "dead" matter subjected for all time to unchangeable nomic laws (p. 114), whence flowed the routine of our perceptions. The Stoics, with greater insight, found these laws replete with reason, but, dogmatic in turn, they postulated a reason akin to man's inherent in matter. The natural theologians, like the materialists, found "dead" matter, but, like the Stoics, they saw strong evidence of reason in its laws; this reason they placed in an external lawgiver. Metaphysician and philosopher filled the measure of obscurity by hypotheses as to mind-stuff and will and consciousness, which had not become consciousness, existing behind the barrier of sense-impression. Science—

refusing to infer wildly where it cannot know, and unwilling to assume new causes where the old have not yet been shown insufficient—treats the " dead matter " of the materialist as a world of sense-impressions. These sense-impressions appear to follow an unchanging routine capable of expression in the brief formulæ of science because the perceptive and reflective faculties are machines of practically the same type in all normal human beings. Like the Stoics, the scientist finds evidence of reason in his examination of natural phenomena, but he is content to think that this reason may be his own till he discovers evidence to the contrary. He recognizes that the so-called law of nature is but a simple *résumé,* a brief description of a wide range of his own perceptions, and that the harmony between his perceptive and reasoning faculties is not incapable of being traced to its origin. Natural law appears to him an intellectual product of man, and not a routine inherent in "dead matter." The progress of science is thus reduced to a more and more complete analysis of the perceptive faculty—an analysis which unconsciously and not unnaturally we project into an analysis of something beyond sense-impression. Thus both the material and the laws of science are inherent in ourselves rather than in an outside world. Our groups of perceptions form for us reality, and the results of our reasoning on these perceptions and the conceptions deduced from them form our only genuine knowledge. Here only we are able to reach truth—to discover similarity and to describe sequence—and we must remorselessly criticize every step we take beyond, if we would avoid the " muddy speculation " which will ever arise when we attempt to extend the field of knowledge by obscure definitions of natural law.

If it should seem to the reader that I have too narrowly circumscribed, not the field of *possible human* knowledge, but the meaning of the word knowledge itself, he must remember the danger which arises when we employ terms without concise meaning and clearly defined limits. The right of science to deal with the beyond of sense-impressions is not the subject of contest, for science confessedly claims no such right. It is within the field of knowledge as we have defined it, especially at points where our knowledge is only in the making, that the right of science has been questioned. It is easy to replace ignorance by hypothesis, and because only the attainment of real knowledge can in many cases demonstrate the falseness of hypothesis, it has come about that many worthy and otherwise excellent persons assert an hypothesis to be true, because science has not yet by positive knowledge demonstrated its falsehood. Here, in the untilled part of the heritage of science, lies the playground of the undisciplined imagination. *Mine*, says science here, as it does not claim of the supersensuous, and it hastens where it can to take effective occupation. Science, we are told, does not explain the origin of life ; science does not explain the development of man's higher faculties ; science does not explain the history of nations.

If by explain[1] is meant "describe in a brief formula," let us admit that science has not yet fully analyzed

[1] No objection can be raised to the words *explain* and *explanation* if they be used in the sense of the descriptive *how*, and not the determinative *why*. The former interpretation is the sole one given to them in this work.

these phenomena. What, then, must follow the admission ? Why, an honest confession of our ignorance and not mistrust in our fundamental principles—no meaningless hunt after unknown origins in the supersensuous, until the known field of perceptions has been shown incapable of yielding the needful basis. To-day our churches still offer up prayers for the weather, and the mystery of Saturn's rings is hardly fully solved ; fifty years ago we could give no account of the origin of species. The mystery of the latter was used as striking evidence of the insufficiency of science and as a valid argument for an anomy, a separate creation of each type of life. Driven from one stronghold of ignorance, those who delight in the undisciplined imagination rather than in positive knowledge, only seek refuge in another. The part played years ago by our ignorance as to the origin of species is now played by our supposed ignorance as to the origin of the higher faculties in man. As well take refuge in the weather or in the mystery of Saturn's rings, for all alike belong to the world of sense-impressions and therefore are material with which the scientific method can and will ultimately cope.

Does science leave no mystery ? On the contrary, it proclaims mystery where others profess knowledge. There is mystery enough in the chaos of sensations and in its capacity for containing those little corners of consciousness which project their own products, of order and law and reason, into an unknown and unknowable world. There is mystery enough here, only let us clearly distinguish it from ignorance within the field of possible knowledge. The one is impenetrable, the other we are daily subduing.

SUMMARY.

1. Scientific law is of a totally different nature from civil law ; it does not involve an intelligent lawgiver, a command and a corresponding duty. It is a brief description in mental shorthand of as wide a range as possible of the sequences of our sense-impressions.

2. There are two distinct meanings to natural law : the mere routine of perception, and the scientific law in the field of nature. The "reason" in natural law is only obvious when we speak of law in the latter sense, and it is then really placed there by the human mind. Thus the supposed reason behind natural law does not enable us to pass from the routine of perceptions to anything of the nature of reason behind the world of sense-impression.

3. The fact that the human reflective faculty is able to express in mental formulæ the routine of perceptions may be due to this routine being a product of the perceptive faculty itself. The perceptive faculty appears to be selective and to have developed in co-ordination with the reflective faculty. Of the world outside sense-impression science can only logically infer chaos, or the absence of the conditions of knowledge ; no human concept, order, reason, or consciousness, can be logically projected into it.

_____

LITERATURE.

AUSTIN, J.—Lectures on Jurisprudence. London, 1879. (Especially Lectures I. to V.)

HUME, D.—Dialogues concerning Natural Religion (pp. 375-468 of vol. ii. of The Philosophical Works, edited by Green and Grose).

STUART, J.—A Chapter of Science ; or, What is a Law of Nature ? London, 1868. (A series of six lectures, of which the first five can still be read with some profit, if read cautiously, while the last forms for the student of logic a useful study in paralogisms.)

CHAPTER IV.

CAUSE AND EFFECT. PROBABILITY

§ 1.—*Mechanism*.

THE discussion of the previous chapter has led us
to see that law in the scientific sense only describes
in mental shorthand the sequences of our perceptions.
It does not explain *why* those perceptions have a
certain order, nor *why* that order repeats itself; the
law discovered by science introduces no element of
necessity into the sequence of our sense-impressions;
it merely gives a concise statement of *how* changes
are taking place. That a certain sequence has occurred
and recurred in the past is a matter of experience to
which we give expression in the concept *causation;*
that it will continue to recur in the future is a matter
of belief to which we give expression in the concept
probability. Science in no case can demonstrate any
inherent necessity in a sequence, nor prove with absolute certainty that it must be repeated. Science for
the past is a description, for the future a belief; it is not,
and has never been, an explanation, if by this word is
meant that science shows the *necessity* of any sequence
of perceptions. Science cannot demonstrate that a
cataclasm will not engulf the universe to-morrow, but
it can prove that past experience, so far from providing
a shred of evidence in favour of any such occur-

rence, does, even in the light of our ignorance of any necessity in the sequence of our perceptions, give an overwhelming probability against such a cataclasm. If the reader has once fully grasped that science is an intellectual *résumé* of past experience and a mental balancing of the probability of future experience, he will be in no danger of contrasting the "mechanical explanation" of science with the "intellectual description" of mythology.

Some years ago (1885) Mr. Gladstone wrote a remarkable article in *The Nineteenth Century* in which he inveighed against the "dead mechanism" to which he asserted men of science reduced the universe. He contrasted the *mechanical* with the *intellectual*, and bravely defended what he termed the "majestic process of creation" described in the first chapter of Genesis against the Darwinian theory of evolution. He has recently repeated several of his arguments in a more elaborate work.[1] Now, when a man of Mr. Gladstone's ability states a paradox of this kind, we may be fairly certain that it arises from some popular confusion in the use of terms, and it befits us to inquire how popular and scientific usage differ as to the word *mechanical*. Unfortunately, some more or less superficial works on natural science give currency to the notion that mechanics is a code of rules which nature of inherent necessity obeys. We are told in books published even within the last few years that mechanics is the science of force, that force is the cause which produces or tends to produce change of motion, and that force is inherent in matter. Force thus appears to the popular mind as an agent inherent in unconscious matter producing change. This agent is very natu-

[1] *The Impregnable Rock of Holy Scripture.* London, 1890.

rally contrasted with the will of a living being, the consciousness of a capacity to produce motion. In matter this consciousness cannot be inferred, and thus force is contrasted as a " dead " agent with will as a "living" agent. The mind which has not probed beyond the surface of physics sympathizes with Mr. Gladstone's revolt against the " dead mechanism " to which, in the imagination of both, science reduces the universe. Now "matter" is for us a group of sense-impressions and "matter in motion " is a sequence of sense-impressions. Hence that which causes change of motion [1] must be that which determines a sequence of sense-impressions, or, in other words, it is the source of a routine of perceptions. But the source of such routine, as we have seen, lies either in the field of the unthinkable beyond sense-impression or else in the nature of the perceptive faculty itself. The " cause of change in motion " thus either lies in the unthinkable or is a factor of perception ; in neither case can it with any intelligible meaning of the words be spoken of as a " dead agent." In the former case the cause of change is unknowable, in the latter it is unknown, and may long remain so, for we are very far at present from understanding how the perceptive faculty can condition a routine of perceptions. Science does not deal with the unknowable, and if force be not unknowable, but unknown, then mechanics as the science of force would as yet have made no progress. The reality is indeed different from this. One of the greatest of German physicists,

[1] We shall see reason in the sequel for asserting that " motion " is a conception, rather than a perception—a scientific mode of representing change of sense-impressions, rather than a sense-impression itself. In this chapter, however, the term " motion " is used in its popular sense for a well-marked class of sequences of sense-impression.

Kirchhoff, thus commences his classical treatise on mechanics [1] :—

" Mechanics is the science of motion ; we define as its object the complete *description* in the *simplest* possible manner of such motions as occur in nature."

In this definition of Kirchhoff's lies, I venture to think, the only consistent view of mechanism and the true conception of scientific law. Mechanics does not differ, as so often has been asserted, from biology or any other branch of science in its essential principles. The laws of motion no more account than the laws of cell-development for the routine of perception ; both solely attempt to describe as completely and simply as possible the repeated sequences of our sense-impressions. Mechanical science no more explains or accounts for the motion of a molecule or a planet than biological science accounts for the growth of a cell. The difference between the two branches of science is rather quantitative than qualitative ; that is, the descriptions of mechanics are simpler and more general than those of biology. So wide-embracing and general are the laws of motion, so completely do they describe our past experience of many forms of change, that with a considerable degree of confidence we believe they will be found to describe all forms of change. It is not a question of reducing the universe to a " dead mechanism," but of measuring the amount of probability that one description of change of a highly generalized and simple kind will ultimately be recognized as capable of replacing another description of a more specialized and complex character. It is not taking biology out of one branch of what might be

[1] *Vorlesungen über mathematische Physik. Bd.* I. *Mechanik,* S. 1. Berlin, 1876.

termed *descriptive* science and removing it into another
—that of *prescriptive* science. Here by *prescriptive*
science we denote an imaginary aspect of science, which
mechanics are too frequently supposed to present,
namely, that of deducing some inherent necessity in
the routine of perceptions, instead of merely describing
that routine in simple statements. When, therefore,
we say that we have reached a "mechanical explanation" of any phenomenon, we only mean that we have
described in the concise language of mechanics a
certain routine of perceptions. We are neither able
to explain why sense-impressions have a definite
sequence, nor to assert that there is really an element
of necessity in the phenomenon. Regarded from this
standpoint the laws of mechanics are seen to be
essentially an intellectual product, and it appears absolutely unreasonable to contrast the mechanical with
the intellectual when once these words are grasped in
their accurate scientific sense.

§ 2.—*Force as a Cause.*

If force be looked upon as the cause of change in the
sense that it necessitates a certain routine of perceptions, then we have no means of dealing with force. It
may be the structure of the perceptive faculty, or it
may be any of the phantasms with which metaphysicians people the beyond of sense-impression. Force
will not, therefore, aid us in our search for a scientific
conception of *cause*. As we have seen that there are
two or even three ideas conveyed by the one term law,
so there are at least two ideas associated with the
word cause, and their confusion has also led to as much
"muddy speculation." Let us first investigate the
popular idea of cause and then see how this is related

to the scientific definition. A very slight amount of observation has shown men that certain sequences of change *apparently* arise from the voluntary action, the will of a living agent. I take up a stone ; no one can predict with certainty what I shall do with it. What follows my picking up the stone is to all appearances a new sequence quite independent of any which preceded it. I can let it fall again ; I can put it into my pocket, or I may throw it into the air in any direction and with any of a great variety of speeds. The result of my action may be a long sequence of physical phenomena to describe which mechanically would require the solution of complex problems in sound, heat, and elasticity. The sequence, however, appears to start in an act of mine, in *my* will. *I* appear to have called it into existence, and in ordinary language I am spoken of as the *cause* of the resulting phenomena. In this sense of the word cause I appear to differ qualitatively from any other stage in the sequence. Had the hand of a stronger man compelled mine to throw the stone, I should at once have sunk into a link in the chain of phenomena ; he, not I, would have been *the* cause of the resulting motion.

It is certainly true that even in popular usage intermediate stages in the sequence will occasionally be spoken of as causes. If the stone from my hand break a window, the cause of the broken window might very likely be spoken of as the moving stone. But although this usage, as we shall see afterwards, is an approach to the scientific usage of the word cause, it yet involves in the popular estimation an idea of enforcement which is not in the latter. That the stone moving with a certain speed *must* produce the destruction of the window is, I think, the idea involved in thus speaking of the moving stone as the cause of the breakage. Were our perceptive organs sufficiently powerful, what science conceives that we should see before the impact would be particles of window and particles of stone moving in a certain manner, and after the impact would be the same particles moving in a very different manner. We might carefully *describe* these motions, but we should be unable to say why one stage would follow another, just as we can describe *how* a stone falls to the earth, but not say *why* it does. Thus, scientifically the idea of *necessity* in the stages of the sequence—stone in motion, broken window—the idea of enforcement would disappear ; we should have a routine of experience, but an unexplained routine. Hence, when we speak of the stages of a sequence in ordinary life as causes, I do not think it is because we are approaching the scientific standpoint, but I fear it arises from our associating, through long usage, the idea of *force* with the stone. The stone is the cause of certain new motions, just as I am looked upon as the cause of certain motions in the stone—that is, both stone and I are supposed to *enforce* subsequent stages in the sequence. Now the reader who has once dismissed the notion of force as a cause, which I think he will probably be prepared to do, will perhaps admit that there is no element of enforcement, but merely a routine of experience in the motions of particles of stone and glass. Still he may say that the will of a living agent does seem to him a cause of motion in the necessarian sense. Nor would he be in this unreasonable, for I must confess that to attribute sequences of motion to will seems at first sight a more scientific hypothesis than to attribute them to an unknown and possibly unknowable source force.

§ 3.—*Will as a Cause.*

It is not unnatural that human beings should be impressed at a very early stage of their mental growth with the real, or at any rate apparent, power which lies in their will of originating "motion." In this manner we find that most primitive peoples attribute all motions to some will behind the moving body; for their first conception of the cause of motion lies in their own will. Thus they consider the sun as carried round by a sun-god, the moon by a moon-god, while rivers flow, trees grow, and winds blow owing to the will of a spirit which dwells within them. It is only in the long course of ages that mankind more or less clearly recognizes will as associated with consciousness and a definite physiological structure; then the spiritualistic explanation of motion is gradually displaced by the scientific description; we eliminate in one case after another the direct action of will in the motion of natural bodies.[1] The idea, however, of enforcement, of some necessity in the order of a sequence remains deeply rooted in men's minds, as a fossil from the spiritualistic explanation of will as the cause of motion. This idea is preserved in association with the scientific description of motion, and in the materialist's notion of force as that which *necessitates* certain changes or sequences of motion, we have the ghost of the old spiritualism. The force of the materialist is the will of the old spiritualist separated

---

[1] The spiritualistic explanation still of course exists where the scientific analysis is incomplete. We continue to appeal to a spirit " at whose command the winds blow and lift up the waves of the sea and who stilleth the waves thereof," or who " sends a plague of rain and waters."

from consciousness. Both carry us into the region beyond our sense-impressions, both are therefore *metaphysical;* but perhaps the inference of the old spiritualist was, if illegitimate, less absurdly so than that of the modern materialist, for the spiritualist did not infer will to exist beyond the sphere of consciousness with which he had always found will associated.

Force as cause of motion [1] is exactly on the same footing as a tree-god as cause of growth—both are but names which hide our ignorance of the *why* in the routine of our perceptions.

§ 4.—*Secondary Causes involve no Enforcement.*

Let us endeavour to see a little more closely how the idea of any inherent necessity in the particular order taken by our perceptions disappears from the scientific conception of a sequence of motions—at least from all but the first stage, if the sequence arise from an apparent act of will. Still speaking in the popular sense, we will term the act of will, if it exists, a first cause, and the successive stages of the sequence secondary causes. Our present proposition is that the scientific description of motion involves no idea of enforcement in the successive stages of motion. We shall see in the sequel that the whole tendency of modern physics has been to describe natural phenomena by reducing them to conceptual motions. From these motions we construct the more complex motions by aid of which we describe actual sequences of sense-impressions. But in no single case have we discovered *why* it is that these motions are

---

[1] Force as a name used for a particular *measure* of motion will be found in our chapter on the " Laws of Motion" to involve no obscurity, and to be in itself a convenient term.

taking place ; science describes how they take place, but the *why* remains a mystery. To term it force might not be so productive of obscurity as it is, were there any suggestion in the elementary text-books that the cause of motion, or of change in motion, may be the nature of the perceptive faculty, or will, or the deity, or any unknowable x amid an unthinkable y and z. The glib transition from force as a cause to force as a measure of motion too often screens the ignorance which it is as much the duty of science to proclaim from the house-tops as it is its duty to assert knowledge on other points. Primitive man placed a sun-god behind the sun (as some of us still place a storm-god behind the storm), because he did not see how and why it moved. The physicist now proceeds to describe *how* the sun moves, by describing how a particle of earth and a particle of sun move in each other's presence. The description of that motion is given by Newton's law of gravitation, but the *why* of that motion is just as mysterious to us as the motion of the sun to the barbarian.[1] No one knows why two ultimate particles influence each other's motion. Even if gravitation be analyzed and described by the motion of some

---

[1] The reader will find it profitable to analyze what is meant by such statements as that the law of gravitation *causes* bodies to fall to the earth. This law really describes how bodies do fall according to our past experience. It tells us that a body at the surface of the earth falls about sixteen feet towards the earth in the first second, and at the distance of the moon about $\frac{1}{3700}$ part of this distance in the same time. The law of gravitation describes the rate at which a body falls, or, better, the rate at which its motion is changed at diverse distances, and the force of gravitation is really a certain measure of this change of motion, and no useful purpose can be served by defining it as the cause of change in motion. Other physical laws ought to be interpreted in the same anti-metaphysical manner.

simpler particle or ether-element, the whole will still be a description, and not an explanation, of motion. Science would still have to content itself with recording the *how*. In what we have termed secondary causes, therefore, science finds no element of enforcement, solely the routine of experience· But the idea of will as a first cause has been over and over again associated with secondary causes. Aristotle, noting the difficulty of explaining why motions take place, introduced not only God as a first cause, but, like primitive man, made God an immediate source of the enforcement in every secondary cause. God, Aristotle held, is continually imparting motion to all the bodies in the universe, and so producing phenomena. Aristotle's doctrine was accepted by the mediæval schoolmen, and for many centuries remained fundamental in philosophical and theological writings. Schopenhauer, the German metaphysician, perceiving that the only known apparent first cause of motion was will, placed will behind all the phenomena of the universe, much like the barbarian who postulates the will of a storm-god behind the storm.[1] But however little logical basis these metaphysical speculations possess—all failing to satisfy our canons of legitimate inference (p. 72)—they still suffice to mark the distinction between the popular or metaphysical conception of cause as enforcement, and the scientific conception of cause as the routine of experience. Every association of inherent necessity with secondary causes is a passage from physics to meta-

---

[1] Sir John Herschel went so far as to identify gravitation and will ! (*Outlines of Astronomy*, arts. 439–40). Other samples of the same animistic tendency will be found in the writings of Dr. J. Martineau and the late Dr. W. B. Carpenter.

physics, from knowledge to fantasy. Historically, I think, the whole association can be traced back through the old spiritualism to the sequences of motion which the will as a first cause can apparently enforce. Here, then, it befits us to ask two questions: Does the will in any way really account for motion? Is there any ground for supposing the will to be an arbitrary first cause?

§ 5.—*Is Will a First Cause?*

Now, in attempting to answer these questions scientifically we must bear in mind that what we term will is only known to us in association with consciousness, and that we can only infer consciousness where we find a certain type of nervous system. Does will as an apparently spontaneous origin of motion throw any light on the mystery of motion? Does it in any way explain the particular sequences motions take? To be consistent we shall have to suppose, with Aristotle, that every phase of motion is the direct product of a conscious being. Let us return to the example of the stone. Apparently, by the arbitrary action of my will, I set the stone in motion. I appear in doing this as a first cause. But a complex sequence of motions now arises. Each stage of this sequence I can conceive myself mechanically *describing*, but I am quite unable to assert the necessity, the *why* of these stages. For example, the stone falls to the ground, and I can say approximately how many feet it will fall in the first and in the following seconds. That is the result of past experience used to predict the future, the result of the classification of phenomena resumed in the law of gravitation; but this law does not explain the *why* of the motion. If I grant that my will set the stone in motion, I cannot suppose it to continue in motion for the same reason, for any amount of willing after the stone has left my hand will not, in the majority of cases, be in the least able to influence its motion. Hence, even in motion started by a conscious being, we have at once a mystery. My will might explain the origin, it cannot explain the continuance of the motion.   If will is to help us at all, we must postulate it as producing motion at every stage.   But clearly this will is not my will ; it must be some other will. Here we are only restating the solutions of primitive man with his spiritualism behind nature, of Schopenhauer with his undefined will behind all phenomena, of Aristotle when he says God moves all things. But this solution involves an extension of the notion of will beyond the sphere where we may legitimately infer its existence. Like the hypothesis of force it postulates an unthinkable x outside sense-impressions. It carries us no-whither. Will cannot, therefore, be looked upon as necessitating a sequence of motion, any more than what we have termed a secondary cause, for in the great majority of cases, if will be supposed to start a motion, it cannot enforce its continuance in a particular sequence, and so far as the will is concerned the motion might cease at its birth.

§ 6.—*Will as a Secondary Cause.*

Will thus appears, like the secondary cause, as a stage in the routine of perceptions. Our experience shows us that in the past an act of will occurred at a certain stage in a routine of perceptions, but we cannot assert that there was anything in the act itself which *enforced* the stages which followed. Does will, however, differ on closer analysis from other secondary causes in being the *first* stage of an observed routine? This leads us to our second question (p. 147), and the answer to it is really involved in the views on consciousness which have been developed in our second chapter.

We have seen that the difference between a voluntary and involuntary exertion lies in the latter being conditioned only by the immediate sense-impression, while the former is conditioned by stored sense-impresses and the conceptions drawn from them. Where consciousness exists, there there may be an interval between sense-impression and exertion, this interval being filled with the " resonance," as it were, of associated but stored sense-impresses and their correlated conceptions. When the exertion is at once determined by the immediate sense-impression (which we associate with a construct projected *outside* ourselves), we do not speak of will, but of reflex action, habit, instinct, &c. In this case both sense-impression and exertion appear as stages in a routine of perceptions, and we do not speak of the exertion as a first cause, but as a direct effect of the sense-impression ; both are secondary causes in a routine of perceptions, and capable of mechanical description. On the other hand, when the exertion is conditioned by the stored sense-impresses, it appears to be conditioned by something *within* ourselves ; by the manner in which memory and past thought have linked together stored sense-impresses and the conceptions drawn from them. No other person can predict with absolute certainty what the exertion will be, for the contents of our mind are not objects to him. None the less the inherited features of our brain, its present physical condition owing to past nurture, exercise, and general health, our past training and experience are all factors determining what sense-impresses will be stored, how they will be associated, and to what conceptions they will give rise. By this we are to understand that, if we could bring into the sphere of perception the processes that intervene in the brain between immediate sense-impression and conscious exertion, we should find them just as much routine changes as what precedes the sense-impression or follows the exertion. In other words, will, when we analyze it, does not appear as the first cause in a routine of perceptions, but merely as a secondary cause or intermediate link in the chain. The " freedom of the will " lies in the fact that exertion is conditioned by our own individuality, that the routine of mental processes which intervenes between sense-impression and exertion is perceived objectively neither by us nor by any one else, and psychically by us alone. Thus will as the first cause of a sequence of motions explains nothing at all; it is only a limit at which very often our power of describing a sequence abruptly terminates.

So much is this recognized by modern science, that special branches of it are entirely devoted to describing the sequences of secondary causes, the routine which precedes special determinations of the will. Science tries to describe how will is influenced by desires and passions, and how these again flow from education, experience, inheritance, physique, disease, all of which are further associated with climate, class, race, or other great factors of evolution. Thus, with the advance of our positive knowledge we come more and more to regard individual acts of will as secondary causes in a long sequence, as stages in a routine which can be described—stages, however, at which the routine changes its at present knowable side from the psychical to the physical. An act of will thus appears as a secondary cause, and no longer as an arbitrary first cause. Evil acts flow indeed from an anti-social will, and as hostile to itself society endeavours to repress them ; but the anti-social will itself is seen as a heritage from a bad stock, or as arising from the conditions of past life and training. Society begins more and more to regard incorrigible criminals as insane, and slight offenders as uneducated children.

§ 7.—*First Causes have no Existence for Science.*

We have now reached some very important conclusions with regard to will as a cause. In the first place, the only will known to us (or the only *like* will that we can logically infer to exist) is seen not to be associated with an arbitrary power to originate, alter, or stop a motion. It appears merely as a secondary cause, as a stage in a routine, but one where the knowable side of the routine changes from the psychical to the physical. Further, there lies in this will no power of enforcing a sequence of motions. The will as first cause is merely a limit arising from some impossibility in our powers of further following the physical side of a routine, or of discovering its further psychical side ; it is merely another way of saying : At this point our ignorance begins. The moment the only will we know or infer ceases to appear as the arbitrary originator or enforcer of a sequence, so soon as it sinks to a stage—if a remarkable stage—in a routine, then it becomes idle to suppose will as the backbone of natural phenomena. Will, as the creator and maintainer of nature, is either an old name used for some unknown and unthinkable existence, or if used in the only sense now intelligible to us, that of a secondary cause or stage in a routine, it gives us no assistance in comprehending routine. We are just as wise if we drop this will behind phenomena, and content ourselves with observing that there is a routine in perceptions. This, in fact, is what science does, not unnecessarily multiplying causes, when no simplification of perceptions arises from postulating their existence.

We have seen that the conception of will as an arbitrary source of motion arose historically, and not unnaturally, from a portion of the routine of which will is a stage being both physically and psychically screened from the observer, owing to its being buried in the individuality of another person. We have further noticed that as will and motion are more carefully analyzed, the conception that will originates motion ceases to have any consistency. But with will as first cause falls to the ground any possible experience of first causes on our part. We can no longer infer even the possibility of the existence of first causes, for there is nothing like them in our experience, and we cannot by the second canon of logical inference (p. 72) pass from the known to something totally unlike it in the unknown. Science knows nothing of first causes. They cannot, as Stanley Jevons has supposed,[1] be inferred from any branch of scientific investigation, and

[1] In the remarkably unscientific chapter entitled, "Reflections on the Results and Limits of Scientific Method," with which his, in so many respects, excellent *Principles of Science* concludes.

where we see them asserted we may be quite sure they mark a permanent or temporary limit to knowledge. We are either inferring something in the beyond of sense-impression, where knowledge and inference are meaningless words, or we are implying ignorance within the sphere of knowledge,[1] in which case it is more honest to say : " Here, for the present, our ignorance begins," than, " Here is a first cause."

§ 8.—*Cause and Effect as the Routine of Experience.*

We are now in a position, I think, to appreciate the scientific value of the word cause. Scientifically, cause, as originating or enforcing a particular sequence of perceptions, is meaningless—we have no experience of anything which originates or enforces something else. Cause, however, used to mark a stage in a routine, is a clear and valuable conception, which throws the idea of cause entirely into the field of sense-impressions, into the sphere where we can reason and reach knowledge. Cause, in this sense, is a stage in a routine of experience, and not one in a routine of inherent necessity. The distinction is, perhaps, a difficult one, but it is all the more needful that the reader should fully grasp it. If I write down a hundred numbers at chance—say by opening carelessly the

---

[1] The latter alternative—the temporary limit to ignorance—has been the chief source of " first causes." So long as the routine of history cannot be traced back more than a few centuries, we find no difficulty in asserting that the world began 6,000 years ago. So long as we do not grasp the evolution of life from its most primitive types, we postulate a first cause creating each type (Paley). So long as we do not observe the various grades of animal intelligence and consciousness, we suppose a soul implanted in every human being at birth. So long as we do not see that the mutual motion of two atoms is as mysterious as the life changes of a cell, we postulate a total difference between the two kinds of motion and a separate creation of life.

pages of a book—there results a sequence of numbers beginning, say—

141, 253, 73, 477, 187, 585, 57, 353 . . . &c.,

in which I cannot predict from any two or three or more numbers those which will follow. The number 477 does not enable me to say that 187 will follow it, the numbers which precede 187 in no way enforce or determine those which follow it. On the other hand, if I take the series—

1, 2, 3, 4, 5, 6, 7, 8 . . .

each individual number leads (by addition of 1) to the immediately following number, or in a certain sense determines it. The first series can, however, be written down so often that we learn it by rote, that it becomes a routine of experience. The analogy must not, of course, be pressed far, but it may still be of service. There is nothing in any scientific cause which compels us of inherent necessity to predict the effect. The effect is associated with the cause simply as a result of past direct or indirect experience. Or again, perhaps the matter may be grasped more clearly from a geometrical analogy. If I form the conception of a circle, it follows of inherent necessity that the angle at the circumference on any diameter is a right-angle. The one conception flows not as a result of experience but as a logical necessity from the other. No sequence of sense-impressions involves in itself a logical necessity. The sequence might be chaotic like our first series of numbers ; it has become for us a routine by repeated experience. The noteworthy fact in a routine of perceptions lies not so much in the particular order of the stages in the sequence, as in the result of experience that this order can exactly repeat itself.

The reader may perhaps wonder how, if the sequences of sense-impressions are really of the chaotic nature represented by our first series of numbers, it is possible to describe such sequences apart from their repetition by those brief formulæ we term scientific laws. As the perceptive faculty presents us, indeed, with the sequence, it is undeniably more like the second than the first series of numbers, for natural phenomena can without doubt be largely described by certain brief laws. We must rather put the actual case in the following form. We observe a person whose motives are quite unknown to us writing down the series—

$$1, 2, 4, 8, 16, 32,$$

and at present he has reached the number 32. A law describing the series is obvious—each number is twice the preceding one. With a great degree of probability we infer that he will now write down 64, especially if· we have seen him write the series up to and beyond 32 before. But there is nothing of logical necessity about his writing 64 after the preceding numbers. Those numbers, when we know the law, suggest his doing so, but do not enforce it.

We are now in a position to scientifically define *cause*. Whenever a sequence of perceptions D, E, F, G is invariably preceded by the perception C, or the perceptions C, D, E, F, G always occur in this order, that is, form a routine of experience, C is said to be a *cause* of D, E, F, G, which are then described as its effects. No phenomenon or stage in a sequence has only one cause, all antecedent stages are successive causes, and, as science has no reason to infer a first cause, the succession of causes can be carried back to the limit of existing knowledge, and beyond that *ad infinitum* in the field of conceivable knowledge. When we scientifically state causes we are really describing the successive stages of a routine of experience. Causation, says John Stuart Mill, is uniform [1] antecedence, and this definition is perfectly in accord with the scientific concept.

§ 9.—*Width of the Term Cause.*

The word cause, even in its scientific sense, is somewhat elastic. It has been used to mark uniform conjunction in space as well as uniform antecedence in time ; while if we take an actually existing group of perceptions, say the particular ash-tree in my garden, the causes of its growth might be widened out into a description of the various past stages of the universe. One of the causes of its growth is the existence of my garden, which is conditioned by the existence of the metropolis ; another cause is the nature of the soil, sand approaching the clay limit, which again is conditioned by the geological structure and past history of the earth. The causes of any *individual* thing thus widen out into the unmanageable history of the universe. The ash-tree is like Tennyson's "flower in the crannied wall" : to know all its causes would be to know the universe. To trace causes in this sense is like tracing back all the lines of ancestry which converge in one individual ; we soon reach a point where we can go no further owing to the bulk of the material. Obviously science in tracing causes attempts no task of this character, but at the same time it is useful to remember how essentially the causes of any finite portions of the universe lead

---

[1] "Uniformity" and "sameness" are, in the perceptual world, however, only relative terms (*see* p. 200).

us irresistibly to the history of the universe as a whole. This thought suggests how closely knit together are in reality the most diverse branches of our positive knowledge. It shows us how difficult it is for the great building of science to advance rapidly and surely unless its various parts keep pace with each other (p. 16). Practically science has to content itself with tracing one line of ancestry, one range of causes at a time, and this not for a special and individual object like the ash-tree in my garden, but for ash-trees or even trees in general. It is because science for its descriptive purposes deals with general notions or conceptions, that the words cause and effect have been withdrawn from the sphere of sense-impressions, from phenomena to which they strictly belong, and applied to the world of conceptions and ideas, where, indeed, there is logical necessity but no true cause and effect. To this point I shall return under § 11.

§ 10.—*The Universe of Sense-Impressions as a Universe of Motions.*

The reader can hardly fail to have been impressed in his past reading and experience with the great burden of explanation which is thrown on that unfortunate metaphysical conception *force*. He will undoubtedly have heard of the " mechanical forces" ruling the universe, of the "vital forces" directing the development of life, and of the "social forces" governing the growth of human societies.[1] He may perhaps

---

[1] A good illustration of the obscurity attaching to the use of the words force and cause may be taken from the recently published *History of Human Marriage*, by E. Westermarck.   The author writes :—"Nothing exists without a cause, but this cause is not sought in an agglomeration of external or internal forces." He thus implies that a cause ought to be sought in this unintelligible "agglomeration of

have concluded, with the present writer, that the word is not infrequently a fetish which symbolizes more or less mental obscurity. But the reason for the repeated occurrence of the word is really not far to seek. Wherever motion, change, growth, were postulated, there in the old metaphysics force as the cause of motion was to be found. The frequent use of the word force was due to the almost invariable association of *motion* with our perceptions, or, in more accurate language, to the analysis of nearly all our sense-impressions by aid of conceptual motions. For example, a coal fire may be said to be a cause of warmth. Here we mean that the group of sense-impressions we term coal, followed by the group we term combustion, has invariably in our experience been accompanied by the sense-impression warmth. We may, if we are chemists, be able to describe the chemical processes, the atomic changes or motions to which the phenomenon of combustion has been reduced ; we may, if we are physicists, describe the motion of the ethereal medium, to which the phenomenon of radiation of heat has been reduced ; we may, if we are physiologists, be able to describe the nerve-motions by aid of which the molecular motion of the finger-tips is interpreted as the sense-impression warmth at the brain. In all these cases we are dealing with the sequences of various types of motion, into which we analyze or reduce a variety of sense-

external and internal forces." Now, what the author attempts to do is to *describe* the various stages through which marriage has passed, and then to express the sequence of these stages by brief formulæ, such as those of natural selection. To use the word *force* hopelessly obscures his method.

impressions. Just as in the special case of gravitation, we can also describe these sequences and can
frequently give a measure to the motions which we
conceive to take place, but we are still wholly unable
to state *why* these motions occur. We may talk,
if we please, about the forces of combustion, the
forces of radiation, or even the forces inherent
in nerve substance ; we might indeed say that
the warmth, of which combustion is the cause, is
due to "an agglomeration of internal and external forces," but in using these phrases we do not
introduce an iota of new knowledge, but too often
a mountain of obscurity. We hide the fact that
all knowledge is concise description, all cause is
routine.

Now, it deserves special note that the sequences
with which we are dealing are all reducible to descriptions of motion, or of change. We need not
start arbitrarily with the combustion of the coal ; its
chemical constitution as an element in the sequence
of causes can, for example, be carried back through
a long past history in the evolution of coal, and we
cannot logically infer (p. 151) any beginning or first
cause in this sequence. Sequences of motion or of
change in natural phenomena go backwards and
forwards through an infinite range of causes, and to
begin or end them anywhere with a first or last cause
is simply to say that at such a point the sphere of
knowledge ends with an unthinkable x. The universe
thus appears to the scientist as a universe of motion,
motion the *why* of which is unknown, but the
sequences of which are, according to our experience,
invariably repeating themselves. The cause of motion
in the scientific sense lying in the sphere of sense-

impressions [1] cannot be the *why* of motions, we must seek it in some *uniform antecedent* of the motion--such, for example, as the past history of the motion, the relative position of the moving bodies, and so forth. How such antecedents are true scientific causes of motion we shall see in our Chapter VIII. devoted to the " Laws of Motion."

§ 11.—*Necessity belongs to the World of Conceptions, not to that of Perceptions.*

At this point the reader may feel inclined to say : " But surely there is as much necessity that a planet describing its elliptic orbit should at a certain time be in a certain position, as that the angles on the diameter of a circle should be right-angles?" With this I entirely agree. The *theory* of planetary motion is in itself as logically necessary as the theory of the circle ; but in both cases the logic and necessity arise from the definitions and axioms with which we mentally start, and do not exist in the sequence of sense-impressions which we hope that they will, at any rate approximately, describe. The necessity lies in the world of conceptions, and is only unconsciously and illogically transferred to the world of perceptions. This difference may be well illustrated by an example due to Mr. James Stuart, formerly Professor of Mechanism in Cambridge. Suppose I were to put a stone on a piece of flat ground and walk round it in that particular curve termed an ellipse, which a planet describes about the sun. We will further suppose the stone to be at that particular point

[1] That the frequently cited " muscular sensation of force" is really only a sense-impression interpreted as one of motion will be shown at a later stage of our work.

termed the focus which in the case of an elliptic orbit is actually occupied by the sun ; and lastly, I will walk round so that a line drawn from the stone to me sweeps out equal areas in equal times, a fundamental characteristic of the laws of planetary motion. Now my motion might be very fairly *described* by the law of gravitation, but it is quite clear that no force from the stone to me, no law of gravitation, could logically be said to cause my motion in the ellipse. We might in *imagination* conceive a point changing its motion according to the law of gravitation and tracing out my ellipse ; it might keep pace with me, and would, of logical necessity, cover equal areas in equal times. This logical necessity would flow from our definition, our conception, namely, that of a gravitating point. This point might be used to describe my elliptic motion, and to predict my positions in the future, but no observer would be logical in inferring that the necessary sequence of positions involved in the concept of a gravitating point could be transferred, or projected into a necessity in the sequence of his perceptions of my motion. I might go round the ellipse a hundred times in the same manner and then stop or go off in an entirely different path. The sole legitimate inference of the observer would then be that the law of gravitation was not a sufficiently wide-embracing formula to describe more than a portion of my motion.[1] This difference between necessity in

[1] The example cited is given by Mr. Stuart on p. 168 of his *A Chapter of Science*. It is there used to support the argument of primitive man ; my will causes me to go round the ellipse, therefore will causes the planets to go round in ellipses, and hence Mr. Stuart passes to Aristotle's God as continual mover of all things. That will is only found associated with certain types of material nervous

conception and routine in perception ought to be carefully borne in mind. The corpuscular, the elastic-solid, and the electro-magnetic theories of light all involve a series of conclusions of logical necessity, and we may use these conclusions as a means of testing our perceptions. So far as they are confirmed, the theory remains valid as a description ; if, on the other hand, our sense-impressions differ from these conclusions, the conclusions have just as much mental necessity, but the theory while valid for the mind is not valid as a description of the routine of perceptions. It is only the very great probability deduced from past experience of routine that enables us to speak of the " invariable order of the universe," or scientists to assert that facts which have hitherto proved obstinate will be ultimately embraced by well-established laws of nature. Not in the field of causation, but in that of conception do we deal with certainties.

§ 12.—*Routine in Perception is a necessary condition of Knowledge.*

While in the nature of perceptions themselves there appears nothing tending to enforce an order D, E, F, G rather than F, G, D, E, there is still a real need, if thought is to be possible, that the perceptive

---

systems is not used by Mr. Stuart, however, to logically infer the material nature of his first cause. He passes by the juggle of a common name from the known to the unthinkable outside the sphere of knowledge and science. The real truth which his *Chapter of Science* contains as to the characteristics of natural law is hopelessly vitiated by his theological standpoint. " I know," he says, "no result of science which could go to discredit any single thing in all the Bible" (p. 184). Mr. Stuart's 'science' is thus incomparably more retrograde than the modern Cambridge theology which discredits Noah's Ark.

faculty should always repeat the sequence in the *same* order. In other words, repetition or routine is an essential condition of thought; the actual order of the sequence is immaterial, but whatever it may be, it must repeat itself if knowledge is to be possible. We express this briefly in the law: *That the same* (p. 200) *set of causes is always accompanied by the same effect.* That the future will be like our experience of the past is the sole condition under which we can predict what is about to happen and so guide our conduct. But thought has been evolved in the struggle for existence as a guide to conduct, and therefore could not have been evolved had this condition been absent. If after the sense-impressions D, E, F, G, the sense-impression H does not uniformly follow, but A, J, or even Z, occur equally often, then knowledge becomes impossible for us, and we must cease to think. The power of thinking, —or of associating groups and sequences of sense-impressions, immediate or stored,—vanishes if these groups and sequences have no permanent elements by which they can be classified and compared.

In the struggle for existence man has won his dictatorship over other forms of life by his power of foreseeing the effects which flow from antecedent causes—not only by his memory of past experience, but by his power of codifying natural law, that is, by his power of generalizing experience in scientific statements. It was not necessary for his success that he should know *why* phenomena take place, but only that he should know *how* they take place, that he should be able to observe in them a routine, a repeated sequence as a basis for his knowledge. We have only to consider in some simple case—say that of the combustion of coal—what would follow for man if the resulting sense-impression were not uniform— if it were, for example, either intense warmth or intense cold—to appreciate that invariable order in the sequence of sense-impressions is an absolute con- dition for man's knowledge, and therefore for the foresight by aid of which he has won his dictatorship. In the chaos of sensations, in the "beyond" of sense- impressions, we cannot infer necessity, order or routine, for these are concepts formed by the mind of man on this side of sense-impressions. Yet if the supremacy of man is due to his reasoning faculty, so the condition for the existence of man as a reasoning being is routine in his perceptions, invariable order in the sequences of his sense-impressions. We can neither assert nor deny that this routine is due to something beyond sense-impression, for in that "beyond" the word routine is meaningless, and we can neither assert nor deny where we are dealing with a field to which the word knowledge cannot be applied. All we can assert is that the reasoning faculty in man connotes a perceptive faculty presenting sense-im- pressions in the same invariable order. That this routine is due to the nature of the perceptive faculty itself—to factors, of which we are unconscious in its constitution, akin to the conscious association and memory of the reasoning faculty—is a plausible if unproven hypothesis. It is one, however, as we have seen, suggested by the contemporaneous growth of perception and reason, and strengthened by the impossibility of any form of perceptive faculty, such as we find in the insane, surviving in the struggle for existence (p. 125).

While invariable order in the sequence of sense-

impressions is thus seen to be an essential characteristic of the perceptive faculty of a rational being, the
power to understand the why and wherefore of any
sequence is not so. It would undoubtedly be of great
intellectual interest to know *why* bodies fall to the
earth, but *how* they invariably fall is the practical
knowledge, which now enables us to build machines
and which enabled our forefathers to throw stones,
and thus helped them as it helps us in the struggle
for existence. Broadly speaking, here as elsewhere,
the perceptive faculty has developed along lines
which strengthen man's powers of self-preservation
and not along those which would merely minister to
his intellectual curiosity.

Anything, be it noted, that tends to weaken our
confidence in the uniform order of phenomena, in
what we have termed the routine of perceptions,
tends also to stultify our reasoning faculty by
destroying the sole basis of knowledge. It decreases
our power of foresight and lessens our strength in
the battle of life. For this reason theosophists and
spiritualists with their modern miracles contradicting
the long-experienced routine of perceptions are very
unlikely to form a society sufficiently stable to
survive in the struggle for existence. Every ecstatic
and mystical state weakens the whole intellectual
character of those who experience it, for it impairs
their belief in the normal routine of perceptions.
The abnormal perceptive faculty, whether that of the
madman or that of the mystic, must ever be a
danger to human society, for it undermines the
efficiency of the reason as a guide to conduct.
Conviction, therefore, of the uniform order of
phenomena is essential to social welfare.

But the reader may object that although this conviction be essential to social welfare, it does not follow that it is well based. Belief in a fetish may be essential to the welfare of a primitive tribe, and he who does not believe in it may be exterminated ; yet this does not demonstrate the rational character of the belief. It is right therefore that we should investigate whether our conviction is well based, and to this point we shall devote the remaining sections of this chapter.

In concluding the present section we may resume the results reached as follows :—

In the order of perceptions (cause and effect) no inherent necessity can be demonstrated.

In the uniformity with which sequences of perceptions are repeated (the routine of perceptions) there is also no inherent necessity, but it is a necessary condition for the existence of thinking beings that there should be a routine in perceptions. The necessity thus lies in the nature of the thinking being and not in the perceptions themselves ; thus it is conceivably a product of the perceptive faculty.

§ 13.—*Probable and Provable.*

Stanley Jevons in his discussion of the theory of probability, which forms one of the most valuable and interesting portions of his *Principles of Science*, remarks that the etymology of the word *probable* does not help us to understand what probability is and where it exists :—

" For, curiously enough, *probable* is ultimately the same word as *provable*—a good instance of one word

becoming differentiated to two opposite meanings "
(p. 197).[1]

Now we have seen that certainty belongs only to
the sphere of conceptions ; that inherent necessity has
a meaning in the mental field of logic, but that we cannot postulate it in the universe of perceptions ; that
the "necessity of natural law" is really an unjustifiable phrase. The word *proof*, therefore, used in the
sense of a demonstrable certainty applies only to
the sphere of conceptions. What are we then to
understand when the word *proof* is applied to natural
phenomena ? Shall we say that it is incorrect to use
the word *prove* at all in such relationship ? Yet our
leading men of science do use it. Here is a passage
from Sir William Thomson's lecture on "The Six
Gateways of Knowledge."[2] He is discussing the
possibility of our having a "magnetic sense," and he
writes :—

"I cannot think that that quality of matter in
space—magnetization—which produces such a prodigious effect upon a piece of metal, can be absolutely
without any—it is certainly not without any—effect
whatever on the matter of a living body ; and that
it can be absolutely without any *perceptible* effect
whatever on the matter of a living body placed there,
seems to me not *proved* even yet, although nothing
has been found."

The word *prove* is here distinctly used of something being demonstrable in the field of perception.

[1] The source of both words must be sought, I think, in the mediæval
Latin *proba*, a sample, test, or trial. Thus *probare* is used in the sense
of extracting a fact by torture, and *probabilis* is that which by aid of the
proba has been attested and approved.
[2] *Popular Lectures and Addresses*, vol. i. p. 261. London, 1889.

There is clearly an inference involved, and this inference is easily seen to be that of the routine of perceptions, namely; that if something has once been perceived, it will under precisely the same circumstances be again perceived. Our conviction of this routine is not a certainty, but, as we have seen, a probability. Hence, when we are speaking of the sphere of perceptions we must remember that provable is ultimately the same word as probable. The association of the two words does not therefore seem without profit ; and the etymology may after all serve to remind us of the character of our knowledge in the field of perception.

The problem before us is the following one : A certain order of perceptions has been experienced in the past, what is the probability that the perceptions will repeat themselves in the same order in the future? The probability is conditioned by two factors, namely : (1) In most cases the order has previously been very often repeated, and (2) past experience shows us that sequences of perceptions are things which have hitherto repeated themselves without fail. Thus there is past experience of repetition in the class, as well as in the individual, strengthening the probability of a future recurrence of the same order. The probability that the sun will rise to-morrow is not only conditioned by men's past experience of the sun's motion, but by their past experience of the uniform order in natural phenomena. There is no need to repeat a cautiously conducted experiment a great number of times to *prove*—that is, to establish an overwhelming probability in favour of—a certain sequence of perceptions. The overwhelming probability drawn from past experience in favour of *all* sequences repeating themselves at once embraces the new sequence. Suppose the solidification of hydrogen to have been *once* accomplished by an experimenter of known probity and caution, and with a method in which criticism fails to detect any flaw. What is the probability that on repetition of the same process the solidification of hydrogen will follow? Now Laplace has asserted that the probability that an event which has occurred p times and has not hitherto failed, will occur again is represented by the fraction $\frac{p+1}{p+2}$. Hence in the case of hydrogen, the probability of repetition would only be $\frac{2}{3}$, or, as we popularly say, the odds would be two to one in its favour. On the other hand, if the sun has risen without fail a million times, the odds in favour of its rising to-morrow would be 1,000,001 to 1. It is clear that on this hypothesis there would be practical certainty with regard to the rising of the sun being repeated, but only some likelihood with regard to the solidification of hydrogen being repeated. The numbers, in fact, do not in the least represent the degrees of belief of the scientist regarding the repetition of the two phenomena. We ought rather to put the problem in this manner: p different sequences of perception have been found to follow the same routine however often repeated, and none have been found to fail, what is the probability that the $(p+1)$th sequence of perceptions will have a routine? Laplace's theorem shows us that the odds are $(p+1)$ to 1 in favour of the new sequence having a routine. In other words, since p represents here the infinite variety of phenomena in which men's past experience has shown that the same causes are on repetition followed by the same effect, there are overwhelming odds that any newly observed phenomena may be classified under this law of causation.[1]   So great and, considering the odds, reasonably great is our belief in this law of causation applying to new phenomena, that when a sequence of perceptions does not appear to repeat itself, we assert with the utmost confidence that the same causes have not been present in the original and in the repeated sequence.

§ 14.—*Probability as to Breaches in the Routine of Perceptions.*

Laplace has even enabled us to take account of possible "miracles," anomies, or breaches of routine in the sequence of perceptions.   He tells us that if an event has happened p times and failed q times, then the probability that it will happen the next time is $\frac{p+1}{p+q+2}$, or the odds in favour of its happening are $p+1$ to $q+1$.   Now if we are as generous as we possibly can be to the reporters of the miraculous, we can hardly assert that a well-authenticated breach of the routine of perceptions has happened *once* in past experience for every 1,000 million cases of routine.   In other words we must take p equal to 1,000 million times q, or the odds against a miracle happening in the next sequence of perceptions would be about 1,000 millions to one.   It is clear from this that any belief that the miraculous will occur in our immediate experience cannot possibly form a factor in the conduct of practical life.   Indeed the odds against a miracle occurring are so great, the percentage of permanently diseased or temporarily dis-

---

[1]   A somewhat greater probability in favour of a new sequence which has repeated itself r times, repeating itself on the $(r+1)$th trial will be given below.

ordered perceptive faculties so large as compared with the percentage of asserted breaches of routine, and the advantage to mankind of evolving an absolutely certain basis of knowledge so great,[1] that we are justified in saying that miracles have been *proved* incredible—the word *proved* being used in the sense in which alone it has meaning when applied to the field of perceptions (p. 168).

§ 15.—*The Bases of Laplace's Theory lie in an Experience as to Ignorance.*

I have said enough, I think, to indicate that if Laplace's theorems be correct and can be *first applied* to measure the probability of the repetition of events, our belief in the routine of perceptions is based upon that high degree of probability, which renders probable and provable practically the same word. Let us consider the basis of Laplace's theory a little more closely. Suppose we take a shilling and toss it, then the chances that head or tail will be uppermost are exactly equal ; unity denoting certainty, we say that the probability of a head equals $\frac{1}{2}$. If we toss it again the chances of a head will not be altered and will again be $\frac{1}{2}$, and so on for each throw, the chance always remaining $\frac{1}{2}$. Since in two throws we might with equal probability have any of the four cases : head, head : tail, tail : head, tail : tail, head, it follows that the recurrence of head has only a probability of $\frac{1}{4}$ or $\frac{1}{2} \times \frac{1}{2}$. Similarly the probability that three heads

---

[1] This refers to the hypothesis (p. 163) that man in the course of evolution has attained a perceptive faculty which in the normal condition can only present sequences of perceptions in the form of routine. Such routine being, as we have seen, the sole basis of knowledge, is of enormous advantage to man.

will be tossed in succession may be easily seen by counting the possible cases to be $\frac{1}{8}$ or $\frac{1}{2} \times \frac{1}{2} \times \frac{1}{2}$; that is, the odds are seven to one against a triple recurrence. Extending this to 20 or 30 recurrences of heads, we soon find that there is an overwhelming probability against a succession of recurrences without a break.

Instead of the shilling, let us take a bag and put into it an equal number of black and white balls. The probability of a random drawing resulting in a white ball will now be $\frac{1}{2}$, and this will at each drawing, provided the balls be · returned to the bag, be the probability in favour of a white ball. Now let us look upon the world of perceptions as a bag containing white and black balls, a white ball representing a routine-order, and a black ball an anomy or breach of routine. Then, since we see no reason why perceptions should have a routine or should not have a routine, may we not assert that each are equally likely, or that there will be the same number of black and white balls in our bag? If this be so, then obviously the odds are seven to one against a routine-order occurring even three times without a single anomy, and are overwhelming against no breach of routine occurring at all. Yet the only supposition that we appear to have made is this : that, knowing nothing of nature, routine and anomy are to be considered as equally likely to occur. Now, we were not really justified in making even this assumption, for it involves a knowledge that we do not possess regarding nature. We use our *experience* of the constitution and action of coins in general to assert that heads and tails are equally probable, but we have no right to assert before experience that, as we know nothing of nature, routine and breach of routine are equally probable. In our ignorance we ought to consider before experience that nature may consist of all routines, all anomies, or a mixture of the two in any proportion whatever, and that all such are equally probable. Which of these constitutions after experience is the most probable must clearly depend on what that experience has been like.

To return to the case of the coin, we must suppose all experience of the action of coins withdrawn from us ; it must be unknown to us, whether coins are so constituted as to have a head on both faces, a tail on both faces, or a head on one and a tail on the other. The probability of any one of these three equally probable constitutions would before experience be $\frac{1}{3}$. Now suppose we had the experience of two tosses both resulting in heads. On the first constitution of the body this would be a certain result, or its probability be represented by 1 ; on the second constitution the result would be impossible, or the probability would be zero, while on the third constitution—that of the customary coin—the probability of the result would be $\frac{1}{4}$. *Experience*, then, shows us that one constitution of the coin is impossible, and that another constitution will certainly give the observed result, while the odds against the remaining possible constitution giving it are 3 : 1. Obviously a double head is a more probable constitution for the coin than head and tail. But in what ratio is this constitution more probable than the other ? This is determined by a principle due to Laplace which we may state as follows :—

" If a result might flow from any one of a certain number of different constitutions, all equally probable before experience, then the several probabilities of each constitution after experience being the real constitution, are proportional to the probabilities that the result would flow from each of these constitutions."

Thus in our case the head-head constitution gives a probability of 1 that the observed result will arise, while head-tail only gives a probability of $\frac{1}{4}$. Hence, on Laplace's principle, the odds are four to one that our coin has a head on both sides. We must be careful to note that this result depends entirely on the assumption that coins may have *any* constitution whatever; it ceases to have application when we have once had the experience that coins usually have a head and a tail. But it may be said, ought we not to have had the actual *experience* that coins may be of any constitution before we can predict that the individual coin which has twice turned up heads is probably a double-headed coin? Can we assume without such experience that, where we are ignorant, all constitutions are *à priori* equally probable? May we for the very reason that we know nothing "distribute our ignorance equally"? The logic of this proceeding has been called in question by more than one writer, notably by the late Professor G. Boole.[1] We may indeed reasonably question whether it is possible to draw knowledge out of complete ignorance. But before we can agree with Boole that Laplace's method is nugatory, we must ask whether, after all, his principle is not based on knowledge, namely, on that derived from experience that in cases where we are ignorant, there in the long run all constitutions will be found to be equally probable.

[1] *An Investigation of the Laws of Thought* (London, 1854), chap. xx. *Problems Relating to the Connexion of Causes and Effects*, especially pp. 363–75.

A good example of this has been given by Professor Edgeworth. Suppose we divide 143,678 by 7 and stop at the fourth figure of the quotient, we have 2,052 as the result. Now we may be supposed ignorant of what the next figure will turn out to be, and in our ignorance *all* the digits from 0 to 9 are equally probable. Why? Because if we divided a very great quantity of numbers of 6 figures by 7, stopping at the fourth digit in the quotient, we should find that the number of times each of the digits from 0 to 9 would occur in the fifth place, were practically equal. In other words, statistics would justify the "equal distribution of our ignorance," or *experience* show us that in our ignorance all constitutions were equally probable. This example may, perhaps, suffice to show that there is an element of human experience at the basis of Laplace's assumption. The reader who wishes to pursue this subject further may be referred in the first place to Professor Edgeworth's article.[1] "I submit," he writes, "the assumption, that any probability-constant about which we know nothing in particular is as likely to have one value as another, is grounded upon the rough but solid experience that such constants do as a matter of fact as often have one value as another."

The reader may, however, ask why may not "nature" change after one set of experiences and before another? The true answer to this question lies in the views expressed partly in earlier chapters of this work, partly in the following chapter on *Space and Time*. Nature, we have seen, is a construct of the human mind (pp. 50, 122–9, 163) ; time and space are not inherent in an outside world, but are modes of dis-

[1] "The Philosophy of Chance," *Mind*, vol. ix. pp. 223–35, 1884.

criminating groups of sense-impresssions (pp. 183, 217).
Thus "nature" is essentially conditioned by our perceptive faculty, and " change " cannot be thought of as
apart from ourselves. That "nature" is identical "before
and after experience" will be admitted, as soon as it is
recognized that time and change relate to perception,
and not to the "beyond" of sense-impressions. The
sameness of the perceptive faculty is the key to the
sameness of the modes of perception. The conditions
for each trial (as in throwing a die or in drawing from
a bag) remaining the same, lie therefore solely in the
identity of the perceptive faculty.

§ 16.—*Nature of Laplace's Investigation.*

We are now in a position to return to our bag of
white and black balls, but we can no longer suppose
an equal number of both kinds, or that routine and
breach of routine are equally probable. We must
assume our " nature bag " to have every possible constitution, or every possible ratio of black to white
balls to be equally likely ; to do this we suppose an
infinitely great number of balls in all. We may then
calculate the probability that with each of these constitutions the observed result, say p white balls and q
black balls (or, p cases of routine, and q anomies)
would arise in $p+q$ drawings.[1] This will determine,
by Laplace's principle, the probability that each hypothetical constitution is the real constitution of the
bag. Let these probabilities be represented by the
letters P_1, P_2, P_3 . . . &c. We may then determine
the probabilities on each of these constitutions that a
white ball will be drawn in the $(p+q+1)$th drawing.
If these probabilities be represented by the letters

---

[1] The reader may suppose the ball returned to the bag after each
drawing.

C_1, C_2, C_3 . . . &c., then by a well-known law for compounding probabilities [1] we shall find that the total probability in favour of a white ball occurring on the $(p+q+1)$th drawing, or of a routine following on p routines and q anomies, is—

$$P_1 \, C_1 + P_2 \, C_2 + P_3 \, C_3 + . \, . \, .$$

Now all this is pure calculation ; it involves no *new* principle, nothing the reader may not take on faith, if he is not an adept in mathematical analysis. We shall therefore suppose the calculation made [2] as Laplace made it, and the result will be found to be that given on our p. 170, namely, the probability that a white ball will be drawn is $\frac{p+1}{p+q+2}$. Or, since q is either zero or vanishingly small as compared with p, we have the overwhelming probability of the routine of perceptions being maintained on the *next* trial.

§ 17.—*The Permanency of Routine for the Future.*

One particular case is worth noting. Suppose we have experienced m sequences of perceptions which have repeated themselves n times without any anomy. Suppose, further, a new sequence to have repeated itself r times also without anomy. Then in all we have had $m (n-1) + r - 1$ repetitions, or cases of routine, and no failures ; hence the probability that the new sequence will repeat itself on the $(r+1)$th occasion is obtained by putting $p = m(n-1) + r - 1$ and $q = o$ in the result of § 16, or the odds in favour of a routine occurring on the next occasion with the new

---

[1] The reader will find this law discussed in any elementary work on algebra. See, for example, Todhunter's *Algebra*, §§ 732 and 746.

[2] See Todhunter's *History of the Theory of Probability*, Arts. 704, 847-8. Boole's *Laws of Thought*, chap. xx. § 23; or T. Galloway, *A Treatise on Probability*, § v., " On the Probability of Future Events deduced from Experience."

sequence are $m(n-1)+r$ to 1. Therefore if m and n are very great, there will be overwhelming odds in favour of the new sequence following routine, although r, or the number of times it has been tested, be very small.[1]

Our discussion of the probability basis for routine in the sequences of perceptions has perforce been brief, and only touched the fringe of a vast and difficult subject. Yet it may perhaps suffice to indicate that the odds in favour of that routine being preserved in the immediate future, or, indeed, for any finite interval, both with regard to old and to new groups of perceptions, are overwhelming.[2] We may be absolutely unable to demonstrate any inherent necessity for routine from our perceptions themselves, but our complete ignorance of such necessity, combined

---

[1] We must be cautious in applying this formula to take a sufficiently comprehensive sequence of perceptions. We must see that the causes are really the same, before we predict on the basis of past experience of routine in perceptions a repetition of sequence in any particular case. That I have twice seen a certain river overflow its banks, and never seen that river without a flood, will not enable me to predict that the flood will always occur when I see the river. I must add to these perceptions, those of the season of the year, of the amount of sun which has acted on the snow-fields and glaciers at its source, of the condition of its banks, &c., &c., before I have a sufficiently wide range of causes to enable me to predict from two repetitions the occurrence of a third. I must indeed show that in my supposed identical sequences there are really the same components. The reader who wishes to study this point more thoroughly must be referred to Mill's " Canons of Induction " (*System of Logic*, book iii.), an elementary discussion of which will be found in the " Lessons on Induction," pp. 210–64, of Stanley Jevons' *Elementary Lessons in Logic*.

[2] The odds in favour of a sequence repeating itself s times, when the past shows p repetitions and no failure are $p+1$ to s. The number of repeated sequences in the universe, or p, is practically infinite, so that the odds are overwhelming so long as s is finite. We cannot, however, argue from this result for an *infinite* future of repetition.

with our past experience, enables us by aid of the theory of probability to gauge roughly how unlikely it is that the possibility of knowledge and the power of thinking will be destroyed in our generation by those breaches of routine, which, in popular language, we term miracles.

So much science can tell us at present ; more we can only hope to *know*, if we admit that routine flows from the nature of our perceptive faculty and not from the sphere beyond sense-impression. If science must at the present stage perforce be content with a *belief* in the immediate permanency of the universe (based on a probability, which in practical life we should term certainty), we must at the same time remember that because a proposition has not yet been proved, we have no right to infer that its converse must be true. It is not a case of balancing contradictory evidence, for not a single valid argument is to be found in the whole range of human experience for inferring a first or last cause. There may be a beginning and an end to life on our planet ; we may term these, if we please, a " first and a last catastrophe." But among the myriad planetary systems we see on a clear night, there surely must be myriad planets which have reached our own stage of development, and teem, or have teemed, with human life. The first and last catastrophe must have occurred a myriad times, and were we able to watch through long thousands of years the changing brilliancy of stars, the first and last catastrophe would appear to us not as a first and last cause, but as much a routine of perceptions as the birth and death of individual men.

SUMMARY.

1. Cause is scientifically used to denote an antecedent stage in a routine of perceptions. In this sense force as a cause is meaningless. First Cause is only a limit, permanent or temporary, to knowledge. No instance, certainly not will, occurs in our experience of an arbitrary first cause in the popular sense of the word.

2. There is no inherent necessity in the routine of perceptions, but the permanent existence of rational beings necessitates a routine of perceptions; with the cessation of routine ceases the possibility of a thinking being. The only necessity we are acquainted with exists in the sphere of conceptions; possibly routine in perceptions is due to the constitution of the perceptive faculty.

3. Proof in the field of perceptions is the demonstration of overwhelming probability. Logically we ought to use the word *know* only of conceptions, and reserve the word *believe* for perceptions. " I know that the angle at the circumference on any diameter of a circle is right," but " I believe that the sun will rise to-morrow." The proof that for no finite future a breach of routine will occur depends upon the solid experience that where we are ignorant, there statistically all constitutions of the unknown are found to be equally probable.

LITERATURE.

BOOLE, G.—An Investigation of the Laws of Thought, chaps. xvi.-xx. London, 1854.

EDGEWORTH, F. Y.—" The Philosophy of Chance," Mind, vol. ix., 1884, pp. 223–35.

GALLOWAY, T.—A Treatise on Probability. Edinburgh, 1839.

JEVONS, W. STANLEY.—The Principles of Science, chaps. x.-xii.

MILL, JOHN STUART.—System of Logic, book iii., Induction. 1st ed., 1843, 8th ed., 1872.

MORGAN, A. DE.—The Theory of Probabilities. London, 1838.

VENN, J.—The Logic of Chance. London, 1866.

The reader who wishes to study Laplace's labours at first-hand will find a guide to his memoirs and some account of the various editions of his Théorie analytique des Probabilités, in Todhunter's History of the Theory of Probability, chap. xx. He may also consult Arts. 841–857 of the same History.

CHAPTER V.

SPACE AND TIME.

§ 1.—*Space as a Mode of Perception.*

IN our second chapter (p. 77) we saw that the distinction between " inside" and " outside" ourselves was not a very real or well-defined one. Certain of the vast complex of our sense-impressions we term inside, others again we term outside. To a savage the beginning of outside, the limit to *self*, is undoubtedly his skin ; although on occasion he may extend the idea of self farther, and be peculiarly careful of what becomes of such outward-lying portions of self as nail-parings and hair-clippings. The skin seems to him to bound off self from an outside world of non-self. The group of sense-impressions which he calls skin, marks off a world which he can see and feel from one which in the normal condition is inaccessible to sight or touch. His first experiences of pain arise, or at least are perpetuated, from something within this invisible and intangible world, and the nerve-vibrations, which he classifies as pain, he postulates as inside self ; his indigestion does not seem immediately associated with the visible and tangible world outside his skin. Thus the sense-impression pain, even when associated later with a group of other sense-impressions classified as those of sight and touch, is still differentiated from them as something especially internal. I receive for a moment, and then they vanish, the feelings of hardness and pain ; both may come to the seat of my consciousness as nerve-vibrations, or even by the same nerve-vibration ; both are associated with stored impresses of past hardnesses and pains, yet I project the sense-impression hardness into something outside self, but the pain I consider as something peculiar to my inside. I speak of *my* pain and *your* pain ; yet not of *my* hardness and *your* hardness, but of hardness as something peculiar to the table-leg. I thus give an objective reality to one group of sense-impressions, which I refuse to another.

Now this distinction seems to me to have arisen from the historical fact that the stored sense-impresses with which we associate hardness have been drawn from the tangible and visible world " outside skin," while those with which we associate pain have been largely drawn from the intangible and invisible world " inside skin." Even as our knowledge develops and " inside skin " becomes less intangible and invisible, even as we learn to associate pain with the stored impresses of various local organs " inside skin," we still feel it a somewhat doubtful use of language to talk of pain as " existing in space." Gradually, however, the skin has ceased to be a well-marked boundary between outside and inside. Self, like the soul of the metaphysicians, has disappeared from body and been concentrated in consciousness. Self, seated (metaphorically, not physically) in the telephonic brain exchange, receives an infinite variety of messages, which we can only assume to reach self in precisely the same manner. Yet self classes some groups of these messages together, and speaks of them as objects existing in space, while to other groups it has denied in the past, or still denies, this spacial existence. How far is this distinction logical, how far historical ? [1]

Now we shall find that the instant we associate a number of sense-impressions in a group, and separate them in perception from other groups, we consider them "to exist in space." Space is thus, in the first place, a mental expression for the fact that the perceptive faculty has separated coexisting sense-impressions into groups of associated impressions. This separation of immediate sense-impressions into groups, this *discriminating* power of the perceptive faculty is, at any rate in the early stages of man's development, most clearly recognized and closely associated with the senses of sight and touch. Hence it comes about that the invisible and intangible "inside skin" is at first not considered as in space. Later, for example, as we localize pain, or associate it with other sense-impressions classified as visible and tangible, we treat "inside skin" as belonging to space. Yet we still frequently consider the presence of visible and tangible members a condition for a *spacial* group of sense-impressions. Space, says Thomas Reid, is known directly by the senses of sight and touch. But probably a like, if less powerful, means of discriminating groups of sense-impressions lies in the senses of sound and smell.[2] We localize

[1] By *historical* I mean that which arises in the natural history of man from imperfect knowledge and illogical inference. Thus the belief in ghosts, witches, and storm-spirits is a perfectly intelligible stage in the natural history of man, but not a logical inference from any natural phenomena in the light of more perfect knowledge.

[2] My baby when three days old was able to distinguish between the snapping of the fingers of the right and left hands, and to follow with

sounds and smells without necessarily associating them with visible and tangible resounding and smelling bodies. It will, I think, be admitted on reflection that whenever we concentrate our attention on a limited group of associated sense-impressions, then we consider them as spacial, or "existing in space." We join together, owing to past experience, certain sense-impressions as a *permanent* group, and we then mentally separate this group from other groups. The actual boundary of the group, however, when we attempt to define it is found in reality to be vague (p. 80). The group, although in the main a permanent association, has a continual flow in and out of junior partners; while some of the partners belong, on closer examination, as much to one association as another. The separation is thus rather practical than real; it arises, in the first place, from the fact that in our perception certain sense-impressions are more or less permanently grouped together, and, in the second place, from the mental habit of concentrating our attention on one of these groups by placing about it in conception an arbitrary boundary separating it from other groups. Such arbitrary boundaries are conceptions drawn doubtless from sense-impressions of sight and touch, but they correspond, as we shall soon see, to nothing real in the world of sense-impression or in phenomena.

The coexistence of more or less permanent and distinct groups of sense-impressions is a fundamental mode of our perception; it is one of the ways in which we perceive things apart. There is nothing in

the ear the direction of the sound. She would turn to a voice long before she paid any attention to bodies moving quite close to her eyes. Difference of position was thus associated with sound.

sense-impressions themselves which involves the notion of space, but whether space be "due" to something behind sense-impression or to the nature of the perceptive faculty itself we are unable at present to decide. Leibniz has defined space as the order of possible coexisting phenomena. This order may "arise" from something behind phenomena, or from the machinery of perception, but in either case the order itself is simply a mode or manner in which we perceive things. The reader must distinguish carefully between the groups of sense-impressions themselves and the order in which we perceive them to coexist. Perhaps the distinction will be best brought out by considering the letters of the alphabet :—

A, B, C, D, E, F, G, ...

The letters may be said to have a real existence like the groups of sense-impressions we term objects. The order of the letters is merely the mode in which we perceive them to coexist as an alphabet. The "existence" we attribute to the order is thus of a totally different character to the "existence" we attribute to the letters. The alphabet has in itself no existence except for the letters it contains, but the letters, on the other hand, could have a real existence if they had never been arranged in any order or alphabet. The alphabet has merely existence as a manner of looking at all the letters together. These results may all be interpreted of coexisting groups of sense-impressions and their order *space*. A single sense-impression might, indeed, exist for us without any coexisting groups being postulated, but space would have no meaning if there were not such coexisting groups. Space is an order or mode of perceiving objects, but it has no existence if objects are withdrawn, no more than the alphabet could have an existence if there were no letters.

If the reader has once grasped this point—and it is undoubtedly a difficult and hard one (for our senses of sight and touch lead us imperceptibly to confuse the reality of sense-impressions with our mode of perceiving them),—then he will cease to look upon space as an enormous void in which objects have been placed by an agency in nowise conditioned by his own perceptive faculty ; he will begin to consider space as an order of things, but not itself a thing. To say, therefore, that a thing "exists in space" is to assert that the perceptive faculty has distinguished it as a group of sense-impressions from other groups of sense-impressions, which actually or possibly coexist. We cannot dogmatically deny that the order of co-existing phenomena "arises" from something behind sense-impressions,[1] but we may feel pretty confident that space, our mode of perceiving these phenomena, is very different from anything in the unknowable world behind sense-impressions. Once recognize space as a mode of the perceptive faculty, and it appears as something peculiar to the *individual* perceptive faculty. Without any perceptive faculty it is conceivable that sensations might exist (see p. 123), but there could not be that mode of perception we term space. The remarkable fact is this : that the order of coexisting phenomena is apparently the same at any rate for the vast majority of human perceptive

[1] Just as little ought we to assert that it does. The word *arise* suggests *causation ;* but the word causation is meaningless as a relation between the unknowable beyond of sense-impression and sense-impression itself (see pp. 82 and 151).

faculties. Why should this mode of perception be
the same for all normal human faculties—or, perhaps
it would be better to say, very approximately the
same ? We express the problem and the mystery
wrongly when we ask "why space seems the same
to you and me ; " we ought more precisely to ask
" why your space and my space are alike."
Because our perceptive faculties are of the normal
type, may be the immediate answer ; but how similar
organizing centres have come to exist in the chaos of
sensations remains still to be described.

Some light perhaps may be thrown on this difficult
problem by considerations which will be more fully
developed in our chapter on *Life*. Man has not
reached his present high stage of development solely
by individualistic tendencies, but also by socialistic
or gregarious tendencies. The struggle of man
against man might suffice to bring about a co-ordination of the individual man's perceptive and reasoning
faculties (p. 124), but in the struggle of group
against group, and of group with its environment, it
is clear that a great advantage would follow to any
group from a close agreement of the perceptive
faculties of its members, and great disadvantage to
any group without this agreement. The survival of
the former would be the natural result.

§ 2.—*The Infinite Bigness of Space.*

" How big is space ? " is a meaningless question as
it stands. " How big is space *for me* ? " admits, however, of an answer. It is just so large as will suffice
to separate all things which coexist for me. Let the
reader try to imagine phenomenal space apart from
groups of sense-impressions and he will quickly

discover how big space is for him.   Space, he will at
once recognize, has no meaning when we cease to
perceive things *apart*—to distinguish between groups
of sense-impressions. We ought constantly to bear
in mind that space is peculiar to ourselves, and that
we ought not reasonably to be stirred to greater
admiration by any one descanting on the "magnitude
of space," than we are wont to be when reflecting on
the complex nature of our own perceptive faculty.
The farthest star and the page of this book are both
for us merely groups of sense-impressions, and the
space which separates them is not in them, but is our
mode of perceiving them.

There is a cheap and, unfortunately, common form
of emotional science which revels in contrasting the
"infinities of space" with the "finite capacities of
man." As instructive samples of this we may take
the following passages from a popular writer on
astronomy :—

"Can it be true that these countless orbs are really
majestic suns, sunk to an appalling depth in the abyss
of unfathomable space ?"

"Yet, after all, how little is all we can see even
with our greatest telescopes, when compared with the
whole extent of infinite space !   No matter how
vast may be the depth which our instruments have
sounded, there is yet a beyond of infinite extent.
Imagine a mighty globe described in space, a globe
of such stupendous dimensions that it shall include
the sun and his system, all the stars and nebulæ, and
even all the objects which our finite capacities can
imagine.   Yet, after all, what must be the relation of
even this great globe to the whole extent of infinite
space ?   The globe will bear to that a ratio infinitely

less than that which the water in a single drop of
dew bears to the water in the whole Atlantic Ocean."[1]

To speak of the mode in which we perceive coexisting phenomena as an abyss of appalling depth is
perhaps rather meaningless phraseology ; but the
statement that infinite space contains more than our
finite capacity can imagine is hopelessly misleading.
In the first place, the space of our perceptions, the
space in which we discriminate phenomena, is
not infinite : it is exactly commensurate with the
contents of that finite capacity we term our perceptive faculty. In the second place, if by "all the
objects which our finite capacities can imagine" the
author means conceptions and not perceptions, he is
confusing two different things—space, as the order of
real coexisting phenomena, what we may term real
space, and the space of our thought, the conceptual
space of geometry, what we may term ideal space.
This latter, as we shall see in the sequel, may be
conceived as either finite or infinite, although a
limited portion of ideal infinite space describes most
easily the real space of our perceptions. Thus the
only infinite space we know of, so far from being a
real immensity overwhelming our finite capacities, is
a product of our own reasoning faculty. On the
other hand cosmical space, the mode of our perception, is finite and limited by the range, not of
what we imagine, but of what we perceive to co-exist.
The mystery of space, whether it be the finite space
of perception or the infinite space of conception, lies
in, and not outside, each human consciousness. We
must seek it either in our power of distinguishing (or
of perceiving apart) so many and varied groups of

[1] Ball's *Story of the Heavens*, pp. 2 and 538.

sense-impressions—or, in our power of drawing conceptions, which enables us to pass from the finite real
to the infinite ideal. Only for us, as perceiving human
beings, has space any meaning ; we cannot infer it
where we do not find psychical machinery similar to
our own.

§ 3.—*The Infinite Divisibility of Space.*

The space of our perceptions, as we have seen, is
finite and varies from individual to individual with
the range and complexity of his perceptions. As it
is just large enough for our perception of phenomena,
so it is just small enough, by which we are to understand that it is not "infinitely divisible." The limit
to its divisibility is the limit to our power of perceiving things apart. Our organs of sense are such
that only sense-impressions of a certain intensity or
amplitude fall within their cognizance. We may
resolve phenomena into smaller and smaller groups
of sense-impressions, but we ultimately reach a limit
at which the sense-impression ceases. We may divide
a piece of paper up into more and more minute
fragments, but ultimately they cease to be sensible
even by the aid of our most powerful microscopes.
We have then reached a limit to our mode of perceiving apart,—in ordinary parlance, to the divisibility
of space. We may possibly *conceive* smaller divisions,
but in doing this we have passed from the sphere of
the real to the ideal—from the space of perception to
the space of geometry. It seems to me that this
transition from perception to conception, often made
quite unconsciously, is the basis of all the difficulties
involved in the paradox as to the infinite divisibility

of space. The point has been referred to by Hume in his *Essay concerning Human Understanding*,[1] where he writes as follows :—

" The chief objection against all abstract reasonings is derived from the ideas of space and time—ideas which, in common life and to a careless view, are very clear and intelligible, but when they pass through the scrutiny of the profound sciences (and they are the chief object of those sciences) afford principles which seem full of absurdity and contradiction. No priestly *dogmas*, invented on purpose to tame and subdue the rebellious reason of mankind, ever shocked common sense more than the doctrine of the infinite divisibility of extension, with its consequences, as they are pompously displayed by all geometricians and metaphysicians with a kind of triumph and exultation. A real quantity, infinitely less than any finite quantity, containing quantities infinitely less than itself, and so on *in infinitum ;* this is an edifice so bold and prodigious that it is too weighty for any pretended demonstration to support, because it shocks the clearest and most natural principles of human reason. But what renders the matter most extraordinary is that these seemingly absurd opinions are supported by a chain of reasoning, the clearest and most natural ; nor is it possible for us to allow the premises without admitting the consequences."

Now the reader should carefully note the unconscious transition in this passage from the *ideas* of space and time to the infinite divisibility of *real* quantities. The transition is even more marked in a

[1] Section xii. part ii. Green and Grose : *Hume's Works*, vol. iv. p. 128.

footnote which accompanies the passage, and which runs thus :—

"Whatever disputes there may be about mathematical points, we must allow that there are physical points—that is, parts of extension, which cannot be divided or lessened either by the eye or imagination. These images, then, which are present to the fancy or senses, are absolutely indivisible, and consequently must be allowed by mathematicians to be infinitely less than any real part of extension ; and yet nothing appears more certain to reason than that an infinite number of them composes an infinite extension. How much more an infinite number of those infinitely small parts of extension, which are still supposed infinitely divisible."

Here the transition from perception to conception and back again is made several times over. A point mathematically defined is a conception and has no real existence in the field of perception. It is true we base this conception on our perceptive experience of things which are not points, but the mathematical point is not a *limit* to any process which could be carried on in the field of perception ; it is the limit to a process which we imagine carried on in the field of thought, in the sphere of conceptions. If Hume means by a physical point the smallest possible groups of sense-impressions which we can perceive apart, then this cannot be divided or lessened by the eye. But this physical point transferred from the field of perception to that of conception can in the imagination be divided over and over again. This remark will be more clearly appreciated when we come to deal with the geometrical conception of space. It suffices for the present to note that Hume passes from the eye to the imagination, from the mathematical to the physical, from the fancy to the senses, as if the geometrical theory of extension, that shorthand method of classifying and describing coexisting phenomena was itself the world of phenomena. Several types of geometry can be elaborated by our rational faculty, and the results, which flow from them, will depend upon the statement of their fundamental axioms. From these types we select that one which will enable us to describe the widest range of phenomena in the briefest possible formula, or which will enable us with the greatest accuracy to classify the differences between groups of sense-impressions. We have no more right to quarrel with the geometrician's conception of the infinite divisibility of space than with his conception of the circle, or with the physicist's conception of the atom. One and all are pure ideals beyond the range of perceptual experience. What we must ask is : How far are these conceptions of service in enabling us to briefly describe and classify our perceptions ; how far do they aid us in mentally storing up past experience as a guide for future action? A point and an ellipse may be absolutely absurd in the world of perceptions, but they are none the less valid and useful conceptions, if they help us to describe and predict the motion of the earth about the sun. The paradoxes which Hume finds in the conclusions of geometry only exist so long as we assert that every conception has a precise counterpart in perception, and forget that science is only a shorthand description of nature and not nature itself.

§ 4.—*The Space of Memory and Thought.*

Before we pass from the subject of real or perceptual

14

space, we ought to note that this mode of perceiving phenomena appears not only in association with immediate sense-impressions, but also with the stored impresses of past experience. To be accurate, we ought perhaps to say that the mode of remembrance is akin to the mode of perception—unless, indeed, we are using the word *perception* to refer to the consciousness alike of an "external" sense-impression and of an "internal" sense-impress. In all probability these processes of what Locke would term external and internal perception are much the same, only the sources from which they draw their material are different. In this case it is sufficient to say that space as a mode of perception applies as much to memory as to phenomena. We certainly gain by this method of regarding the matter new insight into the manner in which space may result from the nature of the psychical machinery. No one can look upon the space whereby the impresses of past experience are grouped and distinguished as a reality apart from internal perceptions ; it is too obviously a mode of the retentive faculty. But the distinction between the world of phenomena and the world of memories lies not in the order and relation of their contents, but in the intensity of the stimulus and the quality of the association in the two cases. The candles, the inkstand, the books and papers on my table have the same order and relation, whether I see and touch them or simply recall them as a memory, but there is a great difference in the vividness [1] of the external

---

[1] Hume's definition of belief, slightly modified, well marks the difference : A group of immediate sense-impressions is a " more vivid, lively, forcible, firm, steady " perception of an object than a group of stored impresses alone is ever able to attain (*Essay Concerning Human Understanding*, sec. v. part ii.).

and internal perceptions, and a considerable change
in the range of stored impresses with which the
contents of perception are associated in the two
cases.

Once recognize space as the mode in which we
perceive coexisting things apart, and we have either
to multiply spaces or to consider that logically all
separation denotes space. Thus our thoughts and
conceptions will be found almost invariably to
involve spacial relationship, while the psychical
processes themselves are, like pain, being more
and more localized or associated with individual
centres of brain-activity. It may fairly be said that
until the spacial relationship is recognized in any field,
until we are able to perceive things apart, we have no
basis for distinction, comparison, classification, and
the resulting scientific knowledge. It is especially
from the localization of psychical processes that we
may hope for great results, for a true science of
psychology in the future. This localization is not a
" materialization " of thought, it is merely an association of " internal " and " external " perceptions,
both equally factors of consciousness. The association is not an association of two totally diverse
and opposed things—matter and mind—but of the
two phases of perception. Groups of sense-impressions in space, being conditioned by the perceptive faculty, are as much a part of the sentient
being as psychical processes themselves.

Logically, then, it seems that whenever we clearly
separate and distinguish coexisting things, we perceive them under the mode space ; and perception
under this mode is what we ought to mean by
" existence in space." Yet historically the notion of

space has arisen from the separation and distinction of groups of sense-impressions, when some one or more members in each group were due to sight or touch ; for these senses are those by which groups have, in the natural history of man, been first perceived apart. Just as these groups of sense-impressions were pro-jected outward from our consciousness, and treated as things unconditioned by our perceptive faculty, as objects independent of the sentient being, so our mode of perception was treated as inherent in them, and given an objective existence, fossils of which are still to be found in the "primeval void" of myth-ology, and the "appalling abyss" of popular astro-nomers. Only gradually have we learnt to recognize that empty space is meaningless, that space is a mode of perception—the order in which our perceptive faculty presents coexistence to us. We are not compelled to postulate a space outside self for pheno-mena, and spaces inside self for memory, thought, and the psychical processes, but rather we must hold that the mode in which we perceive in these different fields is essentially the same, and that this mode is what we term space.

§ 5.—Conceptions and Perceptions.

If such be the space of perception, we have next to ask : How do we scientifically describe it ? What is conceptual space—the space with which we deal in the science of geometry ? We have seen that our perceptive faculty presents sense-impressions to us as separated into groups, and further, that though this separation is most serviceable for practical purposes, it is not very exactly and clearly defined "at the limits" (p. 80). How do we represent in thought, in conception, this separation into groups which results from our mode of perception? The answer is : We *conceive* groups of sense-impressions to be bounded by *surfaces*, to be limited by straight or curved *lines.* Thus our consideration of conceptual space leads us at once to a discussion of surfaces and lines—to a study, in fact, of *Geometry.*

Several important problems at once present themselves for investigation. In the first place, have these surfaces and lines a real existence in the world of perception? Are they phenomena? Or, are they ideal modes whereby we analyze the manner in which we perceive phenomena? In the second place, if they should be only ideals of conception, what is the historical process by which they have been reached? what is their ultimate root in perception?

Now, there is at this stage an important remark to be made, namely, that *what is imperceptible is not therefore inconceivable.* This remark is all the more necessary, for it seems directly opposed to the healthy scepticism of Hume.[1] Yet unless it be true the whole fabric of exact science falls to the ground, neither the concepts of geometry, nor those of mechanics, would be of service ; for example, the circle and the motion of a point would be absurdities if, being imperceptible, they were really inconceivable. The basis of our conceptions doubtless lies in perceptions, but in imagination we can carry on perceptual processes to a limit which is itself not a perception ; we can further associate groups of stored sense-impresses, and form ideas which correspond to nothing in our perceptual experience.

[1] See especially the *Treatise of Human Nature,* part ii. *Of the Ideas of Space and Time.* Green and Grose's *Hume's Works,* vol. i. pp. 334-371.

Here a word of caution is, however, very necessary. Because we conceive a thing, we must not argue that it is either possible or probable as a perception. Indeed, the process or association by which we have reached our conception may in itself suffice to exhibit its perceptual impossibility or improbability. The appeal to experience can alone determine whether a conception is possible as a perception. For example, experience shows me that there is a sensible limit to the visible and tangible ; hence a point, valid as a conception, can never have a real existence as a perception. I reach this conception of a point by carrying to a limit in my imagination a process which cannot be so carried in perception. Exactly of the same character are my conceptions of infinite distance or infinite number ; they are the conceptual limits to processes, which may be *started* in perception, but cannot be carried to a limit except in the imagination. Somewhat different from perceptual impossibility is perceptual improbability. I can conceive Her Majesty Queen Victoria walking down Regent Street, but, tested by my experience of the past actions of royalty, this association of conceptions is hardly a perceptual probability. These instances may be sufficient to indicate that what is improbable or impossible in perception may be valid in conception. But we must ever be careful to bear in mind that the *reality* of the conception, its existence outside thought, can only be demonstrated by an appeal to perceptual experience. The geometrician even asserts the phenomenal impossibility of his points, lines, and surfaces ; the physicist by no means postulates the existence of atoms and molecules as possible perceptions. Science is content for the

present to look upon these concepts as existing only
in the sphere of thought, as purely the product
of man's mind.   It does not, like metaphysics or
theology, demand any existence in or beyond sense-
impression for its conceptions until experience has
shown that the conceptual limit or association can
become a perceptual reality.[1]   The validity of scientific conceptions does not in the first place depend
on their reality as perceptions, but on the means they
provide of classifying and describing perceptions.   If
a circle and a rectangle have no real existence, they
are still invaluable as enabling me to classify my perceptions of form, to describe, however imperfectly, the
difference in shape between the faces of a page of this
book and of my watch.   They are symbols in that
shorthand by means of which science describes the
universe of phenomena.   The atom, if a pure conception, still enables us, by codifying our past experience, to economize thought ; it preserves within
reasonable limits the material upon which we base
our prediction of possible future experience.   If any
one tells us that the storm-god is to some minds as
conceivable as the atom, we must, in the first place,
reply that the conceivable is not the real ; and further,
that the value to man of any ideal of conception
depends upon the extent to which it subsumes the
future in its *résumé* of the past.   The conception
storm-god may, after all, be of some value as a
striking monument to our meteorological ignorance,

[1] Leverrier and Adams *conceived* a planet having a definite orbit as a
method of accounting for the irregularities perceived in the motions of
Uranus.   Their conception might have been valid as a manner of
describing these irregularities, if Neptune itself had never been perceived
—in other words, if their conception had not become a perceptual
reality.

and as a useful reminder that we must "be prepared for all weathers."

What we have at this stage to notice is that the mind is not limited to perceptual association, and that it can carry on in conception a process which may be begun, but cannot be indefinitely continued in the sphere of perception. The scientific value of such conceptions, whether reached by association or as a limit, must in every case be judged by the extent to which they enable us to classify, describe, and predict phenomena.

§ 6.—*Sameness and Continuity.*

Now there are two ideas reached as conceptual limits to perceptual processes which have important bearings on the geometrical representation of space. These may be expressed by the words *sameness* and *continuity*. So far as our perceptual experience goes, probably no two groups of sense-impressions are exactly the same. The sameness in each depends upon the degree of our examination and observation. To a casual observer all the sheep in a flock appear the same, but the shepherd individualizes each. Two coins from one die, or two engravings from one block will always be found to possess some distinguishing marks. We may safely assert that absolute sameness has never occurred in our experience. Not even a "permanent" group of sense-impressions or an object is exactly the same at two different times. Various elements in the group have changed slightly with the time, the light, or the observer. Take a polished piece of metal and note two parts of its surface ; they appear exactly alike, but the microscope reveals their want of sameness. Thus sameness is never a real limit to our experience of phenomena ; the more closely we examine, the less is the sameness. Yet, as a conception, the sameness of two groups of sense-impressions is a very valid idea, and the basis of much of our scientific classification. In the sphere of perceptions sameness denotes the identity for certain practical purposes of two slightly different groups of sense-impressions. In the sphere of conceptions, however, sameness denotes absolute identity of all the members of either group; it is a limit to a process of comparison which cannot be reached in the perceptual world.

The idea of continuity, in the sense in which we are now considering the word, involves that of sameness. If I take a vessel of water, I find a certain permanent group of sense-impressions which leads me to term the contents of the vessel water ; if I take a small quantity of the water out of the vessel I find the " same " group, and this still remains true if I take a smaller and smaller quantity, even to a drop. I may continue to divide the drop, but apparently as long as the portion taken remains sensible at all, there is the same group of sense-impressions, and I term the fraction of the drop water. Now the question arises, if this division could be carried on indefinitely should we at last reach a limit at which the group of sense-impressions would change not only quantitatively, that is in intensity, but also qualitatively ? If we could magnify the sense-impressions due to the infinitesimal fraction of a drop of water up to a sensible intensity, would they so differ from those characteristic of the contents of the original vessel that we should not give them the name water ? Now we cannot test the effects of an indefinitely continued division in the phenomenal world, for we soon reach a stage at which we fail to get, by the means at our disposal, any sense-impressions at all from the divided substances. Our magnifiers of sense-impression have but a limited range.[1] But although in the sphere of perceptions there is no possibility of carrying division to its ultimate limit, we can yet in conception repeat the process indefinitely. If after an infinite number of divisions we conceive that the same group of sense-impressions would be found, then we are said to conceive the substance as *continuous*. We have then to ask how far the conception of continuity applies to the real bodies of our perceptual experience. From the finite process of division which is possible in perception, we might easily conclude that continuity was a property of real substances; and there is small doubt that a slight amount of observation is favourable to the notion that many real substances are continuous, although the infinite division necessary to the conception of continuity fails as a perceptual equivalent. Further observation and wider insight, however, contradict this notion. The physicist and the chemist bring many arguments to show us that the finite process of division which suggests continuity would, if carried to an infinite limit, show bodies to be discontinuous. On a first and untrained inspection we find a continuity and a sameness in perceptions which disappear on closer and more critical examination. The ideas conveyed in these words are found to be no real limits to the actual, but ideal limits to processes which

---

[1] *E.g.*, the microscope, the microphone, the spectroscope, &c. From the spectroscope we obtain, perhaps, positive indications of a qualitative change in many substances as the quantity is diminished.

can only be carried out in the field of conception
Bearing this in mind we may now return to the
geometrical conceptions of space.

§ 7.—*Conceptual Space. Geometrical Boundaries.*

It has been remarked (p. 197) that we conceive
groups of sense-impressions to be limited by surfaces
and lines. We speak of the surface of the table ; the
fly-leaf of this book appears to be separated from the
air above it by a plane surface and that plane to be
bounded at its upper edge by a portion of a straight
line. In the first place we have to ask whether our
geometrical notions of line and plane correspond to
the limits of anything we actually find in perception
or whether they are purely ideal limits to processes
begun in perception, but which it is impossible to
carry to a limit in perception. The answer to these
questions lies in the conceptions of *sameness* and *continuity*. The geometrical ideas of line and plane
involve absolute sameness in all their elements and
absolute continuity. Every element of a straight line
can in conception be made to fit every other element,
and this however it be turned about its terminal
points. Every element of a plane can be made to fit
every other element, and this without regard to side.
Further, every element of a straight line or a plane,
however often divided up, is in conception, when
magnified up, still an element of straight line or
plane.

The geometrical ideas correspond to absolute
sameness and continuity, but do we experience anything like these in our perceptions? The fly-leaf of
this book appears at first sight a plane surface
bounded by a straight line, but a very slight in-

spection with a magnifying lens shows that the surface has hollows and elevations in it, which quite defy all geometrical definition and scientific treatment. The straight line which seems to bound its edge becomes, under a powerful glass, so torn and jagged that its ups and downs are more like a saw edge than a straight line. The sameness and continuity are seen to be wanting on more careful investigation. We take a glass cube skilfully cut and polished, and its faces appear at first as true planes. But we find that a small body placed upon one of its faces does not slide off when the cube is slightly tilted. The face of the cube must, after all, be *rough*, there are hollows and projections in it which catch those of the superposed body ; our plane again appears delusive. Or, we may take one of Whitworth's wonderful metal planes obtained by rubbing the faces of three pieces of metal upon each other. Here again a powerful microscope reveals to us that we are still dealing with a surface having ridges and hollows.

The fact remains, that however great the care we take in the preparation of a plane surface, either a microscope or other means can be found of sufficient power to show that it is not a plane surface. It is precisely the same with a straight line ; however accurate it appears at first to be, exact methods of investigation invariably show it to be widely removed from the conceptual straight line of geometry. It is a race between our power of representing a straight line or plane and our power of creating instruments which demonstrate that the sameness and continuity of the geometrical conceptions are wanting. Absolutely perfect instruments could probably only be constructed if we were already in possession of a true geometrical line or plane, but the instruments we can make appear invariably to win the race. *Our experience gives us no reason to suppose that with any amount of care we could obtain a perceptual straight line or plane, the elements of which would on indefinite magnification satisfy the condition of ultimate sameness involved in the geometrical definitions.* We are thus forced to conclude that the geometrical definitions are the results of processes which may be started, but the limits of which can never be reached in perception ; they are pure conceptions having no correspondence with any possible perceptual experience What we have said of straight lines and planes holds equally of all geometrically defined curves and surfaces. The fundamental conceptions of geometry are only ideal symbols which enable us to form an approximate, but in no sense absolute, analysis of our sense-impressions. They are the scientific shorthand by which we describe, classify, and formulate the characteristics of that mode of perception which we term perceptual space. Their validity, like that of all other conceptions, lies in the power they give us of codifying past and predicting future experience.

We speak of a spherical or cubical body, and say that it is of such and such a capacity. But no perceptual body is ever truly spherical or cubical, and the size we attribute to it is at best an approximate one. Further analysis of our sense-impressions leads us in each case to find variations from the geometrical definition and measurement. Yet the conceptions of sphere and cube are frequently sufficient to enable us to classify and identify various bodies and predict the different types of sense-impression to which these

bodies correspond.[1]   Perhaps no better instance than
geometry can be taken to show how science *describes*
the world of phenomena by aid of conceptions corresponding to no reality in phenomena themselves.
That our geometrical conceptions enable us on the
whole to so effectually describe perceptual space is
only a striking instance of the practically equal
development of our perceptive and reasoning faculties
(p. 125).

§ 8.—*Surfaces as Boundaries.*

Although perceptual boundaries do not, on ultimate
analysis, in any way correspond to any special geometrical definition such as that of plane or sphere, we
have still to inquire whether they answer to our
conception of surface at all.   By surface in this sense
we are to consider, not something of which it would
be possible to analyze the properties by any of the
known processes of geometry, but any *continuous*
boundary between two groups of sense-impressions or
bodies.[2]   Is there a continuous boundary between the

---

[1] Our whole system of measuring size will be found to be based on
geometrical conceptions having no actuality in perception.

[2] " *That which has position, length and breadth but not thickness,* is
called *surface.*

" The word *surface* in ordinary language conveys the idea of extension
in two directions ; for instance, we speak of the surface of the earth, the
surface of the sea, the surface of a sheet of paper.   Although in some
cases the idea of the thickness or the depth of the thing spoken of may
be present in the speaker's mind, yet as a rule no stress is laid on
depth or thickness.   When we speak of a *geometrical surface,* we put
aside the idea of depth and thickness altogether " (H. M. Taylor, *Pitt
Press Euclid,* i.–ii. p. 3).   It seems to me that in ordinary language
there is something more than length and breadth involved—there is an
idea of *continuous boundary*.   It is difficult to say how far this idea is
really involved in the word extension.   A veil may nave extension in
two directions, but it fails to fulfil our idea of surface because it is not a
continuous boundary.

open page of this book and the air above it ? Would it be possible to say at any distinct step of the passage from air to paper, here air ends and paper begins ? At this point we reach one of the most important problems of science. Are we to consider the groups of sense-impressions which we term bodies *continuous* or not ? If bodies are not continuous, then it is clear that boundaries are only mental symbols of separation, and on deeper analysis correspond to no exact reality in the sphere of sense-impression.

Would every element of the surface of a body still appear to us a continuous boundary, however small the element and however much we magnified it up ? If I could take the hundredth part of a square inch of this page and magnify it to a billion times its present size, would there still appear a continuous boundary between air and paper ?

Consider the boundary of still water. It furnishes us with the impression of a continuous surface. On the other hand, examine a heap of sand closely, and it appears to have no continuous boundary at all. Are there any reasons which would lead us to suppose that, if we could sufficiently magnify a small element of this page of paper, it would produce in us sense-impressions not of continuity but of discontinuity ? Would it look, supposing it were still visible, like the surface of water, or rather like a heap of sand, a pile of small shot, or, better still, like a starry patch of the heavens on a clear night ? No group of stars is in perception separated from another by a line or surface. We can *imagine* such boundaries drawn across the heavens, but we do not *perceive* them. We have, then, to ask whether the boundary between paper and air, if immensely magnified, would look

sideways, not indeed like a geometrical line, but
roughly like the first or second of these figures :—

FIGS. 2 AND 3.

Now no direct answer can really be given to this
question, because bodies cease to impress us sensibly
long before we reach the point at which the appearance of continuity might be expected to disappear.
We cannot predict what our sense-impressions would
be, if we could magnify a drop of water up to the size
of the earth. But we may put the question in a
slightly different way. We may ask : Would it enable
us to classify and describe phenomena better if we
conceived bodies to be continuous as in Fig. 2, or
discontinuous as in Fig. 3 ? The physicist promptly
replies : I can only conceive bodies to be discontinuous. Discontinuity is essential to the methods by
which I describe and formulate my sense-impressions
of the phenomenal world.

§ 9.—*Conceptual Discontinuity of Bodies. The Atom.*

Foremost among the physicist's reasons for postulating the discontinuity of bodies is the elasticity
which we notice in all of them. Air can be placed

under a piston in a cylinder and compressed ; a bar of wood can be bent—in other words, a portion of it squeezed and another portion stretched. Even the amounts by which we can squeeze iron or granite are capable of measurement. Now, it is very hard, I think impossible, to *conceive* how we can alter the size of bodies if we suppose them continuous. We feel ourselves compelled to assert that, if the parts of a body move closer together, they must have something free of body into which they can move. If a body were continuous and yet compressible, there appears to be no reason why it should not be indefinitely compressible, or indefinitely extensible, both results repugnant to our experience. Further, our sense-impressions of temperature in both gaseous and solid bodies, and of colour in solid bodies, the phenomena of pressure in gases, and those of the absorption and emission of light, are easily analyzed and described, if we conceive the ultimate parts of bodies to have a capacity for relative motion ; but there is no possibility of conceiving such a motion if all the parts of a body are continuous. A crowd of human beings seen from a great height may look like a turbulent fluid in motion at every point. But we know from experience that this motion is only possible, because there is some void in the crowd. It may become so densely packed that motion is no longer practicable. Thus it is with that relative motion of the parts of bodies upon which so much of modern physics depends ; absolutely close packing, that is continuity, seems to render it impossible. It is only by reducing in conception the complex groups of sense-impressions, which we term bodies, into simple elements directly depending on

15

the motion of discontinuous systems,—of what we may term granular or starlike systems,—that we have been able to resume phenomena in the wide-reaching laws of physics and chemistry. The relative motion of the ultimate parts of bodies, involving the idea of discontinuity is one of the fundamental conceptions of modern science (p. 159). These ultimate parts of bodies we are accustomed to speak of as *atoms ;* groups of atoms which apparently repeat themselves over and over again in the same body,—something like planetary systems in the starry universe,—we term *molecules.* The generally accepted atomic or molecular theory of bodies postulates essentially their discontinuity. Take, for example, a spherical drop of water—to follow Sir William Thomson— suppose it to be as big as a football, then if we could magnify the whole drop up to the size of the earth, the structure, he tells us, would be more coarse-grained than a heap of small shot, but probably less coarse-grained than a heap of footballs. [1]

Now I propose later to return to the atomic hypothesis. At present I will only ask the reader to look upon atom and molecule as *conceptions* which very greatly reduce the complexity of our description of phenomena. But what it is necessary to notice at this stage is : that the conception atom, when applied to our perceptions, is opposed to the conception of surface as the continuous boundary of a body. We have here an important example of what is not an uncommon occurrence in science, namely, two conceptions which cannot both correspond to realities in the perceptual world. Either perceptual bodies have

---

[1] *Popular Lectures and Addresses,* vol. i., " The Size of Atoms," p. 217.

continuous boundaries, and the atomic theory has no
perceptual validity ; or, conversely, bodies have an
atomic structure, and geometrical surfaces are perceptually impossible. At first sight this result might
appear to the reader to involve a contradiction between geometry and physics ; it might seem that
either physical or geometrical conceptions must be
false. But the whole difficulty really lies in the habit
we have formed of considering bodies as objective
realities unconditioned by our perceptive faculty.
We cannot too often recall the fact that bodies are
for us more or less permanent, more or less clearly
defined groups of sense-impressions, and that the
correlations and sequences among the sense-impressions are largely conditioned by the perceptive faculty.
At the present time we have no sense-impressions
corresponding to geometrical surface or to atom; we
may legitimately doubt whether our perceptive
faculty is of such a nature that it could present
impressions in any way corresponding to these conceptions. It is impossible, therefore, to say that one
of these conceptions must be real and the other
unreal, for neither at present has perceptual validity—
that is, exists in the world of real things. As conceptions both are equally valid ; both are equally
ideals, not involved in our sense-impressions themselves, but which the reasoning faculty has discovered and developed as a means of classifying
different types of sense-impressions and of resuming
in brief formulæ their correlations and sequences.

Thus geometrical truths apply with absolute accuracy to no group whatever of our sense-impressions ; but they enable us to classify very wide
ranges of phenomena by aid of the notions of

position, size, and shape. Geometry enables us to predict with absolute certainty a variety of relations between sense-impressions, when these impressions do not involve more than a certain keenness in our senses, more than a certain degree of exactness in our measuring instruments. The absolute sameness and continuity demanded by geometrical conceptions do not exist as *limits* in the world of perceptual experience, but only as approximations or averages.[1] In precisely the same way the theory of atoms treats of ideal conceptions; it enables us to classify another and different range of sense-impressions, and to formulate their mutual relations to a certain degree of keenness again in our senses, or of exactness in our scientific apparatus. Should the atom become a perception as well as a conception, this would not invalidate the usefulness of geometry. Very probably, however, if we could magnify a football up to the size of the earth, so that the perceptual atom, if it existed, would have a size between small shot and a football, we should find that the sense-impressions which the atom was conceived to distinguish and resume, had themselves disappeared under the new conditions.[2] In other words, our scientific conceptions are valid for the world as we know it, but we cannot in the least predict how they would be related to a world which is at present beyond perception.

[1] Geometry might almost be termed a branch of statistics, and the definition of the circle has much the same character as that of Quetelet's *l'homme moyen*.

[2] The visibility and tangibility of bodies may possibly be described by the motion of atoms, but we cannot predict that a *single* atom would be either visible or tangible, still less " bounded by a surface."

§ 10.—*Conceptual Continuity. Ether.*

The reader will now be prepared to appreciate
scientific conceptions, which, if they corresponded to
realities of the phenomenal world, would contradict
each other. Having destroyed the continuity of bodies
by the idea of atom, it might at first sight appear as if
our conceptual space were fundamentally different from
perceptual space. The latter, as we have seen, is our
mode of distinguishing groups of sense-impressions,
and where there is nothing to distinguish, there there
is no space. The perceptive faculty rather than
nature may be said "to abhor a vacuum." On the
other hand, having destroyed the continuity of bodies
by the atomic hypothesis, we seem at first sight to be
postulating a void in conceptual space. But here the
physicist compels us to introduce a new continuity.
This new continuity is that of the *ether*, a medium
which physicists conceive to fill up the interstices
between bodies and between the atoms of bodies.
By aid of this concept, the ether (to which we shall
return later), we are able to classify and resume
other wide groups of sense-impressions. With
regard to the perceptual existence of the ether, it
now stands, some physicists would assert, on a rather
different footing from that of the atom. By the *real*
existence of anything we mean (p. 85) that it forms
a more or less permanent group of sense-impressions.
Now this can hardly be asserted of the ether; we
conceive it rather as a conduit for the motions by
which we interpret sense-impression. The nerves
seem to us conduits of the like kind, but then the
nerves also appear to us as permanent groups of
sense-impressions apart from their function of con-
ductivity. There are no sense-impressions which we

class together and term ether, and on this account it still seems better to consider the ether as a conception rather than a perception. It is true that to some minds the ether may appear as real a perception as the air, and the matter is, perhaps, largely one of definition. Still Hertz's experiments,[1] for example, do not seem to me to have logically demonstrated the perceptual existence of the ether, but to have immensely increased the validity of the scientific concept, ether, by showing that a wider range of perceptual experience may be described in terms of it, than had hitherto been demonstrated by experiment. Further, many of the properties which we associate with the ether are not such as our past experience shows us are likely to become matter for direct sense-impression. I shall therefore continue to speak of the ether as a scientific concept on the same footing as geometrical surface and atom.

§ 11.—*On the General Nature of Scientific Conceptions.*

Our discussion of these special conceptions will the better have enabled the reader to appreciate the nature of scientific conceptions in general. Geometrical surface, atom, ether, exist only in the human mind, and they are " shorthand " methods of distinguishing, classifying, and resuming phases of sense-impression. They do not exist in or beyond the world of sense-impressions, but are the pure product of our reasoning faculty. The universe is not to be

---

[1] *Annalen der Physik,* 1887–9. See also *Nature,* vol. xxxix. pp. 402, 450, 547. An interesting account of Hertz's researches by von Tunzelmann will be found in *The Electrician* for 1888, vol. xxi., pp. 587, 625, 663, 696, 725, 757, 788, and vol. xxii., pp. 16, 41.

thought of as a real complex of atoms floating in ether, both atom and ether being to us unknowable " things-in-themselves," producing or enforcing upon us the world of sense-impressions. This would indeed be for science to repeat the dogmas of the metaphysicians, the crassest paradoxes of a short-sighted materialism. On the contrary, the scientist postulates nothing of the world beyond sense ; for him the atom and the ether are,—like the geometrical surface,—modes by aid of which he resumes the world of sense. The ghostly world of " things-in- themselves" behind sense he leaves as a playground to the metaphysician and the materialist. There these gymnasts, released from the dreary bondage of space and time, can play all sorts of tricks with the unknowable, and explain to the few who can comprehend them how the universe is " created " out of will, or out of atom and ether, how a knowledge of things beyond perception, beyond the knowable, may be attained by the favoured few. The scientist bravely asserts that it is impossible to know what there is behind sense-impression, if indeed there can " be " anything ; [1] he therefore refuses to project his conceptions, atom and ether, into the real world of perception until he has perceived them there. They remain for him valid ideals so long as they continue to economize his thought.

That the conceptions of geometry and physics immensely economize thought is an instance of that wonderful power to which I have previously referred in this work (p. 125), namely, the power the reasoning faculty possesses of resuming in conceptions and

[1] Our notion of "being" is essentially associated with space and time, and it may well be questioned whether it is intelligible to use the word except in association with these modes of perception.

brief formulæ the correlations and sequences it finds in the material presented to it by the perceptive faculty. As our knowledge grows, as our sense becomes keener under the action of evolution and with the guidance of science, so we are compelled to widen our concepts, or to add additional ones. This process does not as a rule signify that the original concepts are invalid, but merely that they form a basis, which is only sufficient for classifying and describing certain phases of sense-impression, certain sides of phenomena. As we grow cognizant of other phases and sides, we are forced to adopt new concepts, or to modify and extend the old. We may ultimately reach perceptions of space which cannot be described by the geometry of Euclid, but none the less that geometry will remain perfectly valid as an analysis and classification of the wide range of perceptions to which it at present applies. If the reader will bear in mind the views here expressed with regard to the concepts of science, he will never consider that science reduces the universe to a "dead mechanism" by asserting a reality for atom or ether or force as the basis of sense-impression. Science, as I have so often reiterated, takes the universe of perceptions as it finds it, and endeavours briefly to describe it. It asserts no perceptual reality for its own shorthand.

One word more before we leave this space of conception, separated by continuous boundaries in the eye of the geometrician, peopled with atoms and ether by the mind of the physicist. How, if geometrical surface, if atom and ether have no perceptual reality, has the mind of man historically reached them? I believe by carrying to a limit in conception processes which have no such limit in perception. Preliminary stages in comparison show apparent sameness and continuity, where more exact and final stages show no such limit ; hence arises the conception of continuous boundaries. The atom again is a conceptual limit to the moving bodies of perception ; while the ether possesses an elasticity, which we have never met with in the elastic bodies of our perceptual experience, but which is a purely conceptual limit to the type of elastic substances with which we are directly acquainted. These concepts themselves are a product of the imagination, but they are suggested, almost insensibly suggested, by what we perceive in the world of phenomena.

§ 12.—*Time as a Mode of Perception.*

I have dealt at greater length with space than it will be necessary to deal with *time*, for much that has been said in the former case as to perception and conception will directly apply to the latter. Space and time are so similar in character, that if space be termed the breadth, time may be termed the length of the field of perception. As space is one mode in which the perceptive faculty distinguishes objects, so time is a second mode. As space marks the co-existence of perceptions at an epoch of time—we measure the breadth of our field—so time marks the progression of perceptions at a position in space—we measure the length of our field. The combination of the two modes, or change of position with change of time, is *motion*, the fundamental manner in which phenomena are in conception presented to us.

If we had solely the power of perceiving coexisting things, our perception might be wide, but it would fall far short of its actuality. The power of "perceiving things apart" by progression or sequence is an essential feature of conscious life, if not of existence. Without this time-mode of perception the only sciences possible would be those which deal with the order or correlation of coexisting things, with number, position, and measurement—in other words, the sciences of Arithmetic, Algebra, and Geometry. Bodies might have size and shape and locality, but science would be unable to deal with colour, warmth, weight, hardness, &c., all of which sense-impressions we conceive to depend upon our appreciation of sequence. In short, the physical, biological, and historical sciences, which have for their essential topics change, or sequence in perception, would be impossible.

I have spoken of certain branches of science being possible or impossible without the time-mode of perception. I ought rather to say that the *material* for these branches of science can or cannot be conceived to exist without time. For in truth all scientific knowledge would be impossible without time ; thought undoubtedly involves an association of immediate and stored sense-impressions (p. 55); every conception, geometrical as well as physical, is ultimately based on perceptual experience, and the very word experience connotes the time-mode of perceiving things. This leads us to what at first sight appears a fundamental distinction between the modes space and time. Space as our method of perceiving coexisting things, of distinguishing groups of immediate sense-impressions, is associated with the world of actual phenomena which we project *outside* ourselves (p. 73). For this reason it has been

termed an *external* mode of perception. On the
other hand, time is the perception of sequence in
stored sense-impresses—the correlation of past perceptions with the immediate perception. Thus time
involves in its essence memory and thought—in other
words, *consciousness.*[1] Consciousness might indeed be
defined as the power of perceiving things apart by
succession. It may perhaps be possible to conceive
consciousness as existing without the space-mode of
perception, but we cannot conceive it to exist without the time-mode. On this account, time has been
termed an *internal* mode of perception. A little consideration, however, soon shows us that this distinction is not a very valid one—as, indeed, no distinction
based on the words *external* and *internal* can ever be
(p. 80). Perception in space is, as a matter of fact,
as largely dependent on the association of immediate
and stored sense-impressions as perception in time. As
we have seen, every object is for us largely a construct (p. 50), and the coexisting objects which we
can perceive apart are indeed very limited. I distinguish the papers, the books, the inkstand, the
candlesticks on my table as separate objects by the
mode space; but at any *instant of time,* it is only a very
small element of this complex of sense-impressions
which is *immediate,* the rest are stored sense-impresses,
capable of becoming immediate sense-impressions in
the next instant, but not so in actuality. Thus in the
case of both time and space the "perceiving apart"

---

[1] For a new-born infant time cannot be said to exist—it is without
consciousness (p. 53). Only as stored sense-impresses result from
immediate sense-impression does the faculty of memory, and so the
time-mode of perception become developed. The rest is reflex action,
the product of inherited and unconscious association.

is the perception of an order existing between a very small element of sense-impression and a much larger range of stored sense-impresses. We do not therefore gain by terming space and time external and internal modes of perception. Both modes of perception are so habitual and yet so difficult of analysis, so commonplace and yet so mysterious, that, although we recognize a distinction between the two, we are often hardly certain whether we are distinguishing things by time or by space. *Why* we perceive things under these modes, the scientist is content to classify with all other *whys* as an idle and irrational question ; but clearer views as to the *how* of these modes of perception will undoubtedly come with the growth of physiological psychology, and with increased observation of the manner in which the lower forms of life and young children discriminate perceptions.

Of time as of space we cannot assert a real existence ; it is not in things, but is our mode of perceiving them. As we cannot postulate anything of the beyond of sense-impression, so we cannot attribute time directly or indirectly to the supersensuous. Like space, it appears to us as one of the plans on which that great sorting-machine, the human perceptive faculty, arranges its material. Through the doorways of perception, through the senses of man, crowd, in our waking state, sense-impression upon sense-impression ; sound and taste, colour and warmth, hardness and weight—all the various elements of an infinite variety of phenomena, all that forms for us reality—crush through the open gateways. The perceptive faculty, sharpened by long centuries of natural selection,[1] sorts

---

[1] We cannot infer the time and space-modes of perception except for perceptive faculties, more or less similar to our own. The order of

and sifts all this mass of sense-impressions, giving to each a place and an instant. Thus the magnitude of space and time depends upon no external world independent of ourselves, but on the complexity of our sense-impressions, immediate and stored. Infinity of space or eternity of time have no meaning in the field of perception, because the correlation and sequence of our perceptions, wide as both undoubtedly are, do not require these enormous frames to exhibit them. Where the senses perceive no object, there there is no space, for there no groups of sense-impressions are to be distinguished. Where I can no longer carry back the sequence of phenomena, there time ceases for me because I no longer require it to distinguish an order of events. Let the reader endeavour to realize empty time, or time with no sequence of events, and he will soon be ready to grant that time is a mode of his own perception and is limited by the contents of his experience.[1] Thus the moments devoted to wonder over the eternities of time are as ill-spent as those consumed in pondering on the immensities of space (p. 188). They are like moments employed in examining the frame of a picture and not its contents, in admiring the constitution of the artist's canvas and not his genius. The

phenomena in both space and time is essentially conditioned by the intensity and quality of the consciousness (p. 101).

[1] It may well be questioned whether anything that falls outside human experience can be said to have existed in *perceptual* time. Such time is essentially the mode by which we distinguish an *immediate* sense-impression from a succession of stored sense-impresses (p. 49). That the world has existed for 60,000,000 years is a *conception*, and the period referred to a conceptual rather than a perceptual one. The *future* also is a notion attaching rather to conceptual than to perceptual time. The full discussion of these points cannot, however, be entered upon at this stage.

frame is just large and strong enough to support the picture, the canvas is just wide and stout enough to sustain the artist's colours. But frame and canvas are only modes by which the artist brings home his idea to us, and our wonder should not be for them, but for the contents of the picture and its author. So it is with time and space—these are but the frame and the canvas by aid of which the perceptive faculty displays our experience. Our admiration is due not to them, but to the complex contents of perception, to the extraordinary discriminating power of the human perceptive faculty. The complexity of nature is conditioned by our perceptive faculty; the comprehensive character of natural law is due to the ingenuity of the human mind. Here, in the human powers of perception and reason, lies the mystery and the grandeur of nature and its laws. Those, whether poets or materialists, who do homage to nature as the sovereign of man, too often forget that the order and complexity they admire are at least as much a product of man's perceptive and reasoning faculties as are their own memories and thoughts.

§ 13.—*Conceptual Time and its Measurement.*

Time as a mode of perception is limited, we have seen, to the extent to which sequences of stored sense-impresses can be carried back; it marks that order of perceptions which is the history of our consciousness. From this it is clear that perceptual time has no future and no eternity in the past. That consciousness in the future will continue as it has done in the past is a conception, but not a perception. We perceive the past, but we only conceive the future. How, then, we may ask, do we pass from perceptual to conceptual time, from our actual sequences of sense-impressions to a scientific mode of describing and measuring them ? Clearly it would he extremely cumbersome to measure time by a detailed account of the changes in our sense-impressions. Imagine the labour of describing all the stages of consciousness between breakfast and dinner as a means of determining the period which has elapsed between the two meals! Yet this method of considering time brings out clearly how time is a relative order of sense-impressions, and how there is no such thing as *absolute* time. Every stage in sense-impression marks in itself an epoch of time, and may form the basis of a measurement of time for an individual. " I am sleepy, it is time to go to bed," says the child ; " I am hungry, it is time to eat," says the savage, and both without thinking of the clock or the sun. Fortunately for us we are not compelled to measure time by a description of the sequence of states of consciousness. There are certain sense-impressions which experience has shown us repeat themselves, and which, on the average, correspond to the same routine of consciousness. In the first place, the recurrence of night and day are observed very early in the natural history of man to mark off approximately like sequences of sense-impressions ; a day and night becomes a measure of a certain interval of consciousness. That the same amount of consciousness can, at any rate approximately, be got into *each* day and night by the normal human being is a matter rather of experience than of demonstration ; it cannot be proved,—it can only be felt.

Very much the same holds for the smaller intervals of time. When we say it is four hours since breakfast, we mean in the first place that the large hand of our clock or watch has gone round the dial-face four times—a repeated sense-impression which we could, if we please, have observed. But how shall we decide whether each of those four hours represents equal amounts of consciousness, and the same amount to-day as yesterday? It may possibly be that our time-keeper has been compared with a standard clock, regulated perhaps from Greenwich Observatory. But what regulates the Greenwich clock? Briefly, without entering into details, it is ultimately regulated by the motion of the earth round its axis, and the motion of the earth round the sun. Assuming, however, as a result of astronomical experience, that the intervals day and year have a constant relation, we can throw back the regulation of our clock on the motion of the earth about its axis. We may regulate what is termed the "mean solar time" of an ordinary clock by "astronomical time" of which the day corresponds to a complete turn of the earth on its axis. Now, if an observer watches a so-called circumpolar star, or one that remains all day and night above the horizon, it will appear, like the end of his astronomical clock-hand, to describe a circle; the star ought to appear to the observer to describe equal parts of its circle in equal times by his clock, or while the end of the clock-hand describes equal parts of its circle. In this manner the hours on the Greenwich astronomical cloch, and ultimately on all ordinary watches and clocks regulated by it, will correspond to the earth turning through equal angles on its axis. We thus throw back our measurement of time on the earth as a time-keeper; we assume that equal turns of the earth on its axis correspond to equal intervals of consciousness. But, all clocks being set by the earth, how shall we be certain that the earth itself is a regular time-keeper? If the earth were gradually to turn more slowly upon its axis, how should we know it was losing time, and how measure the amount? It might be replied that we should find that the year had fewer days in it ; but then how could we settle that it was the day that was growing longer and not the year that was growing shorter? Again, it may be objected that we know a great number of astronomical periods relating to the motion of the planets expressed in terms of days, and that we should be able to tell by comparison with these periods. To this we must answer that the relation of these periods expressed in days, and in terms of each other, appears now indeed invariable ; but what if all these relations are found to have slightly changed a thousand or five thousand years hence? Which body shall we say has been moving uniformly, which bodies have been gaining or losing? Or, what if, the ratios of their periods remaining the same, they were *all* to have lost or gained? How shall we, with such a possibility in view, assert that the hour to-day is the " same " interval as it was a thousand, or better perhaps a million, years back? Now certain investigations with regard to the frictional action of the tides make it highly probable that the earth is not a perfect time-keeper, nor are we able to postulate that regularity of motion, by which alone we could reach absolute time, of any body in our perceptual experience.

Astronomy says it is not in me, nor do we get a more definite answer from physics. Suppose an observer to measure the distance traversed by light in one second ; can this be for all time a permanent record of the length of a second? Another observer a thousand years after measures again the distance for one of *his* seconds, and finds it differs from the old determination. What shall he infer? Is the speed of light really variable, has the planetary system reached a denser portion of the ether, has the second changed its value, or does the fault lie with one or other observer? No more than the astronomer can the physicist provide us with an *absolute* measure of time. So soon as we grasp this we appear to lose our hold on time. The earth, the sole clock by which we can measure millions of years, fails us when we once doubt its regularity. Why should a year now represent the same amount of consciousness as it might have done a few million years back? The absolutely uniform motion by which alone we could reach an absolute measurement of time fails us in perceptual experience. It is, like the geometrical surface, reached in conception, and in conception only, by carrying to a limit there the approximate sameness and uniformity which we observe in certain perceptual motions. Absolute intervals of time are the conceptual means by which we describe the sequence of our sense-impressions, the frame into which we fit the successive stages of the sequence, but in the world of sense-impression itself they have no existence.

Newton, defining what we term here conceptual time, tells us :—

"That absolute, true, and mathematical time is conceived as flowing at a constant rate, unaffected by the speed or slowness of the motions of material things."

Clearly such time is a pure ideal, for how can we

measure it if there be nothing in the sphere of perception which we are certain flows at a constant rate? "Uniform flow," like any other scientific concept, is a limit drawn in imagination—in this case, from the actual "speed or slowness of the motions of material things." But, like other scientific concepts, it is invaluable as a shorthand method of description. Perceptual time is the pure order in succession of our sense-impressions and involves no idea of absolute interval. Conceptual time is like a piece of blank paper ruled with lines at equal distances, upon which we may inscribe the sequence of our perceptions, both the known sequence of the past and the predicted sequence of the future. The fact that upon the ruled lines we have inscribed some standard recurring sense-impression (as the daily transit of a heavenly body over the meridian of Greenwich), must not be taken as signifying that states of consciousness succeed each other uniformly, or that a "uniform flow" of consciousness is in some way a measure of absolute time. It denotes no more than this: that from noon to noon the average human being experiences much the same sequence of sense-impressions, and thus the same space in our conceptual time-log may be conveniently allotted for their inscription. Above all, it must not lead us to project the absolute time of conception into a reality of perception; the blank divisions at the top and bottom of our conceptual time-log are no justification for rhapsodies on the past or future eternities of time, for rhapsodies which, confusing conception and perception claim for these eternities a real meaning in the world of phenomena, in the field of sense-impression.

§ 14.—*Concluding Remarks on Space and Time.*

The reader who has recognized in perceptual space and time the modes in which we distinguish groups of sense-impressions, who has grasped that infinities and eternities are products of conception, not actualities of the real world of phenomena, will be prepared to admit the important conclusions which flow from these views for both practical and mental life. If the individual carries space and time about with him as his modes of perception, we see that the field of miracle is transferred from an external mechanical world of phenomena to the individual perceptive faculty. The knowledge of this in itself is no small gain to clearing up our ideas with regard to such recrudescences of superstition as spiritualism and theosophy. If space and time are to be annihilated, it cannot be done once for all, but it must be done for each individual perceptive faculty. When, for example, theosophists tell us that, putting aside the bondages of space and time, they can communicate with adepts from Central Asia in London drawing-rooms, they are really saying that *their own* perceptive faculties can distinguish groups of sense-impressions in other than those modes of space and time which are characteristic of the normal perceptive faculty. They have not abrogated *our* space and time, only their own. They are merely declaring that their modes of perception are different from ours. If we find from long experience that there is in man a normal perceptive faculty which co-ordinates sense-impressions in space and time in the same uniform manner, then we are justified in classifying the infinitesimal minority who suffer from abnormal modes of perception with the ecstatic and the insane.

Through sickness they have lost, or through atavistic tendencies they have failed to develop, the normal perceptive faculty of a healthy man—the *mens sana in corpore sano*.

No less valuable is the conclusion that it is idle to speak of anything as existing in space or as happening in time which cannot be the material of perception. Whatever by its nature lies beyond sense-impression, beyond the sphere of perception, can neither exist in space nor happen in time. Thus the scientific conception of causation, or that of uniform antecedence, cannot with any meaning be postulated of it—a result we have already reached from a slightly different standpoint (pp. 152 and 186). Indeed, it seems to me that, with a clear appreciation of space and time as modes of perception, most phases of superstition and obscurity fade into nothingness, while the field to which the category of knowledge applies is seen to be sharply defined.

SUMMARY.

1. Space and Time are not realities of the phenomenal world, but the modes under which we perceive things apart. They are not infinitely large nor infinitely divisible, but are essentially limited by the contents of our perception.

2. Scientific concepts are, as a rule, limits drawn in conception to processes which can be started but not carried to a conclusion in perception. The historical origin of the concepts of geometry and physics can thus be traced. Concepts such as geometrical surface, atom, and ether, are not asserted by science to have a real existence in or behind phenomena, but are valid as shorthand methods of describing the correlation and sequence of phenomena. From this standpoint conceptual space and time can be easily appreciated, and the danger avoided of projecting their ideal infinities and eternities into the real world of perceptions.

LITERATURE.

HUME, DAVID.—A Treatise on Human Nature (1739), book i. part ii.
Of the Ideas of Space and Time. Green and Grose : Works of
Hume, vol. i. pp. 334-371.

KANT, IMMANUEL.—Kritik der reinen Vernunft (1781). Elementar-
lehre, i. Theil. Sämmtliche Werke, Ausgabe v. Hartenstein, Bd.
iii. S. 58-80.

A good account of Kant's views will be found in Kuno Fischer's
Geschichte der Philosophie, Bd. iii. S. 312-349. A brief description is
given on pp. 218-20 of Schwegler's Handbook of the History of Philosophy, translated by J. H. Stirling, Edinburgh, 1879.

None of the geometrical or physical text-book writers have hitherto
ventured to discuss how the conceptual space and time which are at the
basis of their investigations are related to perceptual experience. The
reader will, however, find much that is valuable in Clifford's Philosophy
of the Pure Sciences (1873), Lectures and Essays, vol. i. pp. 254-340,
and in his " Of Boundaries in General," Seeing and Thinking (1880),
pp. 127-156.

A criticism of Hume's views will be found on pp. 230-254 of Green's
"General Introduction " to Hume's Works, vol. i., while Kant's doctrines have been attacked by both Trendelenburg and Ueberweg.
References are given in vol. ii. pp. 158, 330, and 525 of the latter
writer's History of Philosophy, London, 1874.

A good deal that is suggestive with regard not only to space and time,
but position and motion, may still with caution be extracted from the
Physics of Aristotle. See especially E. Zeller, Die Philosophie der
Griechen ii. Theil, 2 Abth. S. 384-408, and Ueberweg loc. cit., vol. i.
pp. 163-6. The reader must not be discouraged by the contempt
expressed for Aristotle's ideas of space and motion in George Henry
Lewes's Aristotle : a Chapter from the History of Science, London,
1864 (p. 128 et seq.).

CHAPTER VI.

The Geometry of Motion.

§ 1.—*Motion as the Mixed Mode of Perception.*

WE have seen in the previous chapter that there are two modes under which the perceptive faculty discriminates between groups of perceptions, namely, those of space and time. The combination of these two modes, to which we give the various names of change, motion, growth, evolution, may be said to be the *mixed* mode under which all perception takes place.[1] Science, accordingly, if we except special branches treating of the modes under which we perceive and think, is essentially, as a description of the contents of perception, a description of change or variation. In order to draw a mental picture of the universe, to map out in broad outline its characteristics, science has introduced the conception of geometrical forms ; in order to describe the sequence of perceptions, to form a sort of historical atlas of the universe, science has introduced the conception of geometrical forms changing with absolute time. The

---

[1] Trendelenburg sees in real or constructive motion the basis of all perception and conception. He tries to show that the conception of motion does not require the notions of space and time, which he asserts flow from the conception of motion itself. I do not think he is successful in this, but his attempt is instructive as showing how essentially perception and conception involve motion. (See his *Logische Untersuchungen,*" 2nd edition, Bd. i., chaps. v.–viii., Leipzig, 1862.)

analysis of this conception is what we term the *Geometry of Motion*. The geometry of motion is thus the conceptual mode in which we classify and describe perceptual change. Its validity depends not upon its corresponding absolutely to anything in the real world—a correspondence at once rebutted by the ideal character of geometrical forms—but upon the power it gives us of briefly resuming the facts of perception or of economizing thought.[1] The geometry of motion has been technically termed *kinematics*, from the Greek word κίνημα, signifying a *movement*. It teaches us how to represent and measure motion in the abstract, without reference to those particular types of motion which a long series of experiments, and much careful observation of the world of phenomena, have shown us are best fitted to exhibit the special changes in the sphere of perception. When we apply what we have learnt in the

---

[1] The term *economy of thought*, originally due, I think, to Professor Mach, of Prague, embraces in itself a very important series of ideas. Its value is much more significant, if we remember how thought depends on stored sense-impresses, and that it is difficult to deny to these and to their nexus—association—a physical or kinetic aspect (p. 51). The economy of thought thus becomes closely associated with an economy of energy. The range of perceptions is so wide, their sequences so varied and complex, that no single brain could retain a clear picture of the relationship of the smallest group but for the short-hand descriptions provided by the conceptions of science. Dr. Wallace, in his *Darwinism*, declares that he can find no origin for the existence of pure scientists, especially mathematicians, on the hypothesis of natural selection. If we put aside the fact that great power in theoretical science is correlated with other developments of increasing brain-activity, we may, I think, still account for the existence of pure scientists as Mr. Wallace would himself account for that of worker-bees. Their functions may not fit them individually to survive in the struggle for existence, but they are a source of strength and efficiency to the society which produces them. The solution of Mr. Wallace's difficulty lies, I think, in the social advantage of science as an economy of intellectual energy.

geometry of motion to those particular types of motion—*natural* types as they may be conveniently called—and investigate how they are correlated, then we are led to the so-called *Laws of Motion* and to those conceptions of *Mass* and *Force* [1] upon which our physical description of the universe depends. These will form the topics of succeeding chapters, but, in order to see our way more clearly through that maze of metaphysics which at present obstructs the entry to physics, we must devote some space to a discussion of the elementary notions of kinematics.

§ 2.—*Conceptual Analysis of a Case of Perceptual Motion. Point-Motion.*

We shall, I think, best obtain clear ideas of motion by examining some familiar case of physical change of position and endeavouring to analyze it into simple types which may be easily discussed by the aid of geometrical ideals. Let us take, for instance, the case of a man ascending a staircase which may have several landings and turns in its course. The changes in our sense-impressions during the man's ascent are of an extremely complex character, and we see at once how difficult, if not impossible, it would be to describe all that we perceive. Not only the position of the man on the staircase changes, but his hands and his legs are perpetually varying their position with regard to his trunk, while his trunk itself turns and oscillates, bends and alters its shape. For simplification let us, in the first place, fix our attention on some small element of his person ; let us follow with our eye, for example, the top button of his waistcoat. Now, the

---

[1] Not force as the *cause* of motion, but force as a measure of motion.

first observation that we make is that this button takes up a series of positions which are perfectly continuous from the start to the finish of the ascent. There can be no break in this series of positions anywhere throughout the whole extent of the staircase; for, if there were any, the button must, in accurate language, have ceased to be a permanent group of sense-impressions, and to be distinguished from other groups under the mode space. In ordinary parlance, it must "have left our space and come back to it again"—a phenomenon totally contrary to the experience of the normal human perceptive faculty. If we cut the button off the waistcoat, we could still conceive it to move up the staircase in precisely the same manner as when the man wore it,—carried up, let us suppose, by an invisible spirit hand. It will be obvious that this motion of the button, if fully known to us, would tell us a good deal about the motion of the man. It would not describe, of course, how he moved his legs and arms about, but it would indicate very fairly how long the man took to go from one landing to another, and when he was going quickly, when slowly. But it is still far from clear how we are to describe the motion of the button, so that we could conceive its motion repeated by aid of our description. The button, like the man, has many elements, and the question again arises how we are to describe the motions of them all.

Let us now stretch our imaginations a little further; let us suppose the staircase to be imbedded in a great mass of soft wax, and suppose the button, guided still by the spirit hand, to move up the staircase precisely as it did on the man's waistcoat, but now pushing its way through the wax. The passage of the button

would now form a long tube-like hollow in our mass
of wax extending from the bottom to the top of the
staircase. This tube would not necessarily be of
equal bore throughout, because, owing to the motion
of the man, the button might occasionally move more
or less sideways. Still, the smaller the button the
smaller would be the bore of the tube cut through the
wax. We will now suppose a long piece of stiff wire
passed through the tube and firmly fixed at its ends.
The wax, and even the staircase, may now be removed,
and then, if a small bead be slung on the wire and
move up the wire in the same manner as the button
moved up the tube, we shall be able to describe a
good deal of the motion of the button from that of
the bead. Now in conception we may suppose the
wire to get thinner and thinner, and the bead smaller
and smaller, till in conception the wire ends in a
geometrical line or curve, and the bead in a geometrical point. The motion of the ideal point along
the ideal curve will represent with a great degree of
accuracy the motion of an extremely small button up
a tube through the wax of an extremely small bore.
The reader may feel inclined to ask why we did not
commence by saying : "Consider a point of the man ;
its motion must give a curve passing from top to
bottom of the staircase." The answer lies in this :
that we cannot *perceive* a point. In conception we
reach a point by carrying to a limit the perceptual
process of taking a smaller and smaller element of
the man, and the stages we have indicated from man
to button, bead and geometrical point, indicate how
certain elements of the perceptual motion are dropped
at each stage, till in conception we reach as a limit an
ideal motion capable of being fairly easily described.

The motion of a point along a curve is the simplest ideal motion we can discuss. Obviously, however, it will enable us to classify and describe with considerable exactness a number of our perceptions with regard to the man's motion. Harness the button to the point, and the man to the button ; then if the point move along its path, carrying button and man with it, we shall have a means of describing a good deal of the real motion of the man. When he starts, when he stops, when he goes fast, when he goes slowly, what time he takes from one landing to another will be deducible from the motion of the point. Of course this point-motion does not enable us to *fully* describe the motion of the man. For instance, it is conceivable that he may have turned several somersaults in going upstairs. About such eccentricities in the man's motion the motion of the point may tell us nothing at all. Even had the man been incapable of moving his arms, legs, head, &c.,— had he been a *rigid* body—the point-motion would have been incapable of fully describing his motion. As a rigid body the man might have been turned round and about the point without changing its motion. Did he go upstairs backwards or forwards, head or feet uppermost, or partly in one, partly in another of these modes ? Clearly the motion of the point can tell us nothing of all this. The motion of the point can tell us nothing of how the man as a rigid body might have turned about the point ; we should want to know at each instant of the motion which way the man was facing, what was his *aspect*, and further how he was changing his aspect or rotating about the point. The description of the ideal point-motion would have to be supplemented

by a description of the rotating or spinning motion,
even if the man were supposed to be a rigid body.
The first type of motion, corresponding to change of
position, is termed *motion of translation ;* the second
type, corresponding to the change of aspect of a rigid
body, is termed *motion of rotation.*

§ 3.—*Rigid Bodies as Geometrical Ideals.*

Just as the former motion is described by the purely
ideal conception of a point moving along a curve,
so the latter is also made to depend on geometrical
notions, namely, those of a *rigid* body turning about
a *line* passing through a *point.* What, in the first
place, do we mean by using the term *rigid* body ?
The real man is moving his limbs and bending his
body, and generally changing his form at each instant of the motion.   Now the reader may feel
inclined to say : Replace the man by a wooden table
or chair, and we shall have a rigid body.   But this is
only popular language, and what we are seeking is
an accurate or scientific definition of rigidity.   Such
a definition is usually given in the following words:—

A body is said to remain rigid during any given
motion when the distances between all pairs of its
points remain unaltered throughout the whole dura-
tion of the motion.

But we see at once from this definition that we
have replaced the real body, the group of sense-impressions which forms part of the picture constructed
by our perceptive faculty, by an ideal geometrical
body possessing "points," and that it is a property of
this body—existing only on the ideal map on which
conception plots out perception—that we are defining.   It is quite true that the geometrical ideal of

a rigid body is a better description of a wooden chair than of the flexible body of a man ; yet what is a " point " on the chair, and what is the " distance " between a pair of points ? How, again, am I to ascertain accurately that such distances remain unaltered during the motion ? The very idea of distance, when clearly appreciated, involves the geo- metrical conception of points and does not corre- spond to anything in our perceptual experience. [1] Rigidity is thus seen to be a conceptual limit, which by concentrating our attention on a special group of perceptions forms a valuable method of classification.

Although for the description of some types of motion it may be useful to replace the wooden chair by a body of ideal rigidity in our conceptual map, still the physicist tells us that for the purpose of classifying other phases of sense-impression, he is bound to consider that the chair is *not* rigid, and that he is perceptually able to measure changes in the relative position of its parts. He cannot describe the mechanical action between different parts of the chair without supposing it elastic, and this elasticity in- volves changes of form in its parts. For example,

[1] We speak, for example, of the "distance" from London to Cambridge being fifty-five miles, and this is a practical method of de- scribing the sense-impressions of a journey from one place to the other, and distinguishing it from a journey of fifty-six or fifty-seven miles. But what do we exactly mean ? From Stepney Church to St. Mary's ? If so, from which part of one church to which part of the other ? Or, again, is it from the stone near the gateway of Stepney Church to the last milestone by St. Mary's ? If so, from which side of the one stone to which side of the other ? In the end we find ourselves driven to the conception of a point on either stone—no *perceptual* mark gets over the difficulty of the *where* to the *where*. We are forced to conclude that the idea of distance is a conception, invaluable for classifying our experience but not accurately corresponding to a perceptual reality.

the action between the parts of the chair changes, when it is supported on its back instead of its legs, and thus the chair changes its form in these two positions. A like change of form will take place even if the chair be only rotating. Nor does this variation in shape merely result from the chair being of wood—it would be equally true if the chair were of iron, or any other material. Change of form is in many cases perceptually appreciable, and in most cases we can determine its conceptual value. Thus, so far from the rigid body being a limit which might be reached in perception, our whole perceptual experience seems to indicate that the conception rigidity corresponds to nothing in the real world of phenomena. We perceive that most bodies do change their form, and where we do not perceive it physics compel us to conceive it. Thus rigidity is very much like the spherical surfaces of geometry. The latter correspond accurately to nothing in our perceptual experience, and we cannot even conceive a continuous surface as a limit to be reached in perception. Both, however, are alike valuable bases of classification. By replacing real bodies by ideal rigid bodies we are able, although neglecting their changes of form, to classify and describe a wide range of our perceptions of motion. To classify other perceptions, however, we conceive the same bodies not to be rigid, but varying in form ; we actually measure the very changes in shape, which we purposely neglected in another branch of our survey of the physical universe.

§ 4.—*On Change of Aspect or Rotation.*

Even when we have transferred our moving body from the perceptual to the conceptual sphere by postulating its rigidity, we shall still find the notions of aspect and spin involve further geometrical conceptions. Let us consider our rigid body capable of turning about a point, the question then arises, How can we distinguish one aspect from a second? Clearly, the notion of direction involves that of a line, but the change in direction in *one* line will not be sufficient to describe change of aspect. For if C (Fig. 4) represent the fixed point about which the body rotates, and A be another definite point of the body, the line CA may take up a new position CA'; but the change in position of CA to CA' does not fully determine the aspect of the body, for there is nothing to fix how much the body may have been turned about the line CA while it was moving into the position CA'. We are compelled, therefore, to take a second point B, and a second direction CB; then if we state the new position CB' taken by CB as well as the new position CA' of CA, we shall have absolutely determined the change of aspect of the body. The reader will very easily convince himself that in giving the new positions of two definite points A and B of the rigid body, we have absolutely fixed its position. It is easy to show that this turning of two lines CA and CB into new positions CA' and CB' may also be described as a turning of the body about a certain line of direction CO through a certain angle.[1] Thus the manner in which we conceive

---

[1] This may be proved by the aid of elementary geometry in the following manner :—

Let the triangle CBA be displaced into the position CB'A'. Join the points A, A' and B, B', and let the mid-points of AA' and BB' be M and N respectively. Through C and M draw a plane perpendicular to AA' and through C and N a plane perpendicular to BB'. These two planes meet in a line passing through C, since C is common to

change of aspect to be described and measured is essentially geometrical, or ideal. It depends on the conception of a straight line fixed in the body and fixed in space about which the body turns. It further

them both. Let O be any point in this line, and join it to M and N, then OM and ON are respectively perpendicular to AA' and BB.' In the triangles AOM, A'OM, AM and A'M are equal, OM is common, and the angles at M are right, hence it follows by *Euclid* i. 4 that the third sides OA and OA' are equal. For precisely similar reasons it

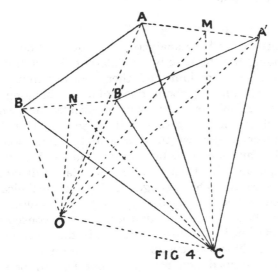

FIG 4.

follows that OB and OB' are equal. Hence the three distances of O from the angles of the triangle ABC are equal to its distances from the three angles of the triangle A'B'C respectively. Thus the two tetrahedrons with summits at O and having bases ABC and A'B'C respectively are equal in every respect, for all their edges are equal each to each. One of them may thus be looked upon as the other in a changed position. They have, however, the same edge OC. Hence one tetrahedron may be moved into the position of the other by rotating it through a certain angle about the edge OC. That is to say, the triangle CBA may be turned into the position CB'A' by rotating it through a certain angle— the angle between the planes BOC and B'OC—about the line OC.

17

involves the conception of the body turning through a certain angle, but an angle Euclid tells us is the inclination of two lines. Thus our description of change of aspect depends upon the conception of lines existing in the rigid body. It is entirely a conceptual description, but like the idea of point-motion, it again serves as a powerful means of discriminating and classifying our experiences of perceptual motion.

§ 5.—*On Change of Form, or Strain.*

Thus far we have analyzed the motion of our man ascending the staircase by considering the motion of an ideal point of him, and then treating him as a rigid body turning about this point, or changing its aspect. It only remains for us to consider how, when the point is in any given position and the man has any given aspect, we may remove the condition of rigidity, and describe how he can move his limbs about, change his form, or alter the relative distances of his parts. This change of form is technically termed *strain*, and its description and measurement forms the third great division in the conceptual motion of bodies. Now we cannot in this work enter into a technical discussion of how strain is scientifically described and measured, but for our present purposes we must ascertain whether the theory of strains deals, like that of the translation of a point and that of the rotation of a rigid body, with conceptual ideals.

There are two fundamental aspects of strain which most of us consciously or unconsciously recognize. These are change of size without change of shape, and change of shape without change of size. Take a thin hollow india-rubber ball and blow more air into its interior. This will increase its size without necessarily changing its shape. It was spherical in shape and remains spherical in shape, only it is larger. We conceive the ball represented by a sphere, and the change in size will depend upon the change in diameter. The ratio of the extension to the original length of the diameter may be taken as a proper basis for the measurement of the strain. Such a ratio is termed a *stretch*, and it may be shown that for a small increase of size the ratio of the increase of volume to the original volume is very nearly three times the stretch of the diameter.[1] This ratio is termed the *dilatation*, and is a proper measure of the change in size. Now it is clear that in order to measure this change of size, we require to measure the diameters in the two conditions of the body. But a diameter, although in the conceptual body definite enough as a straight line terminated by two points, is, in this accurate sense of the word, a meaningless term when we are dealing with a perceptual body. If the body has no continuous boundary, but, according to the physicist, is a mass of discrete atoms (Fig. 5), none or which we can individually feel, and the mutual dis-

---

[1] The volumes of bodies of similar shape are as the cubes of corresponding lengths. Hence if V and V' be the old and new volumes, d and d' the old and new lengths, $V'/V = d'^3/d^3$, but if s be the stretch $(d'-d)/d = s$, or $d' = d(1+s)$. A little elementary algebra gives us for the dilatation δ :—

$$\delta = \frac{V'-V}{V} = \frac{d'^3-d^3}{d^3} = (1+s)^3 - 1 = 3s + 3s^2 + s^3 = 3s, \text{ nearly,}$$

if s, as in most practical cases, be very small. For example, in metal $s = \frac{1}{1000}$ would be a rather large value; but taking $\delta = 3s$, we should only be neglecting about $\frac{1}{1000}$ of the value of δ.

tances of which we cannot measure, it is clear that
the only diameter we can be talking about is that of

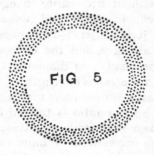

FIG 5

a conceptual sphere by which we have replaced the
perceptual ball.

As it is with change of size, so it is with change of
shape : we are really basing our system of measurement upon conceptions, which enable us to describe
and classify perceptions, but are not real limits to
perception.   Change of shape without change of size
can be realized in the following manner: Take a piece
of woven silk or other slightly elastic material, and
draw a rectangle upon it with sides a few inches long

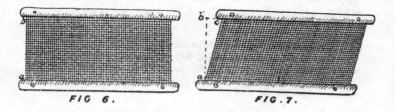

FIG 6.   FIG. 7.

parallel to the warp and woof.   Then if such a
rectangle be held firmly top and bottom between two
pairs of parallel pieces of wood, or even between the

two thumbs and their respective forefingers, a *slide* of the holders parallel to each other will produce a change of form without change of size. Now the extent of such a strain will depend on the amount by which the warp and woof have changed their inclination to each other,—that is to say, on the amount after strain by which the angle between them differs from a right-angle. But this change in angle only becomes of meaning if we suppose the warp and woof to be straight lines. In other words, to get a measure of the strain we replace the perceptual warp and woof by a geometrical network. Such a type of strain is termed a *slide* or *shearing* strain, and all changes of shape without change of size can in conception be analyzed into slides.[1] Further, it may be shown that all changes of form whatever can be analyzed into stretches and slides,[2] or into changes of length and changes of angle. But in the cases of both slide and stretch we are thrown back on geometrical notions, when we come to consider their measurement; in both cases we replace the perceptual body by a conceptual body built up of points, lines, and angles. Thus the whole theory of strain deals with a conceptual means of distinguishing and describing perceptions, and not with something absolutely inherent in the perceptions themselves.

[1] Technically the slide is not measured by the change in angle or by the angle *bac* in Fig. 7, but by the trigonometrical tangent of this angle, or by the ratio of the length *bc* to the length *ba*—in other words, by the ratio of the amount the woof has been slid to the length of the warp.

[2] An elementary discussion of strain will be found in Clifford's *Elements of Dynamic*, part i. pp. 158–90; or in Macgregor's *Kinematics and Dynamics*, pp. 166–84. The reader may also consult §§ 8 and 13, contributed by the present writer to Chapter iii. of Clifford's *Common Sense of the Exact Sciences*.

§ 6.—*Factors of Conceptual Motion.*

We started with a man ascending a staircase, and we have seen by our analysis that the conceptual description of his motion requires us to discuss : (*a*) The Motion of a Point, (*b*) the Motion of a Rigid Body about a Fixed Point, (*c*) the Relative Motion of the Parts of a Body or its Strain. These are the three great divisions of Kinematics, or the Geometry of Motion. But in the case of all these divisions we find that we are thrown back on the ideal conceptions of geometry ; we measure distances between points and angles between lines, which are not true limits to our perceptual experience. Thus our ideas of motion appear as ideal modes, in terms of which we describe and classify the sequences of our sense-impressions : they are purely symbols by aid of which we resume and index the various and continual changes under-gone by the picture our perceptive faculty presents to us. The more fully and clearly the reader grasps this fact, the more readily will he admit that science is a conceptual *description* and classification of our perceptions, a theory of symbols which economizes thought. It is not a final explanation of anything. It is not a *plan* which lies in phenomena themselves. Science may be described as a classified index to the successive pages of sense-impression, but it in nowise accounts for the peculiar structure of that strange book of life.[1]

---

[1] The extremely complex results which flow from the simple basis of the planetary theory have often been taken as an evidence of "design" in the universe. The universe has been with much confusion spoken of as the *conception* of an infinite mind. But the *conceptual* basis of the planetary theory lies in geometrical notions, no ultimate evidence of which can be discovered in the perceptual world. Thus, while the

Of the three types of motion just introduced to the notice of the reader, the first, or point-motion, is that which for our present purposes is most important. The remainder of the present chapter will therefore be devoted to its discussion. The reader will, I trust, pardon its somewhat technical character, for without this investigation of point-motion it would be impossible to analyze the fundamental notions of *Matter* and *Force*, or to rightly interpret the Laws of Motion.

§ 7.—*Point-Motion.* *Relative Character of Position and Motion.*

Motion has been looked upon as change of position, but if we try to represent the position of a point we must do so *with regard to something else.* If space be a mode of distinguishing things, we must have at least two things to distinguish before we can talk about position in space. Position of a point is therefore relative, relative to something else, which for our present purposes we will suppose to be a second point. Absolute position in space, just as absolute space itself (p. 186), is meaningless. Let the letter P (Fig. 8) represent a point, and the letter O a point from which we are to measure P's relative position. Now the distance from O to P would indicate for us

planetary theory answers our purposes of *description*, it could never have been the *conception* upon which the universe was " designed," for the conception is nowhere found perceptually realized. *Starting* with his material endowed with all its peculiar properties, the carpenter makes for us a box according to our geometrical description, but in reality not ultimately geometrical. Starting with nothing but the power of realizing conception in perception, he would have produced from our geometrical plan a geometrical box. Geometrical notions could flow from the material universe, but the latter could not flow from the former.

the position of P relative to O, but in our conceptual space we have in general a variety of other points or geometrical bodies besides O which we wish to distinguish from P, and to do this we must give what is termed direction to the distance OP, we must determine, as it were, whether it runs north and south, south-west and north-east, or upwards and downwards.[1] But even this is not enough. We must be also told the *sense* of this direction, whether, for example, it be *op* or *op'* (Fig. 8), or, say, runs from south-west to north-east or north-east to south-west. Thus, if we want to plot out position in space about

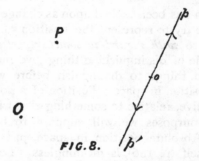

FIG. 8.

a point O, we must do this by measuring distances from O in given directions and with given senses.

[1] In the conceptual space which corresponds most closely to perceptual space—so-called space of *three* dimensions—we require, in order to mark the relative position of all possible bodies, to start from *three* standard points (which must not be in the same straight line) in order to fix direction. Throughout this chapter we shall understand by the position of a point P relative to another point O, the *directed* step OP, and by the motion of P relative to O change in this directed step. A fuller account of *Position* will be found in the chapter under that title contributed by the author to Clifford's *Common Sense of the Exact Sciences.*

We must know distance and *bearing*[1] from O to determine fully a point P. To represent geometrically the position of P with regard to O, we may draw a piece of a straight line (*op*) having as many units of length on our scale as there are units of distance from O to P, the line having the same direction as this distance, and having an arrow-head upon it to mark the sense. Such a line marking the magnitude, direction, and sense of P's position relative to O is termed a *step*. Such a step tells us how to shift our position from O to P. Step so many feet with such and such a bearing, and we shall pass from O to P.

If P be in motion and we know what is the step from O to P at each instant of the motion, we shall have a complete picture of the sequences of positions, the motion of P relative to O. The reader must be careful to notice the relativity of the motion ; absolute motion, like absolute position, is inconceivable: a point P is conceived as describing a path relatively to something else. Thus the button on the man's waistcoat moved relatively to the staircase, but the staircase is rushing perhaps 1,000 miles an hour round the axis of the earth, while the earth itself may be bowling 66,000 miles an hour round the sun. The sun itself is moving towards the constellation of Lyra at some 20,000 miles an hour, while Lyra itself is doubtless in rapid motion with regard to other stars, which, so far from being "fixed," may be travelling thousands of miles an hour relatively to each other. Clearly it is not only impossible to tell how many thousand miles an

---

[1] With the signification in which the words are here used, a line has *direction* but not *bearing*. We must add to direction the conception of *sense* before we form the idea of bearing.

hour we are each one of us to be conceived as speed-ing through space, but the expression itself is mean-ingless. We can only say how fast one thing is moving *relatively* to another, since all things whatsoever are in motion, and no one can be taken as the standard thing, which is definitely " at rest."

Is it correct to say that the earth actually goes round the sun, or that the sun goes round the earth ? Either or neither ; both are conceptions which de-scribe phases of our perception. Relatively to the earth the sun describes approximately an ellipse round the earth in a focus, relatively to the sun the earth describes approximately an ellipse about the sun in a focus. Relatively to Jupiter neither state-ment is correct. Why, then, do we say that it is more scientific to suppose the earth to go round the sun? Simply for this reason : the sun as centre of the planetary system enables us to describe in conception the routine of our perceptions far more clearly and briefly than the earth as centre. Neither of these systems is the description of an absolute motion actually occurring in the world of phenomena. Once realize the relativity of motion and the symmetry of the planetary system is seen to depend largely on the standpoint from which we perceive it : the planetary theory can thus be easily recognized as a mode of description peculiar to an inhabitant of a solar system.

§ 8.—*Position. The Map of the Path.*

Relatively to O, then, our point P describes a con-tinuous curve or path, and its position at any instant of the motion is given by the step OP. In order that the reader may have a clearer conception of what we are considering, we will suppose the motion

to take place in one plane, and conceptualize certain
everyday perceptions. We will suppose O to be a
point taken as the conceptual limit of Charing Cross, P
to be the point which marks the conceptual motion of
translation of a train on the Metropolitan Railway, and
the curve in the above Fig. 9 to be a conceptual map
of the same railway to the scale of about one furlong
to the $\frac{1}{20}$th of an inch. The points P_1, P_2, P_3, . . .
P_{16}, mark the successive stations between Aldgate and
South Kensington. Any step like OP_6 will accurately

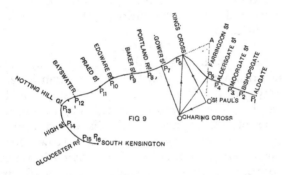

FIG 9

determine a certain position of the train relative to
Charing Cross. The reader must notice an important
result about these steps. Suppose we had been
determining the position of P_6 relative to O'—say St.
Paul's—instead of O. We see at once that there are
two ways of describing the position of P_6 relative to
O'. We might either say, step the directed step
$O'P_6$, or, again, step first from O' to O, and then step
from O to P_6. These two latter steps lead to exactly
the same final position as the former single step.
Now science is not only an economy of thought, but,
what is almost the same thing, an economy of

language. Hence we require a shorthand mode of expressing this equivalence in final result of two stepping operations. This is done as follows :—

$$O'O + OP_6 = O'P_6,$$

which, put into words, reads : Step from O' the directed step O'O, and then take the directed step $O'P_6$, and the spot finally reached will be the same as if the directed step $O'P_6$ had been taken from O'. The reader must be careful not to confuse this geometrical addition with ordinary arithmetical addition. For example, if OO' were eight furlongs, $O'P_6$ ten furlongs, and OP_6 twelve furlongs, then we appear at first sight to have :—

$$8 + 12 = 10,$$

and this is deemed absurd. But it is only absurd to the arithmetician. For the geometrician 8, 12, and 10 may be the lengths of *directed* steps, and he knows that, if he follows a directed step of 8 furlongs by one of 12, he may really have got only ten furlongs from his original position. How, then, is the arithmetician limited ? Why, obviously we must suppose him incapable of stepping out in all directions in space, we must tie him down to motion along one and the same straight line. In this case a step of 8 followed by one of 12 will always make a step of 20, as arithmetic teaches us it should do. Briefly, the freedom of the geometrician consists in his power of *turning corners*.

Let us now go back a little and note that the geometrical addition of steps, $O'O + OP_6 = O'P_6$, may be represented in a slightly different manner. Let us draw the line O'A parallel to OP_6 and P_6A parallel to OO', then we are said to complete the parallelogram on O'O and OP_6, the line $O'P_6$

joining two opposite angles is termed a diagonal, and we have the following rule : Complete the parallelogram on two steps, and its diagonal will measure a single step equivalent to the sum of the other two. This rule is termed addition by the *parallelogram law*, and we see that the steps by which we measure relative position, or displacements, obey this law.   In itself it is the same thing as geometrical addition. Its importance lies in the fact that all the conceptions of the geometry of motion, displacements, velocities, spins, and accelerations may be represented as steps and can be shown to obey the parallelogram law: that is to say, we add together velocities, spins, or accelerations *geometrically* and not arithmetically.   Although our space may not admit of our demonstrating this result for all the conceptions of kinematics,[1] the reader will do well to bear it in mind, as it is an important principle to which we shall have occasion again to refer.

§ 9.—*The Time-Chart.*

Hitherto we have been considering how the position of the point P relative to O might be determined at each instant of time.   We want, however, to know how the position changes, and how this change is to be described and measured.   In order to do this we must consider how the displacement OP_6, for example, changes to the displacement OP_7.   In our geometrical shorthand : $OP_7 = OP_6 + P_6P_7$, and the step P_6P_7 measures the change of position.   We want, then, to ascertain a fitting measure of the

[1] For proofs see Clifford's *Elements of Dynamic*, "Velocities," p. 59, "Spins," pp. 123-4.

manner in which this change varies with the time.
To enable the reader better to conceive our purpose
we will try to turn into geometry a column of
Bradshaw, or, more definitely, a portion of a time-table
of the Metropolitan Railway, corresponding to the
stations marked in Fig. 9. Down the left-hand side of
Fig. 10 are placed the names of the stations represented
in Fig. 9 by the points P_1, P_2, P_3, P_4, . . . P_{16}. These
are placed, as in *Bradshaw*, against a vertical line,
but we will somewhat improve on his arrangement.
He puts the stations at equal distances below each
other, and gives no hint as to the distance between
each pair of them. Now we will place them at such
distances along the vertical from each other that every
$\frac{1}{20}$th of an inch represents a furlong, or $\frac{2}{5}$ths of an inch
represents a mile, so that an inch-scale applied to the
vertical ought theoretically to determine the parliamentary fare between any two stations. In the
next place, we will place off (or *plot off*, as it is termed)
on the horizontal line through P_1 the number of
minutes that the train takes from Aldgate to each of
the other stations. Thus the times of a vertical
column of *Bradshaw* are in our case arranged horizontally. But we will place these times at such
distances that $\frac{1}{8}$th of an inch shall represent a
minute, or the minutes between any pair of stations
may be at once read off by aid of an inch-scale. To
connect each station with its corresponding time we
will draw a horizontal line PQ through the station,
and vertical line tQ through the corresponding time.
These meet in a point Q, and we obtain a series of
points Q_1, Q_2, . . . Q_{16}, in our diagram, corresponding
to the sixteen stations. Now at first sight it may
seem rather an inconvenient form of *Bradshaw*, when

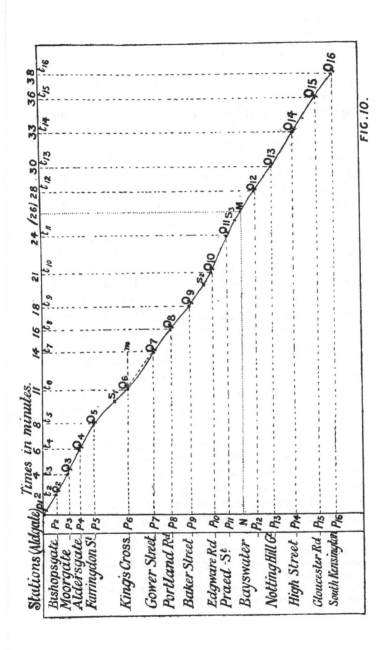

FIG. 10.

each train takes up an entire page.[1] The reader,
however, must wait till we have seen whether our
page may not be made to convey a great deal more
information as to the motion of the train than
Bradshaw's single column.

Now it is clear that what we have done for the
stations may be done for every signal-box, S_1, S_2, S_3,
&c., on the line, and not only for every signal-box,
but for every position along the whole line at which
we choose to observe the time at which the train
passes. We thus obtain a series of points: Q_1, Q_2,
Q_3, Q_4, Q_5, S_1, Q_6, Q_7, Q_8, Q_9, S_2, &c., which are seen to
take more and more the form of a curve as we increase
their number. We will join this series of points by
a continuous line, and to simplify matters we will
suppose our train to run from Aldgate to South Kensington without stopping, otherwise our curve would
have a small straight horizontal piece at each station.
This curve must be carefully distinguished from the
map of the path in Fig. 9 ; it tells us nothing about
the *direction* in which the train is moving at a given
time — that is to say, whether it is going north-
wards, or southwards, or what. But with the help of
Fig. 9 it tells us the exact time the train takes to
reach, not only every station, but every position
whatever between either terminus ; or, on the other
hand, it tells us the exact position for every time
up to 38 minutes after leaving Aldgate. How far
has the train got in 26 minutes, for example ? To

---

[1] Such geometrical *Bradshaws* with, however, many train-curves on a
page, are used by the traffic managers of several French railways. I
possess a fac-simile of that for the Paris–Lyons route containing
between 30 and 40 train-curves, and showing the passing places,
stoppages and speeds of the corresponding trains.

answer this we must scale off along the horizontal
line, or *time-axis*, 26 eighths of an inch ; we must then
draw a vertical line, striking our curve in the point
M; a horizontal through M strikes the vertical line of
stations, or *distance-axis*, at the point N between Praed
Street and Bayswater, and a scale divided into $\frac{2}{5}$ths of
an inch applied to $P_{11}N$ tells us how many furlongs
the train is beyond Praed Street.   An inverse process
will show us the time to any chosen position on the
distance-axis.   Our geometrical time-table, or *time-
chart*, as we shall call it, thus gives us a good deal more
information than *Bradshaw*.   It is further clear that
such a time-chart can be drawn in conception for
every point-motion, and that, taken in conjunction
with a map of the path, it fully describes the most complex point-motion.   Hence the fundamental problem
in such motions is to ascertain the map and the time-
chart.[1]

§ 10.—*Steepness and Slope.*

If we examine the time-chart we see that there
is a considerable difference in its steepness at
different points, and other motions would give us
curves with still greater variations in this respect.
We observe that if we lessen the time between two
stations, say P_{10} and P_{11}, we must shift the line
$Q_{11}t_{11}$ towards $Q_{10}t_{10}$ and the result is that the
curve becomes steeper between Q_{10} and Q_{11}.   On
the other hand, if we lessen the space traversed in a
given time the curve becomes less steep and ultimately quite horizontal if the train stops at a
station.   Thus *the steepness of the time-chart curve*

[1] The time-chart has been generally attributed to Galilei ; I do not
know on what authority.   A *speed-chart* occurs in his *Discorsi* but I
do not think there is anything that could be called a time-chart.

corresponds in some manner to the speed of the train.
We thus reach two new conceptions which need definition and measurement, namely, those of *steepness* and
speed. In Fig. 11 we have a horizontal straight line
AB, and a sloping line AC. Clearly the greater

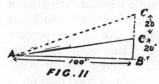

FIG. 11

the angle BAC the steeper
AC will be, and the greater
will be the height we shall
ascend for the horizontal
distance AB. If AB be 100
feet and CB the vertical

through B be 20 feet, we shall have ascended 20 feet
for a horizontal 100, or since the steepness of AC is the
same at all points, we shall ascend 2 feet in 10 feet,
or 200 feet in 2,000 feet, or ⅕ of a foot in 1 foot.[1]
Now, by elementary arithmetic the ratios of 20 to 100,
2 to 10, 200 to 2,000, and ⅕ to 1 are all equal and may
be expressed by the fraction ⅕. This is termed the
slope of the straight line AC, and is a fitting measure
of its steepness. The slope is clearly the number of
units or the fraction of a unit we have risen vertically
for a unit of horizontal distance. If slope be a fit
measure of steepness for a straight line, we have next
to inquire how we can measure the steepness of a
curved line. Let A and C in Fig. 12 be two points
on a curved line, the curve showing no abrupt change
of direction at the point A.[2] Now draw the line,
or so-called chord, AC ; then, whether we go up

[1] This statement depends on the proportionality of the corresponding
sides of similar triangles (see *Euclid*, vi. 4).

[2] A must be in the " middle of continuous curvature," as Newton
expresses it. This condition is important, but for a full discussion of
the steepness of curves we must refer the reader to pp. 44-7 of Clifford's
Elements of Dynamic, part i.

the curve from A to C or along the chord from A to
C, we shall have ascended the same vertical piece
CB for the same hori-
zontal distance AB. The
slope of the chord AC is
then termed the *mean*
slope of the portion AC
of the curve, be-
cause, however the
steepness may vary
from A to C, the
final result CB in

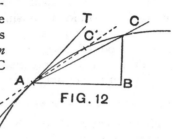

FIG. 12

AB could have been attained by the uniform average
slope of AC.

But this idea of mean slope does not settle the
actual steepness of the curve, say, at the point A.
Now let the reader imagine that the curve AC is a
bent piece of wire, and the chord AC a straight
piece of wire; further, he must suppose small rings
placed about both wires at A and C. In conception
we will suppose the wires to be indefinitely thin, so
that they approach as closely as we please to the
geometrical ideals of curve and line. Then the ring
A, being held firmly at A on the curved wire, let the
ring C be moved along the curved wire towards A.
As it moves the straight wire slips first into the
position AC' and ultimately, when the ring C
reaches A, takes up the position AT. In this
position the straight line is termed the *tangent* to
the curved line at the point A. As the slope of AC
or AC' measures the mean steepness of the curve
from A to C, or from A to C', so does the slope of
the chord in its limiting position of touching line, or
tangent, measure the mean steepness of an indefinitely

small part of the curve about A. The slope of the tangent is then said to measure the steepness of the curve *at* A. It is clear that in this notion of measuring the mean for a vanishingly small length of curve we are dealing with a conception which is invaluable as a method of description. It represents, however, a limit which, no more than a curve or line, can be attained in perceptual experience.

§ 11.—*Speed as a Slope. Velocity.*

Having now reached a conception by aid of which we can measure the steepness of a curve at any point— namely, by the slope of the tangent at that point—we may return to the curve of our time-chart and ask what we are to understand by its slope. Turning to Fig. 10, we observe that the mean slope of the portion Q_6Q_7 of the curve corresponding to the transit from King's Cross to Gower Srteet is Q_7m in Q_6m, or since Q_7m is equal to P_6P_7, and Q_6m to t_6t_7, it is P_6P_7 in t_6t_7. But P_6P_7 is, in a certain scale, the number of miles between the two stations, and $t_6\ t_7$ is, in another scale, the number of minutes between the two stations. Thus the slope, which with one interpretation is a certain rise in a certain horizontal length, is with another interpretation a certain number of miles in a certain number of minutes. Now a certain number of miles in a certain number of minutes is exactly what we understand by the mean or average speed of the train between King's Cross·and Gower Street; the train has increased its distance from Aldgate by so many miles in so many minutes. The manner in which change of distance is taking place during any finite time is thus determined by the slope of the corresponding chord of the time-chart. The average rate of change of distance, or the *mean speed* for any given interval is thus recorded by the slopes of these chords. It is clear, however, that by varying the length of the chord Q_6Q_7—by bringing Q_7 nearer to Q_6 for example—we shall obtain different mean speeds for different lengths of the journey after passing King's Cross. The shorter we take the time the steeper becomes in this case the chord, the greater the mean speed. The conception of a limit to this mean speed is then formed ; namely, the mean speed for a vanishingly small time after leaving King's Cross, and this mean speed is defined as the *actual speed* of passing King's Cross. We see at once that the actual speed will be measured by the slope of the tangent to the time-chart at Q_6, for this tangent is, according to our definition, the limit to the chord. Thus the actual speed at each instant of the motion is determined by the steepness at the corresponding point of the time-chart, and it is measured in miles per minute by the slope of the tangent at that point. We thus find that our time-chart is not only like *Bradshaw*, a time-table, but is also a diagram of the varying speed of the train throughout its journey.

There are one or two points about speed which the reader will find it useful to bear in mind. In the first place speed is a numerical quantity, it is equal to a slope, the unit of which is one vertical unit in or *per* one horizontal unit ; thus the speed unit is one space unit in or *per* one time unit—for example, one mile per minute. Secondly, unless the time-chart has a straight line for its curve, the speed must continually change its magnitude from one point to another of the path. If the curve of the time-chart be a straight line the speed is said to be *uniform*, otherwise it is called *variable*. Lastly, looking back at the map of the path (Fig. 9, p. 251), we see that the *bearing* of the motion as well as the speed varies from point to point of the path. Remembering our definition of tangent we see that the direction of the motion at P is along the tangent at P, and further it has a *sense*—for example, the motion is from P_6 to P_7 and not from P_7 to P_6. Now we see that the change in the motion is of two kinds: change in magnitude, or change in speed, and change in bearing. In order to trace this change still more clearly we form a new conception, namely, that of speed with a certain bearing, and this combination of speed and bearing we term *velocity*. To fully describe the velocity, say at the position P_6 we must therefore combine speed and bearing; the speed is the slope of the tangent at Q_6 (Fig. 10, p. 255), and, when the units of time and space have been chosen, it is solely a number; the bearing is the direction of the tangent to the path at P_6 (Fig. 9) together with the sense, namely, from P_6 to P_7. Like displacement velocity can accordingly be represented by a step, the magnitude of the step measures the speed, the direction of the step shows the direction of the motion, and the arrow-head gives the sense of the motion.

§ 12.—*The Velocity Diagram or Hodograph. Acceleration.*

Now, as it is awkward to have to turn to two different figures—the map of the path and the time-chart—in order to determine velocity, we construct a new figure in the following manner: From any point I we draw a series of rays, IV_1, IV_2, IV_3, IV_4, . . . IV_{16}, parallel to the tangents at the successive points P_1, P_2, P_3, . . . P_{16}, and we measure off along these rays in the sense of the motion as many units of length as there are units of speed in the motion at these points. Each of these rays will, by what precedes, be a step representing the velocity at the corresponding point of the path. If this be done for a very great number of positions the points V_1, V_2, V_3, &c., will be a series approaching more and more closely to a curve. This curve is termed the *hodograph*, from two Greek words signifying a " descrip-

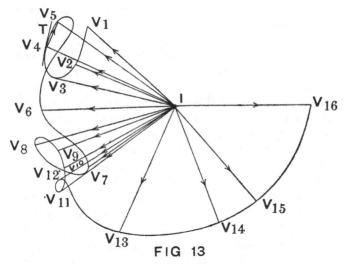

FIG 13

tion of the path." The name has been somewhat unfortunately chosen as the curve is not a " description of the path," but a "description of the motion in the path," rather a *kinesigraph* than a hodograph. Fig. 13 is supposed to represent the hodograph of the motion dealt with in our Figs. 9 and 10.[1]   Thus while

[1] The true hodograph would require a great number of points, such as V, to determine its shape at all accurately.   The constant changes in

the rays of the map of the path (Fig. 9, p. 251) give the position of P relative to O, the rays of the hodograph give the velocities of P relative to O. So soon as we are in possession of the time-chart and the map of the path we can construct this diagram of the velocities. When constructed it forms an accurate picture of how the motion is changing in both magnitude and direction.

Let us now examine this hodograph a little more closely. It consists of a point or *pole* I and rays IV drawn from this pole to a curve V_1 V_2 V_3 . . . V_{16}. Now this is exactly what the map in Fig. 9 consists of. In that figure we have a pole O and rays OP drawn from this pole to a curve P_1 P_2 P_3 . . . P_{16}. In the course of the motion P passes along the whole length of this curve, and in just the same manner we may look upon V as moving along the whole length of the hodograph-curve. The ray IV would in each position be the displacement of V relative to I. The question now arises : Has the motion of V round its curve any meaning for the motion of P in the path? Suppose we were now to treat the hodograph as the map of a new motion, and to construct first the time-chart and then the hodograph of this motion, what would the rays of this second hodograph represent? Now a sort of logical rule-of-three sum will give us the answer to this question. As the rays of the first hodograph are to the map of the path, so are the rays of the second hodograph to the map of V's motion. But we have seen that the rays of the first hodograph measure the velocities of P in its

the direction of the railway (see Fig. 9, p. 251) cause the hodograph curve to bend backwards and forwards, while the slight variations of the speed produce the tangles in the curve.

path, and that these velocities are a fitting measure of how the ray OP, or the position of P relative to O, is changing.   Hence it follows that the rays of the second hodograph would measure the velocities of V in the first hodograph, and that these velocities are a fitting measure of how the ray IV or the velocity of P relative to O is changing.   Thus the velocity of V along the hodograph is the measure of how the velocity of P relative to O is changing.   This velocity of V, or change in the velocity of P, is termed *acceleration*, and we see that a diagram of accelerations may be obtained by drawing the hodograph of the velocity-diagram, treated as if it were itself the map of an independent motion.   Acceleration therefore stands in just the same relation to velocity as velocity stands to the position-step.   As change of position is represented by the steps drawn as rays of the velocity-diagram or first hodograph, so change of velocity is represented by the steps drawn as rays of the acceleration-diagram or second hodograph.[1]   Whatever may be demonstrated of the position-step and velocity will still hold good if the words position-step and velocity be replaced by the words velocity and acceleration respectively.

§ 13.—*Acceleration as a Spurt and a Shunt.*

We must now investigate somewhat more closely this notion of acceleration as a proper measure of the change in velocity.   In a certain interval of time the speed of the point P (Fig. 9, p. 251) changes from a number

---

[1] We might proceed in the same manner to measure the change in acceleration by drawing a third hodograph. Fortunately this third hodograph is rarely, if ever, wanted. The concepts which practically suffice to describe our perceptual experiences of change are position, velocity, and acceleration.

of miles per minute represented by the number of linear units in IV_4 to the number of miles per minute represented by the linear units in IV_5, the speed has in this case (see Fig. 13) quickened, or there has been what we may term a *spurt* in the speed. Further, the bearing of the motion has changed ; instead of the point P moving in the direction IV_4, it now moves in the direction IV_5, that is to say, the direction of the motion has received a *shunt*. Thus the total change in the velocity of P as it moves from P_4 to P_5 consists of a spurt and a shunt. When a train quickens its speed from 40 to 60 miles an hour, and instead of running due north runs north-east, we may describe its motion as spurted and shunted ; technically, we say that its velocity has been *accelerated*. Acceleration has thus two fundamental factors—the spurt and the shunt.[1] If we consider the perceptual world around us, it is clear that the spurting and shunting of motion are conceptions as important for describing our everyday experience as those of the speed and direction of motion itself.

We have seen that the speed changes from the length IV_4 to the length IV_5 in a certain time—that represented by the length t_4t_5 of our time-chart (Fig. 10). The increase of speed per unit of time (or the ratio of the difference of IV_5 and IV_4 to t_4t_5) is termed the *mean speed-acceleration* or the *mean spurt* between P_4 and P_5. Further, the ray IV has been turned from IV_4 to IV_5, or through the angle V_4IV_5 in time t_4t_5. This increase of angle per unit time (or the ratio of the angle V_4IV_5 to t_4t_5) is termed the *mean shunt*, or *mean spin of direction* between

---

[1] Spurt in scientific language includes a retardation or slackening of speed as a negative spurt.

the positions P_4 and P_5. The two combined, or the mean rate of spurting and shunting, form what is termed the *mean acceleration* during the given change of position, or for the given time (t_4t_5). What we measure, therefore, in acceleration is the *rate* at which spurting and shunting take place. Turning to Fig. 13 the reader must notice that there are two processes by aid of which we can conceive the velocity IV_4 converted into IV_5. In the first process we follow the method just discussed : we stretch IV_4 till it is as long as IV_5, that is, we increase the speed from its value in the position P_4 to its value in the position P_5 ; then we spin this stretched length round I till it takes up the position IV_5. This is the spurt and shunt conception of acceleration. In the second process we say add the step V_4V_5 to the step IV_4 and we shall reach the step IV_5 (pp. 252–3)—that is to say, we can consider the new velocity IV_5 obtained from the old velocity IV_4 by adding the step or velocity V_4V_5 by the parallelogram law. The mean acceleration is in this case expressed by the step V_4V_5 added in the given interval t_4t_5. But if we compare Figs. 9 and 13 as maps for the motions of P and V we shall see that adding V_4V_5 in time t_4t_5 corresponds to adding P_4P_5 in time t_4t_5. The latter operation, however, led us, by aid of the time-chart, from the idea of mean speed or mean change in OP to the idea of actual speed or instantaneous change in OP at P_4 ; the instantaneous change in OP_4 was in the direction of the tangent at P_4, and was measured by the slope of the time-chart at Q_4 (see Fig. 10). In precisely the same manner the instantaneous change in IV_4 will be along the tangent at V_4, and will be measured by the

slope of the time-chart *for V's motion* at the corresponding point.   Thus actual acceleration appears, as
in our first discussion of the matter, as the velocity of
V along the hodograph.   Now, however close V_5 is to
V_4, whether we give a stretch and a spin or add the
small step V_4V_5, the final result of the two processes
will be the same.   Hence we can either look upon
actual acceleration as the velocity of V along the
hodograph, or as the combined mode in which IV is
being actually stretched and spun.[1]   Either method
of treating acceleration leads to the same result, and
both possess special advantages for describing various
phases of motion.

In the first case actual acceleration is represented
by a step; the bearing of this step denotes the direction and sense in which V is moving, or the velocity
with which IV is changing ; the number of units of
length in this step denote the number of units of speed
with which V is moving, or the number of units of speed
being actually added per unit of time in the given
direction to the velocity IV of P.   By " added in the
given direction " we are to understand that the incre-
ments of velocity are to be added geometrically or by
the parallelogram law (*e.g.*, $IV_5 = IV_4 + V_4V_5$ and this
however small conception V_4V_5 may be in).

§ 14.—*Curvature.*

In the spurt and shunt method of regarding acceleration, on the other hand, actual acceleration will be
specified by two factors : (i.) the rate at which velocity
is being spurted or IV being stretched ; (ii.) the

---

[1] What we have here stated of acceleration applies just as much to
change of position. Turning to Fig. 9, we may look upon the change
of position of OP as measured by the velocity of P along its path or by
the manner in which OP is being actually stretched and spun.

rate at which velocity is being shunted, or IV being spun about I (Fig. 13, p. 263).   As in the first case the direction of actual acceleration at V_4 is that of V_4T or the tangent at V_4, it is clear that as a rule acceleration will not be in the direction of velocity,[1] but will act partly in the direction of velocity and partly at right-angles to it.   This result is so important that the reader will, I hope, pardon me for considering it from a slightly different standpoint.   Let us imagine the acceleration to be always such that it never stretches IV, and let us try to analyze this case a little more closely.

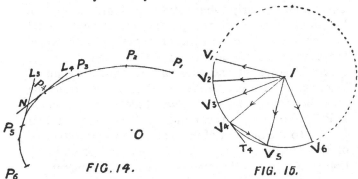

FIG. 14.   FIG. 15.

Obviously if IV is never stretched, if the speed is never spurted, the point V can only describe a circle, for IV remains uniform in length.   Uniform speed can, however, be conceived associated with a point moving in any curved path whatever.   Let Fig. 14 represent this path, and let Fig. 15 be the circular hodograph, corresponding points of both curves being denoted by the same subscript numerals attached to the letters P and V.

---

[1] At V_3, for example, IV_3 appears to coincide with the direction of the tangent at V_3.   In this case the whole effect of acceleration is instantaneously to spurt without shunting.

Now, since all the acceleration in this case depends upon the change in the direction of motion, or the change in the direction of the tangent to the path, we must stay for a moment to consider how this change in direction, or the *bending* of the path, may be scientifically described and measured. Now if we pass, for example, from the point P_4 to P_5 on the path, and P_4L_4, P_5L_5 be the tangents (p. 259) at P_4, P_5 respectively, then the direction of the curve has continuously altered from P_4L_4 to P_5L_5 as we traverse the length P_4P_5 of the curve. The angle between these directions is L_4NL_5, and clearly the greater this angle for a given length of curve P_4P_5, the greater will be the amount of bending.[1] The amount of angle through which the tangent has been turned for a given length of curve forms a fit measure of the total amount of bending in that length. Accordingly we define the mean bending or *mean curvature* of the element of curve P_4P_5 as the ratio of the number of units of angle in L_4NL_5 to the number of units of length in the element of curve P_4P_5. Thus the mean curvature of any portion of a curve is the average turn of its tangent per unit length of the curve. From the mean curvature we can reach a conception of *actual curvature* as a limit when the element of arc P_4P_5 is very small in just the same manner as from mean speed we reached a conception of actual speed. This process of reaching a limit in conception, which cannot be really attained in perception, is so important that we will again repeat it for this special case, in order that the reader may have

[1] We are supposing here that the direction of bending between P_4 and P_5 does not change, that the curve is not like this : $\backsim$. We can always insure that no such change takes place by taking a sufficiently small length of arc.

little difficulty henceforth in discovering and discussing such limits for himself. Let us accordingly suppose the distances between the points P_1, P_2, P_3, ... P_6 plotted off (Fig. 16) down a vertical line as in the time-chart of Fig. 10 (p. 255). Along the horizontal line P_1M_6 instead of assuming units of length to represent units of time, let them represent units of angle,[1] and let the

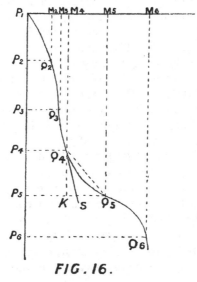

FIG. 16.

number of units taken from P_1 represent successively the number of units of angle between the tangents P_2L_2, P_3L_3, P_4L_4, &c., in Fig. 14 (p. 269), and the tangent

---

[1] According to *Euclid* iii. 29, and vi. 33, the angles at the centre of a circle which stand on equal arcs are themselves equal; if we double or treble the arc we must double or treble the angle; the arc is thus seen to be a fit measure of the angle. Further (Clifford's *Common Sense of the Exact Sciences*, pp. 123–5), the arcs of different circles subtending equal angles at their respective centres are easily shown to be in the

to the curve at P_1. Thus let P_1M_4 represent the angle between the tangents at P_1 and at P_4; P_1M_5 that between the tangents at P_1 and at P_5 and so on. Now draw in Fig. 16 vertical lines through the points M_2, M_3, &c., and horizontal lines through the points P_2, P_3, &c., and suppose these lines pair and pair to meet in the points Q_2, Q_3, &c. We have then a series of points Q, which increase in number as we increase the points P in Fig. 14, and in conception ultimately give us the curve marked in Fig. 16 by the continuous line. The diagram thus obtained is a chart of the bending or curvature in Fig. 14. For, the mean curvature in the length P_4P_5 is the ratio of the angle L_4NL_5 to the length P_4P_5 in Fig. 14, or, what is the same thing, the ratio of the number of units in M_4M_5 to the number in P_4P_5 in Fig. 16. But if Q_4K be drawn parallel to M_5Q_5 to meet P_5Q_5 in K, this ratio is that of KQ_5 to Q_4K, or is the *slope* of the chord Q_4Q_5 to the *vertical* line P_1P_6. Thus the slope of any chord of the curvative-chart to the vertical measures the mean curvature of the corresponding portion of the curve in Fig. 14. When we make the chord Q_4Q_5 smaller and smaller by causing Q_5 to move towards Q_4, the mean curvature becomes more and more nearly the mean curvature at and about P_4;

---

ratio of their radii. If, therefore, we take as our standard circle for measuring angles the circle whose radius is the unit of length, its arc c for any given angle will be to the arc a of a circle of radius r subtending the same angle in the ratio of 1 to r, or in the form of a proportion, $c : a :: 1 : r$, whence it follows that $c = a/r$, or the *circular measure* c of any angle is the ratio of the arc a subtended by this angle at the centre of any circle to the radius r of this circle. The unit of angle in circular measure will therefore be one for which a equals r, or which subtends an arc equal to the radius. This unit is termed a *radian*, and is generally used in theoretical investigations.

but as on p. 259 the chord becomes more and more nearly the tangent at Q_4. As we have defined actual curvature to be the limit to the mean curvature in a vanishingly small length of curve beyond P_4 (see Fig. 14), we see that the actual curvature at P_4 is the slope to the vertical of the tangent Q_4S at the corresponding point Q_4 of the curvature-chart. This slope, and accordingly the actual curvature, is therefore a measurable quantity at each point of any curve.[1]

§ 15.—*The Relation between Curvature and Normal Acceleration.*

Returning again to Figs. 14 and 15, we note that the mean curvature over the length P_4P_5 is the ratio of the number of angle units in L_4NL_5 to the number of length units in the element of curve P_4P_5. Now the speed in the length P_4P_5 is constant and equal to IV_4; hence if the point P traverse this length in

[1] The mean curvature over any arc *ab* of a circle centre O is the ratio of the angle between the tangents at its extremities, or—what is the same thing, since the tangents are perpendicular to the radii O*a* and O*b*—of the angle *aOb* at the centre to the arc *ab*. But we have seen in the footnote, p. 271, that the measure of this angle in *radians* is the ratio of the arc *ab* to the radius. Hence it follows that the mean curvature of a circle is equal to the inverse of the radius (or unity divided by the radius). As this mean curvature is therefore independent of the length of the arc, it follows that the

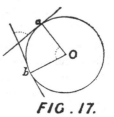

FIG . 17.

actual curvature at each point must be the same and be equal to the inverse of the radius. Since the radius of a circle can take every value from zero to infinity, a circle can always be found which has the same amount of bending as a curve at a given point, and thus fits it more closely at that point than a circle of any other radius. The radius of this circle is termed the *radius of curvature* of the curve at the given point. Hence the curvature of a curve is the inverse of its radius of curvature.

a number of minutes, which we will represent by the letter t, we must have, since speed is the number of units of length per minute, the length P_4P_5 equal to the product of IV_4 and t (or in symbols $P_4P_5 = IV_4 \times t$). Further, since the angle L_4NL_5 is turned through by the tangent also in time t, the ratio of the angle L_4NL_5 to t is the mean rate at which the tangent is turning round in the time t, or is the *mean spin* of the tangent (or, if the mean spin be denoted by the letter S, we have in symbols $L_4NL_5 = S \times t$). From these results it follows at once that the mean curvature which is the ratio of L_4NL_5 to P_4P_5 must be equally the ratio of the mean spin, S, to the mean speed IV_4. Thus we have directly connected motion with curvature.

Proceeding in conception to the limit we have the important kinematic result that : *If a point moves along a curve the ratio of the spin of the tangent to the speed of the point is the actual curvature at each situation of the point.*

It remains to connect this result with the acceleration. The acceleration in the case we are dealing with is the velocity of V along its circle (Fig. 15). This acceleration at V_4, for example, is along the tangent V_4T_4 to the circle, or at right-angles to IV_4 the direction of the velocity of P (Fig. 14) ; it has thus, as we have seen, purely a shunting and no spurting effect. Now, since IV_4 and IV_5 were drawn parallel to the directions of motion L_4P_4, L_5P_5 at P_4 and P_5 respectively, it follows that the angles L_4NL_5 and V_4IV_5—between two pairs of parallel lines—must be equal. Hence the mean spin of the tangent from P_4 to P_5 must be the ratio of the angle V_4IV_5 to the time t in which P passes from P_4 to P_5, or, what is

the same thing, in which V passes from V_4 to V_5.
But the magnitude of the angle V_4IV_5 is (see the
footnote, p. 271) the ratio of the arc V_4V_5 to the
radius IV_4. Further, the ratio of the arc V_4V_5 to
the time t is the mean speed of V from V_4 to V_5
(p. 260). Thus it follows that the mean spin of the
tangent (Fig. 14) is the ratio of the mean speed of V to
the radius IV_4. Taking P_5 closer and closer to P_4, and
therefore V_5 to V_4, mean values become the actual
values at P_4 and V_4; we therefore conclude that the
actual spin of the tangent at P_4 is the ratio of the
actual speed of V at V_4 to IV_4, or, in other words,
to the speed of P. Thus the spin of the tangent is
the ratio of the speed of V to the speed of P. But
the speed of V is the magnitude of the acceleration,
which in this case is all shunt. Hence we conclude that
the rate of shunting at P is properly measured by the
product of the spin of the tangent and the speed of P
(or in symbols, shunt acceleration $= S \times U$, U being the
speed of P). But we have seen above that the curvature is the ratio of the spin of the tangent to the speed
of P (or in symbols curvature $= S/U$). Combining,
accordingly, these two results we see that the shunt
acceleration in this case is properly measured by the
product of curvature and the square of the speed.[1]
This acceleration takes place in the direction V_4T_4, or
is perpendicular to the direction of motion at P.

A little consideration will show the reader that the
expression we have deduced for the acceleration per-

[1] If r be the radius of curvature (see the footnote, p. 273), then $1/r$ will
be the curvature, and if we term this element of acceleration *normal
acceleration*, we have, by the above results, the three equivalent values :

normal acceleration $= \dfrac{U^2}{r} = S \times U = r\, S^2.$

pendicular to the motion would not be altered were the speed to vary between P_4 and P_5. For, returning to Fig. 13, we note that IV_4 is to be changed to IV_5. This can be conceived as accomplished in the following two stages (p. 267) : (i.) rotate IV_4 round I without changing its length into the position IV_5 ; (ii.) stretch IV_4 in its new position into IV_5. The first stage corresponds to the type of motion we have just dealt with, or shunt acceleration without spurt ; the second stage to the case of spurt acceleration without shunt. In the limit when IV_5 is indefinitely close to IV_4, the first stage gives us the element of acceleration *perpendicular* to the direction of motion, and the second stage the element of acceleration in the direction of motion. By the above reasoning the former is seen to be measured by the product of the square of the speed and the curvature.

§ 16.—*Fundamental Propositions in the Geometry of Motion.*

We are now in a position, after restating our results, to draw one or two important conclusions.

Acceleration has spurt and shunt components.

The spurt acceleration takes place in the direction of motion, and is measured by the rate at which speed is being increased (or, it may be, decreased).

The shunt acceleration takes place perpendicular to the direction of motion, and is measured by the product of the curvature and the square of the speed.

These two kinds of acceleration are usually spoken of as *speed acceleration* and *normal acceleration*.

From these results we conclude that :—

1. If a point be not accelerated it will describe a straight line with uniform speed. For there will be no spurt, and therefore the speed must be uniform,

and there will be no shunt, and therefore the path must have zero curvature, but the only path without bending is a straight line. Neither uniform speed nor zero curvature *alone* denote an absence of acceleration.

2. When a point is constrained to move in a given path the normal acceleration may be determined in each position from the speed and the form of the path, *i.e.*, from its curvature or bending. In this case the problem is to find the speed from the speed acceleration.

3. When a point is free to move in a given plane, then its motion can be theoretically determined, if we know its velocity in any one position, and its acceleration for all positions. For from the normal acceleration and the speed we can calculate the initial amount of bending of the path; thus the initial form of the path is known. For a closely adjacent position on this initial form, we can determine from the speed acceleration the change in speed due to this change of position. Hence we obtain the speed in the new position. From the speed in the new position and the normal acceleration in this position, the bending in the next little element of path may be deduced. This process may be repeated as often as we please, till the whole path of the motion is constructed. The succession of positions may be taken so close together that we obtain the form of the path to any degree of accuracy required. Knowing the path and the speed at each point of it we are able to construct a time-chart like that of our Fig. 10 (p. 255). For we know from the speeds the slope at each point of the Q-curve. Hence we commence by drawing a little element, say P_1Q_2, at the slope given by the initial speed; this element by aid of the horizontal Q_2P_2,

through its terminal Q_2, gives a new position at distance P_1P_2 from the initial position ; the speed in this new position determines the slope of the next little element Q_2Q_3 of the curve ; Q_3 by aid of the horizontal Q_3P_3 gives a third position with a third speed a ıd so a slope for the third element, and this process can be continued till we have constructed the time-chart by a succession of little elements. By taking these elements sufficiently small, we make the resulting polygonal line differ as little from the true curve of the time-chart as we please. Now we have seen that when the map of the path and the time-chart are known, the motion has been fully described. Thus we conclude that : *Given the velocity of a point in any position and the acceleration of the point in all positions, the motion of the point is fully determined.*[1]

This proposition is the basis of the whole of our mechanical description of the universe. Rightly interpreted, it contains all that we can assert of the " mechanical determinism " of nature ; wrongly interpreted, it is the basis of that crude materialism which pictures the universe as an aggregate of objective material bodies, enforcing for all eternity certain motions on each other, and a perception of those motions upon us. What the proposition exactly tells us is this : that a motion is fully determined, that is, can be described, either by giving the path and the

[1] The methods by which we have shown that the initial velocity and position, together with the acceleration in all positions, determine the map of the path and the time-chart, are only theoretical methods of construction. The practical methods of constructing these curves involve the highest refinement of mathematical analysis. Our object here is only to show that the motion is theoretically determined by a knowledge of the above quantities.

time to each position of the path, or by giving the velocity in any one position and the acceleration in all positions. We are really dealing with two different modes of *describing* motion, either of which can be deduced from the other, but neither of which *explains* why the motion takes place, or can be said to "determine" it in the sense of the materialists.

§ 17.—*The Relativity of Motion. Its Synthesis from Simple Components.*

There still remains a point to which it is needful to draw the reader's attention. The whole motion of our point P (Fig. 9, p. 251) has been considered relative to a point O. We started with a position relative to O, and it follows that the velocity and acccleration we have been discussing describe changes of motion relative to O also. Thus *absolute* velocity and *absolute* acceleration are seen to be as meaningless as absolute position. If the points O and P were *both* to have their motions accelerated in the same manner the relative path would not be changed—any more than the map (Fig. 9) is changed by our moving about, in any manner we please, the page on which it is printed. But the fact that all motion is relative leads us at once to the very natural question : How are we to pass from the motion of a point relative to one pole O to motion relative to a second pole O'? We must look at this point somewhat closely, for it involves some important consequences.

Let us suppose the motion of P relative to O known, and the motion of O' relative to O known, we require to find the motion of P relative to O'. Let P_1, P_2 (Fig. 18) be two successive positions of P relative to O, and O'_1, O'_2 the corresponding positions of O'. Then O'_1P_1 is the first and O'_2P_2 is the second step, measuring the position of P relative to O'. From O'_1 draw $O'_1P'_2$ parallel and equal to O'_2P_2, then O'_1P_1 and $O'_1P'_2$ give the relative motion of P with regard to O_1, and the relative displacement in the given interval is $P_1P'_2$. Now draw O'_1O_2 parallel and equal to O'_2O, then O'_1O, and O'_2O, or O'_1O_2, give the relative positions of O with regard to O. But by the equality of opposite sides of parallelograms OO_2 equals $O'_2O'_1$, equals $P_2P'_2$. Hence

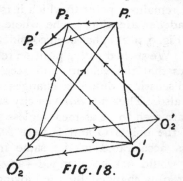

FIG. 18.

$P_2P'_2$ is equal to the displacement of O relative to O'. But in the geometry of steps (p. 252):—

$$P_1P'_2 = P_1P_2 + P_2P'_2,$$

or in words : the displacement of P relative to O' is equal to the displacement of P relative to O added *geometrically* to the displacement of O relative to O'. Now this result is true, however large or small these displacements may be, and these displacements divided by the number of units in the interval of time which is the same for all of them, represent the mean velocities in this interval. Hence we conclude that: the mean velocity of P relative to O' is equal to the mean velocity of P relative to O added *geometrically* to the mean velocity of O relative to O'. If we take the interval of time, and consequently the displacements smaller and smaller, mean velocities become in the limit the actual velocities. These actual velocities have always the direction of the displacements $P_1P'_2$, P_1P_2 and OO_2 which ultimately from chords become tangents to the corresponding paths; further, since the interval of time is the same for all the displacements, the magnitudes or speeds of these velocities are always proportional to the sides $P_1P'_2$, P_1P_2, and $P_2P'_2$, (or OO_2) of the triangle $P_1P'_2P_2$. Hence the mean velocities and ultimately the actual velocities always form the three sides of a triangle which has its sides parallel and proportional to the sides of the triangle $P_1P'_2P_2$, and this however small the latter triangle becomes. The actual velocity of P relative to O' thus forms one side of a triangle of which the actual velocities of P relative to O and of O relative to O' form the other two sides. In other words, the actual velocity of P relative to O' is obtained from the actual velocities of P relative to O, and of O relative to O' by adding them geometrically, or by the *parallelogram law*. Just as the position of P relative to O' was found by applying the parallelogram law to the steps O'O and OP (p. 253), so we obtain the velocity of P relative to O' by applying the same law to the velocities of P relative to O, and of O relative to O'. A very similar proof shows us that the acceleration of P relative to O' may be obtained in the same way from the accelerations of P relative to O and O relative to O'. We thus obtain an easy rule —that of the parallelogram law—for passing from the motion of P relative to O to that of P relative to O'.

The whole of this discussion may be looked at from a somewhat different standpoint. We may suppose the plane of the paper in which the motion of P about O takes place to be always moved as a whole so that the point O' remains stationary. In order to do this we must always be shifting the paper so that O'_2 falls back on O'_1, and $O'_2O'_1$ will measure the fitting shift of the paper. This carries P_2 clearly forward to P'_2 and O to O_2. Thus the motion of P relative to O' may be looked at as the motion of P due to two sources—a movement of P about O, and a movement of the plane containing P and O; this later motion is the motion of O about O', or is equal and opposite to the perfectly arbitrary motion of O' about O. Thus we conclude that if a point P has two independent velocities (corresponding to the limits of the displacements P_1P_2 and $P_2P'_2$) then the actual velocity of P will be found by adding these velocities geometrically. This statement is usually termed the *parallelogram of velocities.* A precisely similar statement holds for independent accelerations (p. 253), and is called the *parallelogram of accelerations.* To these important results we shall have occasion again to refer. We conclude, therefore, with the general statement that the independent displacements, the independent velocities, and the independent accelerations of a moving point are respectively added geometrically as we add steps, or by the so-called parallelogram law.

The value of this rule of combination lies in the power it gives us of building up complex cases of motion from simple cases. If we find as a result of experience that the perceptual antecedents [1] of one acceleration may be superposed on the perceptual antecedents of a second acceleration—without these accelerations altering their value to our degree of refinement in measuring them,—then the parallelogram of accelerations will be invaluable as a mode of *synthesis*, or of constructing the complex from the simple. The law of gravitation applied to the planetary theory is a striking example of the value of such a synthesis.

In this chapter we have seen how the relative position, velocity, and acceleration of points may be defined, described, and measured. We have been gleaning wholly in the conceptual field of geometrical ideals. We have next to ask how these conceptions may be applied to describe our perceptual, experience of change in the world of phenomena. How are these three factors, position, velocity, and acceleration, related to each other in that ideal dance of corpuscles to which we reduce the physical universe, in that atomic gallop by aid of which we describe and resume our sense-impressions? How do we conceive the relative position of these corpuscles to change? How are their speeds and directions of motion varying? Does experience show us that relative position produces a definite speed, or a definite spurt and shunt? The answer to these questions lies in the so-called properties of matter and in the laws of motion which will be the topics of our two following chapters.

---

[1] By "perceptual antecedents" we are to understand *cause* in the scientific sense, but the word has not been used in the above paragraph, because the reader might have supposed the cause of acceleration to be the metaphysical (and imperceptible) entity *force*, whereas it really lies in *perceptible* relative position (p. 345).

SUMMARY.

1. All the notions by aid of which we describe and measure change are *geometrical*, and thus are not real perceptual limits. They are forms distinguishing and classifying the contents of our perceptual experience under the mixed mode of motion. The principal of these forms are point-motion, spin of a rigid body and strain. Motion is found to be relative, never absolute; for example, it is meaningless to speak of the motion of a point without reference to what system the motion of the point is considered with regard to.

2. An analysis of point-motion leads us to the conceptions of velocity and acceleration, the first as a proper measure of the manner in which position is instantaneous changing, the second as a proper measure of how velocity itself is changing. It is found that a motion is fully determined, or theoretically a complete description of its path and position at each instant of time may be deduced, when the velocity in any one position and the acceleration for all positions are given.

3. The parallelogram law as the general rule for combining motions is the foundation of the synthesis by which complex motions are constructed out of simple motions.

LITERATURE.

CLERK-MAXWELL, J.—Matter and Motion, chaps. i. and ii. London, 1876.

CLIFFORD, W. K.—The Common Sense of the Exact Sciences, chap. iv. "Position," and chap. v. "Motion"; London, 1885. Also for a more advanced treatment the same writer's Elements of Dynamic, part i., book i. chaps. i. and ii. ; bcok ii. chaps. i. and ii. ; book iii. chap. i. ; London, 1878.

MACGREGOR, J. G.—An Elementary Treatise on Kinematics and Dynamics, part i., "Kinematics," chaps. i.-iii., v. and vii. London, 1887.

CHAPTER VII.

MATTER.

§ 1.—"*All things move*"—*but only in Conception.*

AN old Greek philosopher, who lived perhaps some five hundred years B.C., chose as the dictum in which he summed up his teaching the phrase : "*All things flow.*" After ages, not understanding what Heraclitus meant—it is doubtful whether he understood himself —dubbed him "Heraclitus the Obscure." But to-day we find modern science almost repeating Heraclitus' dictum when it says : "*All things are in motion.*" Like all dicta which briefly resume wide truths, this dictum of modern science requires expanding and explaining if it is not to be misinterpreted. By the words "All things are in motion" we are to understand that, step by step, science has found it possible to describe our experience of perceptual changes by types of relative motion : this motion being that of the ideal points, the ideal rigid bodies or the ideal strainable media which stand for us as the signs or symbols of the real world of sense-impressions. We interpret, describe, and resume the sequences of this real world of sense-impressions by discussing the relative positions, velocities, accelerations, rotations, spins, and strains of an ideal geometrical world which stands for us as a conceptual representation of the perceptual world. In our Chapter V. we saw that space and time did not themselves correspond to actual perceptions, but were *modes* under which we perceived, and by which we discriminated, groups of sense-impressions. So motion as the combination of space with time is essentially a *mode* of perception, and not in itself a perception (p. 231). The more clearly this is realized the better able the reader will be to appreciate that the " motion of bodies " is not a reality of perception, but is the conceptual manner in which we represent this mode of perception and by aid of which we describe changes in groups of sense-impressions ; the perceptual reality is the complexity and variety of the sense-impressions which crowd into the telephonic brain-exchange. That the results which flow from the conceptual world of geometrical motions agree so closely with our perceptual experience of the outside world of phenomena (p. 77) is a phase of that accordance between the perceptive and reasoning faculties upon which we have laid stress in an earlier part of this volume (p. 124).

Wherein lies the advance from Heraclitus to the modern scientist ? Why was the dictum of one not unjustly termed obscure, while the other claims— and rightly claims—to find in the development of his dictum the sole basis for our knowledge of the physical universe ? The difference lies in this : Heraclitus left his flow undescribed and unmeasured, while modern science devotes its best energies to the accurate investigation and analysis of each and every type of motion which can possibly be used as a means of describing and resuming any sequence of sense-impressions. The whole object of physical science is the discovery of ideal elementary motions which will enable us to describe in the simplest language the

widest ranges of phenomena ; it lies in the symbolization of the physical universe by aid of the
geometrical motions of a group of geometrical forms.
To do this is to construct the world mechanically [1] ;
but this mechanism, be it noted, is a product of conception, and does not lie in our perceptions themselves
(p. 139). Startling as it may, when first stated, appear
to the reader, it is nevertheless true that the mind
struggles in vain to clearly realize the motion of anything which is neither a geometrical point nor a body
bounded by continuous surfaces; the mind absolutely
rebels against the notion of anything moving but these
conceptual creations, which are limits, unrealizable,
as we have seen, in the field of perception. If the
world of phenomena be, as the materialists would
have us to believe, a world of moving bodies like
the conceptual world by which science symbolizes it,
if we are to assert the perceptual existence of atom
and ether, then in bqth cases we are incapable of considering the ultimate element which moves as anything but a perceptual realization of geometrical
ideals. Yet so far as our *sensible* experience goes
these geometrical ideals have no phenomenal
existence ! We have clearly, then, no right to infer
as a basis of perception things which our whole
experience up to the present shows us exist solely in
the field of conception. It is absolutely illogical to
fill up a void in our perceptual experience by projecting into it a load of conceptions utterly unlike
the adjacent perceptual strata. It is " a profound
psychological mistake," says George Henry Lewes,
" to assert that whenever we can form clear ideas, not

[1] This word is here used in the scientific sense of Kirchhoff, and not
in the popular sense of Mr. Gladstone : see pp. 137 and 139.

in themselves contradictory, these ideas must of necessity represent truths of nature."[1]   The reader will, we feel certain, find it impossible to conceive anything other than geometrical ideals as the moving element at the basis of phenomena. The attempt, however, to conceive something else is worth the making for it inevitably leads us to the conclusion that the term "moving body" is not scientific when applied to perceptual experience. In external perception (p. 219) we have sense-impressions and more or less permanent groups of sense-impressions. These sense-impressions vary, dissolve, form new groups—that is, they *change*. Of these messages received at the brain telephonic exchange, or of groups of them, we cannot say they move—they appear, disappear, and reappear. Change is the right word to apply to them rather than motion. It is in the field of conception solely that we can properly talk of the motion of bodies ; it is there, and there only, that geometrical forms change their position in absolute time—that is, *move*. In the field of perception motion is but a popular expression to describe the mixed mode in which we discriminate and distinguish groups of sense-impressions.

§ 2.—*The Three Problems.*

That we speak of the motion of bodies as a fact of perceptual experience is largely due to the constructive elements associated with immediate sense-impression [2] (p. 49). These constructive elements are

---

[1] See especially §§ 69, 69a, and 108 of his *Aristotle : a Chapter from the History of Science.* London, 1864.

[2] The writer is not objecting to the current use of such expressions as "the sun moves," or "the train moves." Both do move—in concep-

drawn from our conceptual notions of change, which
again flow very naturally from a limited perception ;
a deeper perceptual experience is required to
demonstrate their purely ideal character (p. 203). But
the reader will, perhaps, hardly be prepared to accept
the conclusion that change is perceptual, motion conceptual, without closer analysis. This analysis may
be summed up in the three questions : *What is it that
moves ?   Why does it move ?   How does it move ?*
In the first place we must settle whether we are
asking these questions of the conceptual or perceptual spheres.   If it be of the former, the world of
symbolic motions by aid of which science describes
the sequences of our sense-impressions, then these
questions are easy to answer.   The things which
move are points, rigid bodies and strainable media,
geometrical concepts one and all.   To ask why they
move is to ask why we form conceptions at all, and
ultimately to question why science exists.   Finally,
the manner in which they move is that which enables
us most effectually to describe the results of our perceptual experience.

If we turn to the perceptual sphere and ask what
it is that moves and why it moves, we are compelled
to confess ourselves utterly incapable of finding any
answers whatever. *Ignorabimus*, we shall always be
ignorant, say some scientists. That we are really
ignorant will be the theme of the present chapter, but
I believe that this ignorance does not arise from the
limitation of our perceptive or reasoning faculties. It
is rather due to our having asked unanswerable

tion ; in perception there is a change of sense-impressions. So soon as
space is recognized as a mode of perception, and not itself a phenomenon,
this conclusion cannot be avoided.

questions. We may legitimately ask why the complex of our sense-impressions changes, but, according to the views expressed above, motion is not a reality of perception, and it is therefore, for the sphere of perception, idle to ask what moves and why it moves. With the growth of more accurate insight into the conceptual nature of motion these questions will, I believe, be dismissed like the older questions as to the blue milk of the witches and the influence of the stars (p. 27). With their dismissal, however, physical science will be for ever relieved of the metaphysical difficulties as to matter and force which it has inherited from scholastic traditions. *Ignorabimus*, therefore, does not seem the true answer to the first two questions ; it may be a true answer to the problem of changes in sense-impression (see our pp. 129 and 288). The third question—How do things move ?—also wants restating to be of any real value, and when restated it merges in the same question asked of the conceptual sphere. What, we must ask, are the conceptual types of motion best suited to describe the stages of our perceptual experience ? The answer to this question forms the subject-matter of our next chapter.

Some of my readers may feel inclined to consider that in this discussion we are entirely deserting the plane of common sense. What moves? Why, natural bodies move, they will say, is the common-sense answer. But common-sense is often a name for intellectual apathy. Being inquisitive, we naturally ask what these bodies consist in, and probably shall be told that they are quantities of *matter*. Still persisting with our questions we ask : What, then, is matter? It will not do to put us off with the reply that matter is that which moves. All we should, then, have done would be to give a name to the moving thing, but in doing so we should not have succeeded in defining or describing it.   The reader may, perhaps, imagine that insight into the nature of matter will be gained by consulting the accepted text-books of science.   Let us accordingly examine the statements of one or two.

§ 3.—*How the Physicists define Matter.*

A first writer says : "*Matter is a primary conception of the human mind,*" and more than one elementary text-book provides us with practically the same definition.   Now, the obscurity and paralogism of this statement could only be equalled by the perversities of metaphysics.[1]   Matter, we are told, is what moves in the phenomenal world, and if it were asserted that matter is a primary *perception* of the human mind we might be no wiser, but at any rate the statement would not be without sense.   But perhaps the phrase is not to be taken literally as signifying that a primary conception actually moves among perceptions, but only that we can form intuitively a conception of what moves perceptually — that the perceptual actually corresponds to the conceptual.   In this case

[1] "Matter," says Hegel, "is the mere abstract or indeterminate reflection-into-something-else, or reflection-into-self at the same time as determinate; it is consequently Thinghood which then and there is,—the subsistence or substratum of the thing.   By this means the thing finds in the matters its reflection-into-self ; it subsists not in its own self, but in the matters, and is only a superficial association between them, or an external bond over them " (*The Logic of Hegel,* translated by W. Wallace, Oxford, 1874, p. 202).   We may smile over such absurdities, but that they should be taught in the last decade of the nineteenth century in our universities, and this to immature minds, and largely at the public expense, is a cause for sorrow rather than amusement.   The much-abused schoolmen never rivalled these Hegelian quagmires even before they were transferred to English soil.

we are again thrown back on the fact that conceptual motion is a motion of geometrical ideals, and that these correspond in no accurate sense to our perceptions. Indeed, if matter be a conception at all, like the conception of a circle it ought to be a clear and definite idea, whereas the reader who will honestly ask himself what he *conceives* by matter will find that an answer is impossible, or that in attempting one he is sinking deeper and deeper into the metaphysical quagmire.

Proceeding further, we naturally turn to the little work termed *Matter and Motion*, by Clerk-Maxwell, one of the greatest British physicists of our generation. This is what he writes of matter :—

"*We are acquainted with matter only as that which may have energy communicated to it from other matter, and which may in its turn communicate energy to other matter.*"

Now this appears something definite ; the only way in which we can understand matter is through the energy which it transfers. What, then, is energy ? Here is Clerk-Maxwell's answer :—

"*Energy, on the other hand, we know only as that which in all natural phenomena is continually passing from one portion of matter to another.*"

All our hopes are shattered ! The only way to understand energy is through matter. Matter has been defined in terms of energy, and energy again in terms of matter. Now Clerk-Maxwell's statements are extremely valuable as expressing concisely the nature of certain conceptual processes, by aid of which we describe certain phases of our perceptual experience, but as defining matter they carry us no further than the statement that matter is that which moves.

We will now turn to the famous *Treatise on Natural Philosophy* of Sir William Thomson and Professor Tait —the standard work in the English language on its own branches of physical science. These writers, in § 207, tell us :—

"We cannot, of course, give a definition of *matter* which will satisfy the metaphysician, but the naturalist may be content to know matter as *that which can be perceived by the senses*, or as *that which can be acted upon by, or can exert, force*. The latter, and indeed the former also, of these definitions involves the idea of *force*, which, in point of fact, is a direct object of sense ; probably of all our senses, and certainly of the 'muscular sense.' To our chapter on 'Properties of Matter' we must refer for further discussion of the question, *What is matter?*"

That the naturalist nowadays is not bound to satisfy the metaphysician—any more than he is bound to satisfy the theologian—will be admitted at once by the sympathetic reader of our own volume. But the naturalist is bound in the spirit of science to probe and question every statement, however high the authority on which it is made; and he is further bound to inquire whether a statement as to a physical fact is also in accord with his psychological experience. Science cannot be separated into compartments which have no mutual relationship, no mutual dependence, and no intercommunication. Science and its method form a whole, and if a physical definition be not psychologically true, it is not physically true. Now we have seen that the contents of perception are sense-impressions and stored sense-impresses, and that which can be perceived by the senses are these and these only. Do our authors mean to define all sense-impressions as matter ? Would they call colour, hardness, pain, matter ? We think this is hardly likely ; they would probably tell us that the *source* of certain groups of sense-impressions is what they term matter; but this is not what they say. Had they said it they must themselves have recognized that they were passing beyond the veil of sense-impression and postulating a "thing-in-itself" (p. 87) behind the world of phenomena. They would then have seen that they were unconsciously endeavouring to satisfy the metaphysician, whom they had so properly disowned. This unconscious attempt to satisfy the "metaphysician within themselves" is further evidenced by their second statement, which throws back matter upon *force*. But *force* for these authors is the cause of motion (§ 217), not in the import of an antecedent or accompanying sense-impression—as, for example, relative position—but in the metaphysical sense of a moving agent. They do not, indeed, place this moving agent behind sense-impression ; they even describe it as a "direct object of sense," but from the psychological standpoint force must either be a sense-impression or a group of sense-impressions, for as source or object of sense-impressions it would be purely metaphysical. But as a group of sense-impressions in us, force cannot be that which *causes* motion in an objective world. As to our muscular appreciation of force, that is a point to which we shall find occasion to return later. We ought not, however, to lay much stress on these authors' remarks as to matter, for they expressly tell us that what matter is will be further discussed in another chapter of their work. Unfortunately, this portion of their great treatise has never been published, although they wrote the above remarks more than twenty years ago. Perhaps, had they returned to the subject, they would have recognized that, if the word matter had not appeared more frequently in their text than it does in their index, their volumes would have lost not an iota of their inestimable value to the physicist.

One of the two authors of the *Treatise on Natural Philosophy* has, however, published a separate work, entitled, *The Properties of Matter*. On pp. 12–13 of this work we have no less than nine, and on pp. 287–91 we have no less than twenty-five definitions or descriptions of matter, yet so far from matter being rendered intelligible by all these statements with regard to it, Professor Tait himself writes :—

"*We do not know, and are probably incapable of discovering, what matter is.*" And again : " *The discovery of the ultimate nature of matter is probably beyond the range of human intelligence.*"

Now these statements mark a considerable advance on the standpoint of the *Treatise on Natural Philosophy*. They will at least suggest to the reader that it is no mere whim on my part to question the right of matter to appear *at all* in scientific treatises. When one author tells us it is a primary conception of the human mind, and another that it is probably beyond the range of human intelligence, we feel an uncomfortable sense of the metaphysician smiling somewhere round the corner. If our leading scientists either fail to tell us what matter is, or even go as far as to assert that we are probably incapable of knowing, it is surely time to question whether this fetish of the metaphysicians need be preserved in the temple of science.

§ 4.—*Does Matter occupy Space?*

But to return to Professor Tait; he has called his book *The Properties of Matter*, and this the reader will say means something, and something very definite. Now, for the purposes of classifying our sense-impressions, it is undoubtedly useful to term particular groups of them which have certain distinguishing characteristics " material sense-impressions," and these material sense-impressions are what Professor Tait deals with under the properties of matter. It is Professor Tait, the unconscious metaphysician, who groups this class of sense-impressions together and supposes them to flow as properties from something beyond the sphere of perception, namely, matter.[1] As a working definition of matter, Professor Tait considers that we may say : "*Matter is whatever can occupy space.*" Now this definition will lead us to a number of ideas which it is instructive to follow up. In the first place, is it perceptual or conceptual space to which the definition applies ? If the latter, then matter must be a geometrical form—a result which we think our author does not intend. We think it more probable that Professor Tait looks upon space as itself objective, although he avoids any definite statement on this really important issue (p. 47). From the standpoint of our present volume, however, space is the mode by which we distinguish

---

[1] The unconscious metaphysics of Professor Tait occur on nearly every page of his treatment of the fundamental concepts of physical science. Thus he asserts the "objectivity of matter," while force is not objective, we are told, but subjective. Notwithstanding this assertion, "matter is, as it were, the plaything of force." How this nothing, this "mere phantom suggestion of our muscular sense," this force, can have an objective plaything it would puzzle a metaphysician to explain.

coexisting groups of sense-impressions, and therefore
only groups of sense-impressions can be said to
" occupy " space. This definition would therefore
lead us to identify matter with groups of sense-
impressions, and in practical everyday life the things
which we term matter are certainly more or less
permanent groups of sense-impressions, not unknowable "things-in-themselves" beyond sense-impression.
Now there can be no scientific objection to our classifying certain more or less permanent groups of sense-
impressions together and terming them matter,—to
do so indeed leads us very near to John Stuart Mill's
definition of matter as a "permanent possibility of
sensation " [1]—but this definition of matter then leads
us entirely away from matter as the thing which
moves. It can hardly be said that weight, hardness,
impenetrability *move ;* these are sense-impressions in
the brain telephonic exchange ; their grouping, their
variation and succession may lead us to the *conception*
of motion, but a sense-impression in itself cannot be
said to move ; it is there at the brain terminal or not
there. In order to bring motion into the sphere of
sense-impression, we are compelled to associate colour,
hardness, weight, &c., with geometrical forms, and in
making such constructs (p. 49) we pass from the
plane of perception to that of conception. I move
my hand ; my power to realize this motion depends
on my conceiving my hand bounded by a continuous
surface. If the physicist tells me that my hand is an

---

[1] *System of Logic,* bk. i. chap. iii. That groups of sense-impressions
recur in a more or less permanent state is an experience we have every
moment of our lives. There is a " permanent possibility of sense-
impressions." We are not forced to assert anything about this possibility residing in a *supersensuous* entity matter.

aggregation of discrete molecules, then my idea of
the motion of the hand is thrown back on the motion
of the swarm of molecules. But the same difficulty
arises about the individual molecule. I may surmount
it by supposing the molecule to be in itself a corpora-
tion of atoms, but I cannot conceive the atom's motion
unless it be bounded by a continuous surface or else be
a point. The only other way out of the difficulty is to
construct the atom of still smaller atoms—(and there
are certain phenomena presented by the spectrum
analysis of the gaseous elements that might well
induce us to believe that the atom cannot be conceived as the ultimate or " prime element of
matter ")—but what about these smaller atoms, are
they geometrical ideals or are they built up of tinier
atoms still, and if so where are we to stop ? The
process reminds us of the lines of Swift :—

> " So naturalists observe, a flea
> Has smaller fleas that on him prey ;
> And these have smaller still to bite 'em,
> And so proceed *ad infinitum*."

I am unable to verify Swift's statement as to the fleas,
but I feel quite sure that to assert the real existence
in the world of phenomena of all the concepts by aid
of which we scientifically describe phenomena—
molecule, atom, prime-atom—even if it be *ad infinitum*,
will not save us from having ultimately to consider
the moving thing to be a geometrical ideal, from
having to postulate the phenomenal existence of what
is contrary to our perceptual experience. This point
brings out very clearly what the present writer holds
to be a fundamental canon of scientific method,
namely : *To no concept, however invaluable it may be*

as a means of describing the routine of perceptions, ought phenomenal existence to be ascribed until its perceptual equivalent has been actually disclosed.

Whenever we disregard this canon, when, for example, we assert reality for the mechanisms by aid of which we describe our physical experience, then we are more likely than not to conclude with an *antinomy*, or a conflict of rules. For such mechanisms are constructs largely based on conceptual limits, which are unattainable in the field of perception. When we consider space as objective and matter as that which occupies it, we are forming a construct largely based on the geometrical symbols by aid of which we analyze motion conceptually. We are projecting the form and volume of conception into perception, and so accustomed have we got to this conceptual element in the construct that we confuse it with a reality of perception itself. When we go a stage further in the phenomenalizing of conceptions, and postulate the reality of atoms, the antinomy becomes clear. If bodies are made up of swarms of atoms, how can they have a real volume or form? What is the volume or form of a swarm of bees or a cloud of dust? Obviously we can only give them shape and size by enclosing them conceptually in an ideal geometrical surface. Just as in a swarm of bees or a cloud of dust odd members of the community near this imaginary surface are continually passing in and out, so—if we phenomenalize conception—we must assert that at the surface of water or of iron odd molecules or atoms are perpetually leaving or, it may be, re-entering the swarm. Condensation and evaporation go on at the surface of the water and iron has a metallic smell. Now if the swarm be in this continual state of flow at the surface we can only speak of it as having volume or form *ideally*, or as a mode of conceptually distinguishing one group of sens?-impressions from another (p. 197). It is the conceptual volume or form which occupies space, and it is this form, and not the sense-impressions, which we conceive to move. If we throw back the occupancy of space on the individual members of the swarm, it is certainly not the volumes or forms of the individuals, which we consider as the volume or form of the material body, for the former we treat as imperceptible and the latter as perceptible. Further, we must then infer that the unknown is ultimately unlike the known, that geometrical ideals can be realized in the imperceptible. This, however, is a distinct breach of the canon of logical inference (p. 72).

So far, then, our analysis of the physicist's definitions of matter irresistibly forces upon us the following conclusions : That matter as the unknowable cause of sense-impression is a metaphysical entity[1] as meaningless for science as any other postulating of causation in the beyond of sense-impression ; it is as idle as any other *thing-in-itself*, as any other projection into the supersensuous, be it the force of the materialists, or the infinite mind of the philosophers. The classification of certain groups of sense-impressions as material groups is, on the other hand, scientifically of value ; it throws no light, however, on matter as that which perceptually moves.

Conceptually all motion is the motion of geometrical ideals, which are so chosen as best to describe those

---

[1] The scientific reader must for the present have at least sufficient confidence in the author not to believe that *mass* is thrown overboard with the fetish matter.

changes of sense-impression which in ordinary language we term perceptual motion.

§ 5.—*The " Common-sense" View of Matter as Impenetrable and Hard.*

Now the reader may feel inclined, on the basis of his daily experience, to assert that both the physicists above referred to and the author are really quibbling about words, and that we can sufficiently describe matter by saying that it is *impenetrable* and *hard*. Now these terms describe important classes of sense-impressions, and the sense-impressions of impenetrability and hardness are very frequently factors of what we have called material groups of sense-impressions. But it is very doubtful whether we can consider them as invariably associated with these material groups. At any rate if we do we shall find ourselves again involved in the antinomies which result when we pass incautiously to and fro from the field of perception to that of conception. When we say a thing is impenetrable, we can only mean that something else will not pass through it, or that there are two groups of sense-impressions which, in our perceptual experience, we have always been able to distinguish under the mode space. Impenetrability, therefore, can only be a relative term ; one thing is impenetrable for a second. When we say that matter is impenetrable we cannot mean that nothing whatever can pass through it. A bird cannot fly through a sheet of plate glass, but a ray of light does penetrate it perfectly easily. A ray of light cannot pass through a brick wall, but a wave of electric oscillations can. In order to describe the motion of these luminous and electric waves the physicist conceives ether to penetrate all bodies and to act as a medium for the transit of energy through them. Matter cannot therefore be looked upon as the thing which is *absolutely* impenetrable.

Or, are we missing the point of what is meant, when it is asserted that matter is that which is impenetrable? Are we to postulate the real existence of atoms and then to suppose the individual members of the swarm impenetrable? Here again a difficulty arises. There is much that tends to convince physicists that the atom cannot be conceived as the simplest element of the conceptual analysis of material groups. Just as a bell when struck sets the air in motion and gives a note, so we conceive an atom capable of being struck, and of setting not the air but the ether in motion, of giving, as we might express it, an ether note. These notes produce in us certain optical sense-impressions— for example, the bright lines of the spectrum of an attenuated gas. As without seeing two bells we might, and indeed often do, distinguish them by their notes,[1] so the physicist distinguishes an atom of hydrogen from an atom of oxygen, although he has never seen either, by the different light notes which he conceives to arise from them. But as the bell to give a note must be considered as vibrating—changing its shape or undergoing strain—so the physicist practically finds himself compelled to conceive the atom as undergoing strain, or changing its shape. This conception forces us to suppose the atom built up of distinct parts capable of changing their relative

---

[1] The householder is generally able to distinguish the sound of the back-door from that of the front-door bell, although, probably, in ninety-nine cases out of a hundred he may never have seen the bells in his house.

position. What are these ultimate parts of the atom, by the relative motion of which we describe our sense-impressions of the bright lines in the spectrum? We have as yet formed no conception. Does the ether or anything else penetrate between these ultimate parts of the atom? We cannot say. In the present state of our knowledge it is impossible to tell whether it would or would not simplify things to conceive the atom as penetrable or impenetrable. Hence, even if we go so far as to give the concept atom a phenomenal existence, it will not help us to understand what is meant by the assertion that matter is impenetrable.

§ 6.—*Individuality does not denote Sameness in Substratum.*

Shall we, however, be more dogmatic still, and, denying that ether is matter, assert that matter is impenetrable *relative* to matter? In order to give any definite answer to this question we have again to pass from the perceptible material group to its supposed elementary basis, the atom, and to ask whether we have any reason for conceiving atoms as incapable of penetrating each other. In the first place the physicist, although he has never caught an atom, yet conceives it as something which is incapable of disappearing—*it continues to be.* In the next place, if we conceive it as entering into combination with a second atom, although we have no reason for asserting that the two atoms do not mutually penetrate, we are still compelled, in order to describe by aid of atoms our perceptual experience, to conceive that, out of the combination, two separate atoms can again be obtained with the same individual characteristics as the original two possessed. What right have we to postulate these laws with regard to atoms when atoms are, even if real, still absolutely imperceptible to us, when we are absolutely unable to observe their mutual action? We have exactly the same logical right as we have to lay down any scientific law whatever. Namely, we find that these laws as to the action of single atoms, when applied to large groups of atoms, enable us to describe with very great accuracy what occurs in those phenomenal bodies, which we scientifically symbolize by groups of atoms; they enable us to construct without contradiction by perceptual experience, those routines of sense-impression which we term chemical reactions.

The hypotheses that the individual atom is both indestructible and impenetrable suffice to elucidate certain physical and chemical properties of the bodies we construct from atoms. But the continued existence of atoms under physical changes and the reproduction of their individuality on the dissolution of chemical combination might possibly be deduced from other hypotheses than those of the indestructibility and impenetrability of the individual atom. It does not follow of logical necessity that because we experience the same group of sense-impressions at different times and in different places, or even continuously, that there must be one and the same thing at the basis of these sense-impressions. An example will clearly show the reader what we mean and at the same time demonstrate that however useful as hypotheses the indestructibility and impenetrability of the atom may be, they are still not absolutely necessary conceptions; so that even if we do project our atom into an imperceptible of the phenomenal world, it will not follow that there must be an unchangeable individual something at all times and in all positions as the basal element of a permanent group of sense-impressions. The permanency and sameness of the phenomenal body may lie in the individual grouping of the sense-impressions and not in the sameness of an imperceptible something projected from conception into phenomena.

The example we will take is that of a wave on the

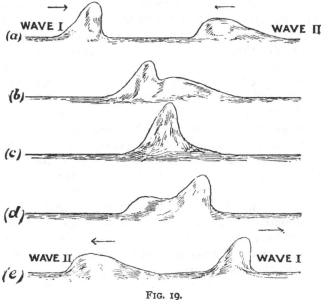

FIG. 19.

surface of the sea. The wave forms for us a group of sense-impressions, and we look upon it, and speak of it, as if it were an individual thing. But we are compelled to conceive the wave when it is fifty yards off as consisting of quite different moving things to what it does when it reaches our feet—the substratum of the wave has changed. Throw a cork in; it rises

21

and falls as the wave passes it, but is not carried along by it. The wave may retain its form and be for us exactly the same group of sense-impressions in different positions and at different times, and yet its substratum may be continually changing. We might even push the illustration further ; we might send two waves of different individual shapes (Ffg. 19) along the surface of still water in opposite directions (*a*), or in the same direction if the pursuing wave had the greater speed. One of these waves would meet or overtake the other (*b*) ; they would coalesce or combine (*c*), producing in us for a time (which depends entirely on their relative speeds), a new group of sense-impressions differing totally from either individual group ; but they would ultimately pass each other (*d*) and emerge with their distinct individualities the same as of old (*e*). Throughout the whole of this sequence the substrata of the two individual waves are changing and for the time of the combination their substratum is identical, and yet the waves are able to preserve their individual characteristics, so far as reappearing with them after combination is concerned.[1] Thus sameness of sense-impressions before and after a combination is seen from a perceptual example not to involve of necessity a sameness of substratum.

Now we have cited this example of the wave for two reasons. In the first place it shows us that it is possible to conceive atoms as penetrable by atoms, and as varying from moment to moment

---

[1] If analogy were to be sought to the sameness of total weight before, during, and after combination, it might be found in the sameness of the volume of fluid raised above the sea-level, before, during, and after coalition. Thus sameness of weight does not in conception necessarily involve sameness of substratum.

in their substratum, without at the same time denying the possibility of their physical permanency and individual reproduction after chemical combination. To consider an atom as consisting always of the same substratum, and as impenetrable by other atoms, may help us to describe easily certain physical and chemical phenomena; but it is quite conceivable that other hypotheses may equally well account for these phenomena, and this being so we have clearly no right first to project special conceptions into the world of real phenomena, and then to assert on the strength of this that matter, penetrable in itself, is impenetrable in its ultimate element, the atom. Clearly impenetrability is neither in perception nor conception a necessary factor of material groups of sense-impressions. Further, the permanence and sameness of such a group do not necessarily involve the conception of a permanent and same substratum for the group.

My second reason for citing this wave example lies in the light it throws on the possibilities involved in the statement: "*Matter is that which moves.*" The wave consists of a particular form of motion in the substratum which for the time constitutes the wave. This form of motion itself moves along the surface of the water. Hence we see that besides the substratum something else can be conceived as moving, namely, *forms of motion.* What if, after all, matter as the moving thing could be best expressed in conception by a form of motion moving, and this whether the substratum remain the same or not? To this suggestion we shall return later, as it is one extremely fruitful in its results.

§ 7.—*Hardness not Characteristic of Matter.*

It remains for us now to deal with the other characteristic, hardness, which is popularly attributed to
matter. There are certain persons who, when men's
ignorance as to the nature of matter is suggested to
them, are content to remark that one has only to knock
one's head against a stone wall in order to have a
valid demonstration of the existence and the nature of
matter. Now if this statement be of any value, it can
only mean that the sense-impression of hardness is the
essential test of the presence of matter in these persons'
opinion. But none of us doubt the existence of the
sense-impression hardness associated with other sense-
impressions in certain permanent groups ; we have
been aware of it from childhood's days, and do not
require its existence to be experimentally demonstrated now. It is one of those muscular sense-
impressions which we shall see are conceived by
science to be describable in terms of the relative
acceleration of certain parts of our body and of
external bodies. But it is difficult to grasp how the
sense-impression of hardness can tell us more of the
nature of matter than the sense-impression of softness might be supposed to do. There are clearly
many things which are popularly termed matter and
are certainly not hard. Further, there are things
which satisfy the definitions of matter as that which
moves or as that which fills space, but which are very
far indeed from producing any sense-impression of the
nature of hardness or softness; nor would they even
satisfy our definition if we said that matter is that
which is heavy, heaviness being certainly a more widely-
spread factor of material groups of sense-impressions
than hardness. Between the sun and planets, between

the atoms of bodies, physicists conceive the ether to exist, a medium whose vibrations constitute the channel by means of which electro-magnetic and optical energy is transferred from one body to another. In the first place, the ether is a pure conception by aid of which we correlate in conceptual space various motions. These motions are the symbols by which we briefly describe the sequences and relationships we perceive between various groups of phenomena. The ether is thus a mode of resuming our perceptual experience; but like a good many other conceptions of which we have no direct perception, physicists project it into the phenomenal world and assert its real existence. There seems to be just as much, or little, logic in this assertion as in the postulate that there is a real substratum, matter, at the back of groups of sense-impressions; both at present are metaphysical statements. Now there is no evidence forthcoming that the ether must be conceived as either hard or heavy,[1] and yet it can be strained or its parts put in relative motion. Further, from Professor Tait's standpoint, it occupies space. Hence those who associate matter with hardness and weight must be prepared to deny that the ether is matter, or be content to call it non-matter. It is worth noting, at the same time, that the metaphysicians—whether they be materialists asserting the phenomenal existence both of space and of a permanent substratum of sense-impression, or " common-sense " philosophers asking us to knock our

[1] I venture to think Sir William Thomson's attempt to *weigh* ether a retrograde step (see his *Lectures on Molecular Dynamics*, pp. 206-8, Baltimore, 1884). If the ether be a sufficiently wide-embracing conception, gravitation should flow from it, and this certainly was Sir William's view when he propounded the vortex atom.

heads against stone walls—reach hopelessly divergent
results when they say that matter is that which moves,
that matter occupies space, and that matter is that
which is heavy and hard.

§ 8.—*Matter as non-Matter in Motion.*

There is, however, a still greater dilemma in store
for the "common-sense" philosophers. We have not
yet reached a clear conception of what the ether, the
non-matter of our philosophers, consists in. There
are in fact two, at first sight, completely divergent
ways in which the ether is reached as a conceptual
limit to our perceptual experience (see p. 217), but it is
the great hope of science at the present day that "hard
and heavy matter" will be shown to be ether in motion.
In other words, it is well within the range of possibility
that during the next quarter of a century science will
have discovered that our symbolic description of the
phenomenal universe will be immensely simplified, if
we take as our symbolic basis for material groups of
sense-impressions a type of motion of the conceptual
ether; in other, more expressive if less accurate,
language, if we treat our friends' matter as their non-
matter in motion. We shall then find that our sense-
impressions of hardness, weight, colour, temperature,
cohesion, and chemical constitution, may all be
described by aid of the motions of a single medium,
which itself is conceived to have no hardness, weight,
colour, temperature, nor indeed elasticity of the
ordinary perceptual type. This would mean an
immeasurably great advance in our scientific power
of description. Yet if physicists even then persist in
projecting the conceptual into the sphere of sense-
impression, and in asserting a phenomenal existence

for the ether, we should still be ignorant of what it is that moves, of what ether-matter may really consist in.

Our analysis, therefore, of the various statements made by physicists and common-sense philosophers with regard to the nature of matter, shows us that they are one and all *metaphysical*—that is, they attempt to describe something beyond sense-impression, beyond perception, and appear, therefore, at best as dogmas, at worst as inconsistencies. If we confine ourselves to the field of logical inference, we see in the phenomenal universe not matter in motion, but sense-impressions and changes of sense-impressions, coexistence and sequence, correlation and routine. This world of sense-impression science symbolizes in conception by an infinitely extended medium, whose various types of motion correspond to diverse groups of sense-impressions, and enable us to describe the correlations and sequences of these groups. The moving elements of this medium can in thought be conceived of only as geometrical ideals, as points or continuous surfaces. To make our symbolic chart or picture agree the better with perceptual experience, we find it necessary to endow these geometrical ideals with certain relative positions, velocities, and accelerations, the correlations of which are expressible in certain simple laws termed the laws of motion (see the following Chapter). If we choose to term the moving things of the conceptual chart *matter*, there can be no objection to the term, provided we carefully distinguish this conceptual matter from any metaphysical ideas of matter as the substratum of sense-impression, as that which perceptually moves, as that which fills space, or as that which can be defined as heavy, hard, and impenetrable. Conceptual matter is thus merely a name for the geometrical ideals endowed with certain correlated motions by aid of which we describe the routine of our external perceptions. It is in this sense that we shall use the term matter for the remainder of this work, unless we are expressly referring to the matter of the metaphysicians. " Heavy " matter will be a name for the conceptual symbol by which we represent what we have termed material groups of sense-impressions, while ether-matter will be a name for the symbol by which we describe other phases of sense-impression, especially the correlation in space and time of sense-impressions belonging to diverse material groups. We shall not project our conceptions into imperceptibles in the field of perception (!) [1]—except in so far as it may be necessary in order to criticize current physical notions. We shall try and preserve throughout the standpoint that science is a description of perceptual experience by aid of conceptual shorthand, the symbols of this shorthand being in general *ideal* limits to perceptual processes, and as such having no exact perceptual equivalents.

The reduction of " matter to non-matter in motion," of heavy-matter to ether-matter in motion, is so important as a possible simplification of our scientific analysis of phenomena that we must devote a few pages to its discussion. We will term the fundamental element of heavy matter, the element out of which,

---

[1] The reader may perhaps expect the words "unperceived things " rather than "imperceptibles." But as every external perception is a group of sense-impressions, and as our senses are limited, the atom, if a real phenomena, could only appear sensible by colour, hardness, temperature, &c., the very sense-impressions it is conceived to describe. Hence, if the atom is to be *not* these things but their source, it may be truly termed *imperceptible.*

perhaps, chemical atoms themselves are to be conceived as built up, the *prime-atom*. We have, then, to ask what types of motion in the ether have been suggested as possible forms for the prime-atom. There are two suggestions to which reference may be made, both of which depend upon our postulating the same constitution for the ether. We must here make a brief digression in order to throw some light on this constitution of the ether.

§ 9.—*The Ether as " Perfect Fluid" and " Perfect Jelly."*

The reader is certainly acquainted with two types of perceptual bodies which may be roughly described as liquid and elastic. As specimens of these two types we will take water and jelly. As substances water and jelly have a remarkable agreement in one respect and a remarkable divergence in another. If we put either water or jelly into a cylinder closed at the bottom and attempt to compress them by aid of a heavily-loaded piston, we shall find that the compression is either insensible or of very small amount indeed. Careful experiments with elaborate apparatus show that these substances are compressible, but the amount of compression, although measurable, is exceedingly minute as compared, for example, with the amount that air would be compressed by the same load. We express this result by saying that both water and jelly, offer great resistance to one form of strain, namely, change of size (p. 243). But this resistance is only relative, relative to other substances, such as gases, and to the machinery of compression at our disposal. So far as our perceptive experience goes there is no substance which resists absolutely all change of size, and for which change of size is impossible. Hence an incompressible substance is merely a conceptual limit which has not its equivalent in the world of phenomena, but which is reached in conception by carrying on indefinitely a process (or a classification of compressible bodies) starting in perception.

Turning from this agreement to the divergence between water and jelly, we remark that if a lath of wood or even a knife-blade be pressed downwards on a jelly it requires considerable effort to shear or separate the jelly into two parts ; on the other hand, the water is separated by the lath without any sensible resistance. Now the change of shape we are in this case concerned with is of the nature of a slide (p. 245), and we say that the water offers little and the jelly considerable resistance to sliding strain. Here, again, the question of the amount of resistance is relative. So far as our perceptual experience goes, all fluids offer some, however small, resistance to the sliding of their parts over each other. The fluid which offers absolute resistance to compression and no resistance at all to slide of its parts,—or the parts of which slip over each other without anything of the nature of frictional action,—is only a conceptual limit. Such a fluid is termed a *perfect fluid*. On the other hand, by proceeding to the opposite limit in the case of an incompressible jelly, that is, by supposing it to resist absolutely change of shape by sliding, we should obtain a body incapable of changing its form by either compression or slide, and thus reach that conceptual limit, the *rigid body*. If we suppose absolute resistance to compression and partial resistance to slide, we have in conception a medium which might perhaps be described as a *perfect jelly*.

Returning now to our ether, we note that physicists

conceive it incompressible, but that for some purposes they appear to treat it as a *perfect fluid,* for other purposes as a *perfect jelly.*[1] This might at first sight appear a contradiction or conflict of conceptions, and it does undoubtedly involve difficulties which physicists are at present far from having thoroughly mastered. If we consider the ether as purely conceptual, then, in order to describe different phases of phenomena, we are certainly at liberty to first consider it as of one nature and then as of another. But in doing so it is evident that we are leaving room for a wider conception which will resume both phases of phenomena at once, and will not lead us into logical contradictions if both phases have to be dealt with in the same investigation. Thus, if the ether as a perfect fluid enables us to describe atoms by its types of motion, and the ether as a perfect jelly enables us to describe the radiation of light, it is clear that when we treat the atom as a source of light-radiations, we may get into serious confusion by the conception that the ether is at the same time a perfect fluid *and* a perfect jelly. We are compelled, indeed, to try and find some reconciliation between these two conceptions. If we turn to perceptual experience for a suggestion, we may note that water is the principal component of jelly, and may, by the addition of more or less gelatinous material, be stiffened to a jelly of any consistency. In the like manner we can conceive a series of perfect jellies formed, ranging in their resistance to slide, from the perfect fluid, through all stages of viscosity, up to the perfectly rigid body. We might, then, out of this series of jellies choose one which, for sliding strains of a certain magnitude, was sensibly

[1] For further purposes again scarcely as either.

a perfect fluid, while for smaller **strains**, such as are involved in the theory of light-radiation, it would act as a perfect jelly. This is the solution propounded in 1845 by Sir George G. Stokes,[1] and it may be termed the jelly-theory of the ether. The jelly-theory of the ether has undoubtedly been of value in simplifying many of our conceptions of physical phenomena, but how far it can be reconciled with any system of ether-motion as a basis for the prime-atom yet awaits investigation.[2]

There is another possibility to which I can only briefly refer here—namely, that the ether is to be conceived as a perfect fluid, but that just as a certain type of motion of this ether corresponds to the atom, so types of motion may be used to stiffen the ether, or to give it elastic rigidity. The ether may be a perfect fluid, but, owing to the turbulence of its motion, it may act for certain purposes as a perfect jelly. This hypothesis will be better appreciated when I have said a few words as to the ether-motions which may constitute the prime-atom.

§ 10.—*The Vortex-Ring Atom and the Ether-Squirt Atom.*

In constructing an atom out of an ether-motion we have first to gain some idea of how it is possible that ether, not being itself hard or resisting change

---

[1] *Mathematical and Physical Papers*, vol. i. pp. 125–29, and vol. ii. pp. 12–13. The present writer considers, however, that there is a difference in quality as well as in degree between a viscous fluid and an elastic medium. The complete difference in type between the equations of a plastic solid and a viscous fluid is sufficient evidence of this. In the former case, any shear *above a certain magnitude* produces set; in the latter any shear *whatever*, *if continued long enough*.

[2] For example, Sir William Thomson's vortex atom would hardly be a possibility.

of shape, can yet be conceived to produce the sensations of hardness and resistance by its motion. Some general idea can easily be got of the sort of resistance produced by particular types of motion in the following manner : Take an ordinary spinning-top, and suppose we succeed by great care in balancing it on its peg. Clearly the least touch of the hand will upset it ; it offers no resistance to the motion of the hand. The same remark applies if the peg of the top were fixed by a ball-and-socket joint to the table. But, on the other hand, if the top be set spinning, we shall find the case entirely altered ; it will now present considerable resistance to being upset, and, if partially turned round its ball-and-socket joint, will tend to return to the old vertical position. A considerable number of such spinning-tops would offer a large amount of resistance to a hand passed over the table at a less distance than their height. This example may perhaps bring home to the reader how a certain type of motion may suffice to stiffen a body not otherwise stiff. Another example of motion stiffening a body is the smoke-ring, with which most devotees of tobacco are well acquainted. Two such smoke-rings will not coalesce ; they pass through or wriggle round each other, and round solid corners which come in their way, and, furthermore, their relative motion is easily seen to closely depend upon their relative position. Now we see smoke-rings because the moist particles in the smoke render the gaseous mixture visible, as similar particles render steam visible ; but we might blow air-rings in air, which would act precisely as the smoke-rings do, only they would be invisible. Such rings are termed *vortex-rings ;* and if we study the action of such

rings not in air or water but in our conceptual perfect fluid, we shall find that, like atoms, they retain
their own individuality ; they enter into combination,
but cannot be created or destroyed. This is the basis of
Sir William Thomson's vortex-ring theory of matter
—a prime atom, according to his theory, is an ether
vortex-ring.[1] By the aid of vortex-motion, or spinning elements of liquid in a liquid, we are also able
to conceive a liquid stiffened up to a required degree
of resistance to sliding strain, and thus to replace the
ether as a perfect jelly by the ether as a perfect fluid
in a turbulent condition.[2] We can then dispense with
Sir George Stokes' hypothesis of slight viscosity. But
however suggestive these ideas may be for the lines
upon which we may in future work out our conceptions of ether and atom, they are very far indeed from
being at present worked out, and there are many
difficulties in the vortex-atom theory—notably that
of deducing gravitation—which the present writer is
not very hopeful will ever be surmounted.

While Sir William Thomson's theory supposes
that the substratum of an atom always consists of the
same elements of moving ether, the author has ventured to put forward a theory in which, while the
ether is still looked upon as a perfect fluid, the individual atom does not always consist of the same elements of ether. In this theory an atom is conceived
to be a point at which ether flows in all directions

---

[1] For a fuller account of this theory see Clerk-Maxwell's article
"Atom," in the *Encyclopædia Britannica*, or his *Scientific Papers*,
vol. ii. pp. 445–84. See also as to spin producing elastic resistance Sir
William Thomson's *Popular Lectures and Addresses*, vol. i. pp. 142–46
and 235–52.

[2] See G. F. Fitzgerald : "On an Electro-magnetic Interpretation of
Turbulent Fluid Motion," *Nature*, vol. xl. pp. 32–4.

into space ; such a point is termed an *ether-squirt*. An ether-squirt in the ether is thus something like a tap turned on under water, except that the machinery of the tap is dispensed with in the case of the squirt. Two · such squirts, if placed in ether, move relatively to each other, exactly like two gravitating particles, the mass of either corresponding to the mean ·rate at which ether is poured in at the squirt. From periodic variations of the rate of squirting, as influenced by the mutual action of groups of squirts, we are able to deduce many of the phenomena of chemical action, cohesion, light, and electro-magnetism. Indeed the ether-squirt seems a conceptual mechanism capable of describing a very considerable range of phenomena. It involves, of course, the conception of negative matter, or ether-*sinks ;* for the amount squirted into an incompressible fluid must be at least equalled by the amount which passes out. As, however, an ether-squirt and an ether-sink must be conceived to repel each other, there need be no surprise that we are compelled to consider our portion of the universe as built up of positive matter ; the negative matter, or ether-sinks, would long ago have passed out of the range of ether-squirts.[1]

11.—*A Material Loophole into the Supersensuous.*

Now the reader may naturally ask : Where can we

[1] Carnelley, however, demanded an element of negative atomic weight, and a substance of negative weight is by no means inconceivable. Should the reader be interested in a mathematical account of this theory he may consult : " Ether-squirts; Being an Attempt to Specialize the Form of Ether-Motion which forms an Atom in a Theory propounded in former Papers," *American Journal of Mathematics,* vol. xiii. pp. 309–62. See also *Camb. Phil. Trans.* vol. xiv. p. 71 ; *London Math. Society,* vol. xx. pp. 38 and 297.

conceive the ether to come from when it pours in at the squirt or prime-atom ?   In taking the ether-squirt as a model dynamical system for the atom, we are not bound to answer this question in order to demonstrate its validity, any more than we are bound to explain why ether and atom themselves come to be. From our standpoint, they are justified as conceptions if they enable us to resume our perceptual experience. But as there are many who will insist on projecting the conceptual into the phenomenal field, I will endeavour to answer the question by suggestion.

Suppose we had two opaque horizontal plane surfaces placed close together, and containing between them water in which lived a flat fish, say a flounder. Now it is clear that the perceptions of our fish would be limited to motion forwards or backwards, to right or to left, but vertically upwards or downwards would be an imperceptible, and therefore probably inconceivable, motion for him.   Now let us pass in conception to a limit unrealizable in perception ; let us suppose our flounder to get flatter and flatter, and the film of water thinner and thinner, as the planes are pressed closer together.   The motion of the flounder and the motion of the water may then, for conceptual purposes, be supposed to take place in one horizontal plane.   Now if we were to make a hole in one of the planes and squirt water in, it is clear that our flounder would experience new sense-impressions when he came into the neighbourhood of the squirt.   Indeed the pressure produced by the flow of water might compel the flounder to circumnavigate the squirt—that is, the squirt might be for him hard and impenetrable.   Such squirts, although only water in motion, might form very *material* groups of sense-impressions for our fish. If, however, he were told that matter was formed of squirts, he would be quite unable to conceive where the squirt came from. It could be from neither forwards nor backwards, neither from right nor left, for it flows *in* in all these directions. The flounder would presume we were quite mad did we suggest that the water came vertically upwards or downwards ; that there was another direction in space—" upward and outward in the direction of his stomach," as the author of *Flatland*[1] felicitously expresses it. Could the flounder get out of his space through the squirt—*through and out in the direction of matter*—he would reach a new

Fig. 20.

world, wherein he would perceive what squirts were, what his matter really consisted in. Through the eye of the needle, out through the matter of flatland, the flounder would reach the heaven of our three-dimensioned space, where we go up and down, as well as forward and backward, and to right and left. But for the flounder this " out through matter " would remain inconceivable, not to say ridiculous ; it would be to penetrate behind the surface of sense-impressions.

Now this parable of the flounder is specially in-

[1] *Flatland: a Romance of Many Dimensions*, by A Square. London, 1884.

tended for those minds which, strive as they will, cannot wholly repress their metaphysical tendencies, which *must* project their conceptions into realities beyond perception. The danger of this metaphysical speculation lies in the frequency with which it contradicts our perceptual experience when it passes from the " beyond " of sense-impression to the world of phenomena. Now a happy conception as to how the prime-atom is to be constructed, fitting in with all our perceptual experience (that is, enabling us to describe it symbolically with great accuracy), *might* leave a loophole for the metaphysical mind to pass to something which does not symbolize the perceptual, and therefore might *dogmatically* be assumed to belong to the supersensuous. Out from our space through the ether-squirt, out through matter we in conception pass, like the flounder, to another dimensioned space. This space has for a number of years past formed the subject of elaborate investigations by some of our best mathematicians,[1] and it possesses this great advantage : that when we pass from the conclusions drawn for this higher space to the space of our perceptual experience, then we are not involved in the contradictions which abound in the transition from the older metaphysics to our physical experience. Here in this new playroom, entered, perhaps, by the doorway of matter, metaphysician and theologian can for the present safely spin beyond the sensible the cobwebs, which have been swept away by the scientific broom whenever they encumbered the habitable. apartments of knowledge. The necessary mathematical equipment required for genuine research in the field of higher-dimensioned space will

---

[1] Riemann, Helmholtz, Beltrami, and Clifford.

at any rate act as a safeguard against over light-hearted expeditions "beyond the sensible"! Should a time ever come, which may, perhaps, be doubted, when a happy conception as to the structure of the prime-atom is discovered to be a *perceptual* fact, then if such a conception involves the existence of four-dimensioned space,[1] our friends will have done yeoman service in preparing a way for a scientific theory of the supersensuous—*out through the doorway of matter!*

§ 12.—*The Difficulties of a Perceptual Ether.*

But we have romanced enough for the sake of the metaphysically-minded. Returning to the solid ground of fact, we have to remember that no hypothesis as to the structure of the prime-atom from ether in motion is at present scientifically accepted ; no model dynamical system for the atom has as yet been shown to have such a wide-reaching power of describing our perceptual experience that it has passed from the field of imagination and become a current symbol of scientific shorthand. Nor is the reason far to seek ; we desire to construct, if possible, the prime-atom from an ether-motion, but our conceptions of the ether are at present very ill-defined.

---

[1] The ether-squirt is not the only atomic theory which suggests a space beyond our own. Clifford imagined matter to be a *wrinkle* in our space, which suggests the idea of another space to bend it in. This notion of Clifford's may, perhaps, be brought home to our reader by imagining the flounder rigidly flat and a crumple or wrinkle in his plane of motion. The wrinkle would, like matter, be impenetrable to the fish ; he could not *fit* it ; either the wrinkle or he would have to get out of the way. This non-fitting of two kinds of space has not hitherto, however, been developed as a mode of describing any of our fundamental physical experiences.

We are agreed that it must be conceived as a medium which resists strain, but we are not certain how to represent best the relative motions that follow on relative change in the position of the ether-elements. We are not yet satisfied with a perfect fluid, a perfect jelly, or even a turbulent perfect fluid conception of the ether.

Treating the ether not as a conception but as a phenomenon, we find it difficult to realize how a *continuous* and *same* medium could offer any resistance to a sliding motion of its parts, for the continuity and sameness would involve, after any displacement, everything being the same as before displacement. The idea of a perfect jelly appears to involve some change in structure as we magnify smaller and smaller elements larger and larger. Finally, any relative motion of translation as distinct from one of rotation seems excluded by the idea of absolute incompressibility.[1] It is not a metaphysical quibble when we demand that two things shall not occupy the same space, but that when motion begins there shall be *somewhere* unoccupied for *something* to move into. The obvious fact is that while in conception we can represent the moving parts of the ether *as points*, and we can endow these points with such relative velocities and accelerations as will best describe our perceptual experience, yet when we project the ether into the phenomenal world it is at once recognized as a conceptual limit unparalleled in perceptual experience, and we do not feel at home with it. The old problems as to "heavy matter" recur. What is the ultimate element of the ether which moves? and why does it

---

[1] For absolutely incompressible elements (other than points) motion round any closed curve other than a circle seems inconceivable.

move? Build a perceptual matter out of a phenomenal ether, and we have again thrust upon us the question as to ether-matter's nature. Is it also to be a *terra incognita nunc et in æternum?* The mind again fails to rest in peace until it reaches somewhere the motion of a point, the sizeless ultimate element of matter postulated by Boscovich. We find ourselves again involved in the contradictions which flow from asserting a reality for motion in the phenomenal field. We are again forced to the conclusion that motion is a pure conception, which may describe perceptual changes, but cannot be projected into the phenomenal world without involving us in inexplicable difficulties.

§ 13.—*Why do Bodies Move?*

We have left but little space for the discussion of our second question : Why do bodies move? But the answer to this question must be clear after what precedes. If we mean : Why do sense-impressions change in a certain manner?—then we have already seen what are the possibilities of knowledge on this point when considering consciousness, the nature of the perceptive faculty and the routine of perceptions (pp. 122–9). If we mean : Why do the geometrical symbols by which we conceptualize material groups of sense-impressions move in a certain fashion?—then the answer is, that after many guesses we have found these types of motion to be best capable of describing the past and predicting the future routine of our perceptions. If, however, any one persists in phenomenalizing our conceptual symbols of motion, then science can only reply to this question : Why does matter move? *We don't know.* Let us suppose that the earth actually moves in an ellipse round the sun in a focus, and then let us attempt to analyze the *why* of it. Well, conceptually we construct this motion out of a certain relative motion of the elementary parts of sun and earth. We say that if these elementary parts have certain relative accelerations when in each other's presence, then the earth will describe an ellipse about the sun. These elementary parts may be looked upon as atoms or groups of atoms, but to save any hypothesis let us simply term them *particles* of matter. Now, why do two particles when in each other's presence move relative to each other in a certain fashion? It will not do to answer: Owing to the *law* of gravitation. That merely describes how they move. Nor can we say: Owing to the *force* of gravitation. That is merely throwing the answer on the beyond of sense-impression—it is the metaphysical method of avoiding saying: We don't know.

When we see two persons dancing round each other we assume that they do it because they wish to, because they *will* to. They cannot be said, if one is not holding the other, to enforce each other's motion. To attribute the dance to their common will is the sole explanation we can give of it.[1] When we find the ultimate particles of matter dancing about each other, we can hardly, like Schopenhauer, attribute it to their common will to dance thus, because will denotes the presence of consciousness, and consciousness we cannot logically infer unless there be certain types of material sense-impressions associated with it. Thus will, if it had any meaning as a cause of motion— which we have seen it has not (p. 150)—could not

[1] See Appendix, *Note V.*

help us with regard to our dance of material particles. All we can scientifically say is, that the *cause* of their motion is their relative position ; but this is no explanation of why they move when in that position. The difficulty cannot be surmounted by appealing to the notion of force. Of the metaphysical conception of force we have said enough (pp. 140 *et seq.*), and we need not reconsider it here. But force is sometimes said to be a sense-impression—we are said to have a "muscular sensation" of force. I *will* to push a thing with my hand, and on the will becoming action a "muscular sensation" occurs which is termed the exertion of force. But why is this more a sense-impression of force than a sense-impression of changes in the motion, or of relative accelerations in the particles of my finger-tips ? Add to this that the so-called "muscular sensation" of force is associated with a conscious being, or is a subjective side of some changes of motion in his person, and we see that it can throw absolutely no light on the reason why material particles move. "Force is a direct *object* of sense," write Sir William Thomson and Professor Tait.[1] Force "is not a term for anything objective," writes Professor Tait.[2] In the face of such contradictions, is it not better to cease supposing that any lucid explanation of the why of motion can be abstracted from the idea of force?

But may not our particles, like two dancers, *hold hands*, and so the one "enforce" the other's motion ? We must not say that this holding hands is impossible, although they be 90,000,000 miles apart. We conceive

[1] *A Treatise on Natural Philosophy*, part i. p. 220. Cambridge, 1879.
[2] *The Properties of Matter*. Edinburgh, 1885.

light as easily traversing those 90,000,000 miles by aid of the ether, and may not our particles hold hands by means of the ether? All scientists hope that this may be so, at any rate conceptually, although they have not yet conceived how it can be so. But if we phenomenalized the ether and were able to describe by aid of it action at a distance of millions of miles, we should still be left with the problem : Why does the relative position of two adjacent parts of ether influence the motion of those parts? It might seem at first sight easier to explain why two adjacent ether elements "move each other" than why two distant particles of matter do. The common-sense philosopher is ready at once with an explanation : They *pull* or *push* each other. But what do we mean by these words? A tendency when a body is strained to resume its original form ; a tendency in a certain relative position of its parts to a certain relative motion of its parts. But why does this motion follow on a particular position? It is the old problem over again, with the difference that relative position now involves small instead of large distances. It will not do to attribute it to the *elasticity* of the medium ; this is merely giving the fact a *name*. We do indeed try to describe the phenomenon of elasticity conceptually, but this is solely by constructing elastic bodies out of *non-adjacent* particles, the changes of position of which we associate with certain relative motions. In other words, to appeal to the conception of elasticity is only to "explain" one "action at a distance" by a second "action at a distance." If the ether-elements owe their elasticity to such an arrangement, we shall want another ether to "explain" the motion of the first, and the process will have to be continued *ad infinitum.* Clearly the phenomenalization of the ether is absolutely useless as a means of explaining why matter moves. It still leaves us with the same problem in another form : Why does ether-matter move ? And here no answer can be given. We cannot proceed for ever "explaining" mechanism by mechanism. Those who insist on phenomenalizing mechanism must ultimately say : "*Here we are ignorant,*" or, what is the same thing, must take refuge in matter and force. According to Paul du Bois-Reymond, the problem of action at a distance is the third *Ignorabimus,*[1] but the problem is really identical with that of Emil du Bois-Reymond's first *Ignorabimus,* the nature of matter and force.

It seems to me that we are ignorant and shall be ignorant just as long as we project our conceptual chart, which symbolizes but is not the world of phenomena, into that world ; just as long as we try to find realities corresponding to geometrical ideals and other purely conceptual limits.. So long as we do this we mistake the object of science, which is not to explain but to describe by conceptual shorthand our perceptual experience. When we once clearly recognize that change of sense-impression is the reality, motion and mechanism the descriptive ideal, then the Brothers du Bois-Reymonds' first and third problems, and their cry of *Ignorabimus* become meaningless. Matter and force and "action at a distance" are witch - and - blue - milk problems (p. 27), if mechanism be purely a conceptual description. What moves in conception is a geometrical ideal, and it moves because we conceive it to move. *How* it

[1] See the work cited on our p. 46.

moves becomes the all-important question, for it is the means by which we regulate our mechanism so as to describe our past and predict our future experience. This *how* of motion is the point to which we must next turn. The laws of motion in the widest sense embrace all physical science—perhaps it were not too much to say all science whatever. All laws, von Helmholtz tells us, must ultimately be merged in laws of motion. Even such a complex phenomenon as that of heredity is at bottom, Haeckel holds, a transference of motion. Strong in her power of describing *how* changes take place, Science can well afford to neglect the *why*. She may not go so far as to fully accept even Emil du Bois-Reymond's second *Ignorabimus*, so long at least as psychology stands where it does ; but as to what consciousness is and why there is a routine of sense-impressions she is content for the present to say, "*Ignoramus.*"

SUMMARY.

The notion of matter is found to be equally obscure whether we seek for definition in the writings of physicists or of "common-sense" philosophers. The difficulties with regard to it appear to arise from asserting the phenomenal but imperceptible existence of conceptual symbols. Change of sense-impression is the proper term for external perception, motion for our conceptual symbolization of this change. Of perception the questions "what moves" and "why it moves" are seen to be idle. In the field of conception the moving bodies are geometrical ideals.

Of the du Bois-Reymonds' three cries of *Ignorabimus*, only the second in a modified sense is scientifically valuable, the others are unintelligible, because we find that matter, force, and "action at a distance" are not terms which express real problems of the phenomenal world.

LITERATURE.

Bois-Reymond, Emil du.—Ueber die Grenzen des Naturerkennens. Leipzig, 1876.

Clerk-Maxwell, J.—Articles " Atom " and " Ether " in the Encyclopædia Britannica, reprinted in the Scientific Papers, vol. ii. pp. 445 and 763. The article on the " Constitution of Bodies " may also be consulted with advantage.

Clifford, W. K.—Lectures and Essays, vol. i. (" Atoms " and " The Unseen Universe "). London, 1879.

Tait, P. G.—Properties of Matter (especially chaps. i.–v.). Edinburgh, 1885.

Thomson, Sir William.—Popular Lectures and Addresses, vol. i. (especially pp. 142–52). London, 1889.

CHAPTER VIII.

THE LAWS OF MOTION.

§ 1.—*Corpuscles and their Structure.*

IN the last chapter we have seen how the physicist conceptually constructs the universe by aid of a vast atomic dance. I use the word *atom* although it is very probably the ultimate element of the ether, which we ought to talk about as the fundamental unit of the dance. Let us term this latter unit the *ether-element*, without intending to assert by the use of this word that the ether is necessarily discontinuous.[1] Two adjacent ether-elements will be the symbols, necessarily geometrical, by which we represent the relative motion of the parts of the ether. On the basis of the ether-element let us try and conceive how the physicist imagines his mechanical model of the universe constructed. Perceptual experience gives us no hint as to what we ought to conceive the ether-element to consist of, or how we ought to imagine it to act, if it could be isolated. But we are compelled to consider ether-elements when in each other's presence as moving in certain definite modes, as taking part in a regulated dance. Perceptually

[1] If we suppose the ether to be a conceptual limit to a perceptual fluid or jelly (pp. 313 and 328), then to conceptualize at all its transmission of stress or its elasticity we are, I think, compelled to suppose it discontinuous.

there is no reason for this dance, conceptually it enables us to describe the world of sense-impressions.

Probably, although this point is far from being definitely settled, one type of motion among the ether-elements may be conceived as constituting the prime-atom. These prime-atoms, the *protyle* of Crookes, are to be taken as symbols of the ultimate basis of material groups of sense-impressions, or, in ordinary language, of gross or sensible "matter." Prime-atoms in themselves, or, what is more likely, in groups, form the atom of the chemist, the conceptual substratum of the so-called simple elements such as hydrogen, oxygen, iron, carbon, &c., by aid of which the chemist classifies all the known heavy matter of the physical universe. If the prime-atom of the physicist is really the atom of the chemist, then the prime-atom must be conceived as having variations either in its structure or in its type of motion corresponding to the different chemical elements. There are certain perceptual facts, however, which suggest that we should describe phenomena best by conceiving the atom of the simple chemical element to be constructed from groups of prime-atoms, the disassociation of which corresponds to no definite perceptual results which the chemist has hitherto succeeded in attaining. Out of the atoms of the simple elements the chemist constructs *compounds ;* that is, by combining conceptually these atoms in certain groupings he forms the *molecule* of the compound. Thus two atoms of hydrogen and one of oxygen are united to form the molecule of water. Any portion of the compound substance itself is conceived as composed of an immense number of molecules. In order to describe the sense-impressions which we physically associate with a "piece of a given substance" we are bound to postulate that the smallest physical element of it is to be considered as containing millions of molecules.[1]

If we take a piece of any substance, say a bit of chalk, and divide it into small fragments, these still possess the properties of chalk. Divide any fragment again and again, and so long as a divided fragment is perceptible by aid of the microscope it still appears chalk. Now the physicist is in the habit of defining the smallest portion of a substance which, he conceives, could possess the physical properties of the original substance as a *particle*. The par-

---

[1] The reasons for this statement are chiefly drawn from the Kinetic Theory of Gases. Clerk-Maxwell in his article "Atom" (*Encyclopædia Britannica*) considers that the *minimum visibile* of the present day may be conceived as containing sixty to one hundred million atoms of oxygen or nitrogen. He proceeds to draw from this result conclusions, which I think quite unwarranted, as to our power of describing by aid of molecular structure the physiological facts of heredity. He remarks that : "Since the molecules of organised substances contain on an average fifty of the more elementary atoms, we may assume that the smallest particle visible under the microscope contains about two million molecules of organic matter. At least half of every living organism consists of water, so that the smallest living being visible under the microscope does not contain more than about a million organic molecules. Some exceedingly simple organism may be supposed built up of not more than a million similar molecules. It is impossible, however, to conceive so small a number sufficient to form a being furnished with a whole system of specialized organs."

This reasoning is simply a form of special pleading based on the assumption that variations in physiological organs depend *solely* on chemical constitution and not on physical structure. Why are we to put on one side the facts that there are upwards of fifty atoms in the organic molecule, that there *is* a certain proportion of water, and that these organic molecules must be conceived as *closely* packed into a scarce visible germ? Why are these one hundred million atoms not to be conceived as physically influencing each other's *motion* ? If this be so,

ticle is thus a purely conceptual notion, for we cannot say when we should reach the exact limit of subdivision at which the physical properties of the substance would cease to be. But the particle is of great value in our conceptual model of the universe, for we represent its motion by the motion of a geometrical point. In other words, we suppose it to have solely a motion of translation (pp. 237 and 246) ; we neglect its motions of rotation and of strain. The physicist has here reached a purely conceptual limit to perceptual experience ; he takes a smaller and smaller element of gross " matter," and supposing it always to be of the same substance (*i.e.*, to produce the same sense-impressions although it becomes imperceptible), he deals with it as a moving point. What right has the physicist to invent this ideal particle ? He has never perceived the limiting quantity, the *minimum esse* of a substance, and therefore cannot assert that it would not produce in him sense-impressions which could only be described by aid of the concepts spin and strain. The logical right of the physicist is, however, exactly that on which all scientific conceptions are based. We have to ask whether postulating an ideal

then their relative position, the structure of the germ as a dynamical system, may be shown to involve no less than 10,000 million million periodic motions, having various relative positions in space, and apart from this relative position having in amplitude, phase, and "note," three hundred million variables at the disposal of the physiologist ! Whether heredity can or cannot be described by the influence of such a molecular structure on other molecules is quite beyond our present scientific knowledge to determine ; but we certainly cannot dogmatically assert with Maxwell that : " Molecular science sets us face to face with physiological theories. It forbids the physiologist from imagining that structural details of infinitely small dimensions can furnish an explanation of the infinite variety which exists in the properties and functions of the most minute organism ."

of this sort enables us to construct out of the motion of groups of particles those more complex motions by aid of which we describe the physical universe. Is the particle a symbol by aid of which we can describe our past and predict our future sequences of sense-impressions with a great and uniform degree of accuracy? If it be, then its use is justified as a scientific method of simplifying our ideas and economizing thought.

The reader must note that this hypothesis of the particle is made use of by Newton in the statement of his law of gravitation : " Every *particle* of matter in the universe attracts every other *particle*," he tells us, in such and such a manner. Yet Newton is here dealing with conceptual notions, for he never saw, nor has any physicist since his time ever seen, individual particles, or been able to examine how the motion of two such particles is related to their position. The justification of the law of gravitation lies in the power it gives us of constructing the motion of the groups of particles by aid of which we symbolize physical bodies and ultimately describe and predict the routine of our sense-impressions. The particle, therefore, as the symbolic unit of physical substance with its simple motion of translation is as valid as the law of gravitation, in the statement of which it is indeed involved.

Lastly, groups of particles bounded in conception by continuous surfaces are the symbols by which we represent those material groups of sense-impressions that are currently spoken of as physical bodies or objects. To find the simplest possible types of relative motion for these various concepts, and thence to construct the motion of the geometrical forms by which we symbolize physical bodies, so that the motion describes to any required degree of accuracy our routine of sense-impressions, is the scope of physical science. We find that by assuming certain laws for the relative motion of these conceptual symbols—the laws of motion in their widest sense—we are able to construct a world of geometrical forms moving in conceptual space and time, which describe with wonderful exactness the complex phases of our perceptual experience.

§ 2.—*The Limits to Mechanism.*

Let us now resume the elements of our conceptual model of the physical universe in a purely *diagrammatic* manner.[1] An asterisk shall represent the ether-

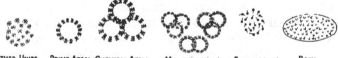

ETHER-UNITS. PRIME ATOM. CHEMICAL ATOM MOLECULE (= *ι*) PARTICLE (= *ν*) BODY

FIG. 21.

element, a ring of asterisks will suggest the prime-atom probably constructed from a special ether-element motion—for example, a vortex-ring. One, two, or more-prime atoms form the chemical atom, and for its symbol we will take three interlaced rings. Combinations of chemical atoms form the molecule, in our diagram represented by two chemical atoms of three and one of two prime atoms. Millions of these molecules, of which we can only represent a few by the shorthand symbol *ι*, would form the particle (shorthand symbol *ν*), while millions of particles, here

---

[1] The diagram is only to suggest the physical relationships to the reader, and has no meaning from the standpoint of relative size or form.

merely suggested, conceptually enclosed by a continuous surface, symbolize the physical bodies of our
perceptual experience. These concepts, it must be
borne in mind, from ether-element to particle, have no
perceptual equivalents, and it is only by experiments
on the perceptual equivalent of the last of the series,
the conceptual body, that the physicist is able to
test the truth of the laws of motion he propounds.

In the first place he postulated these laws for particles, and demonstrated their validity by showing
that they enabled him to describe the routine of his
sense-impressions with regard to physical "bodies."
But with the growth of our ideas as to the nature of
ether and gross "matter," we naturally begin to question
whether the laws which describe the relative motion
of two particles are to be conceived as holding for
two molecules, two chemical atoms, two prime-atoms,
and ultimately for two ether-elements. Or, what may
possibly be still more important, are they to hold for
the relative motion of a prime-atom and adjacent
ether-elements? How far are we to consider the laws
of motion as applied to particles of gross "matter" to
result from the manner in which particles are built up
from molecules, molecules from atoms, and ultimately
atoms probably from ether-elements? Now this is a
very important issue, and one which does not appear
to have always been sufficiently regarded. If we
assume that the particle is ultimately based on a
certain type of ether-motion, then we must admit the
existence of other types of ether-motion which do not
constitute gross "matter." In this case it will by no
means follow that the relative motion of two particles,
or of two prime-atoms, will follow the same laws as
the relative motion of two ether-elements. It is quite

clear, of course, that modes of motion peculiar to gross " matter " must arise from its special structure, and not be assumed to flow from laws applying to *all* moving things. For example, gravitation, magnetization, electrification, the absorption and emission of heat and light are all phases of sense-impression which we associate with gross " matter," and therefore they must be described by modes of motion characteristic of gross "matter," or modes which flow from its peculiar constitution. As kinetic formulæ or special laws of motion they cannot be extended to the ether in general. But there are still more general laws of motion, which we may describe as the Newtonian laws, and which certainly when applied to particles are confirmed by our perceptual experience of bodies. Ought we to assert that these laws hold in their entirety for all the scale from particle to ether-element ? Shall we find our conceptual description of the universe simplified, or the reverse, by sup posing complete *mechanism* to extend from particle to ether-element ? Or will it be more advantageous to postulate that mechanism in whole or part flows from the ascending complexity of our structures, that the ether-element is largely the *source* of mechanism, but is not completely mechanical [1] in the sense of obeying the laws of motion as given in dynamical text-books ? The question is undoubtedly an important one, but one which cannot be answered offhand. Nor, indeed, till we have much clearer conceptions of the structure of the prime-atom than we have at present reached, will it be possible to say how

[1] For example, as will be shown in the sequel, the " mass " of a particle must be considered as in all probability very different from the " mass " of an ether-element (p. 368).

far the mechanism we postulate of particles may be conceived to flow from its structure.

In order to remind the reader that the general laws of motion we are about to discuss may either entirely or only in part hold for the whole series of physical concepts from particle to ether-element, we will class the whole series together as *corpuscles*, a word simply signifying little elementary bodies. We shall then have to ask in each case to which of the ideal corpuscles we are to suppose our laws to apply. The test will always be the same, namely : How far is the assumption necessary in order to obtain a model which will enable us to describe briefly the routine of perception ?

§ 3.—*The First Law of Motion.*

Let us now return to our conception of the universe as the regulated dance of the elemental groups which we have termed prime-atoms, chemical atoms, molecules, and particles. Individual corpuscles dance in groups, groups dance round groups, and groups of groups dance relatively to each other. *How*, we have next to ask, do two corpuscles dance with regard to each other ? In the first place we must observe that, at least in the case of gross " matter ", a corpuscle which is conceived as forming part of the sun must be considered as regulating its dance with due regard to a corpuscle forming part of the earth. We cannot assert that it would not be best to conceive this as really done through a chain of partners, namely, ether-elements intervening between the sun and earth corpuscles, but as we have not yet settled how this chain of partners is to act, we must content ourselves at present by the statement that sun and earth corpuscles do regard each other's presence. But if they can do this at 90 million miles, there is every reason for inferring no breach in continuity and supposing they would also do it at 90 billion miles. We note, however, at once that it is necessary to conceive a particle at the surface of the earth paying more attention in its dance to an earth particle than to a sun particle, and again the phenomenon of cohesion tells us that two adjacent particles of the same piece of substance pay more heed to each other than particles of different pieces. Hence we conclude that : (1) in general terms corpuscles must be conceived as moving with greater regard to their immediate partners in the dance than to their near neighbours, and with greater regard to near neighbours than to still more distant corpuscles ; but, (2) there is no limit to the distance at which we conceive corpuscles can influence each other's motion. . This influence may, however, be so small that even when summed for the bodies that we construct from corpuscles, there is no perceptual equivalent to be found for it by aid of any instrument at our disposal. We can now state a first general law of motion :—

Every corpuscle in the conceptual model of the universe must be conceived as moving with due regard to the presence of every other corpuscle, although for very distant corpuscles the regard paid is extremely small as compared with that paid to immediate neighbours.

If the reader once grasps that every corpuscle in the universe must be conceived as influencing the motion of every other corpuscle, he will then fully appreciate the complexity of the corpuscular dance by aid of which we symbolize the world of sense-impressions. The law of motion just stated probably applies to prime-atoms, and through them to chemical atoms, molecules, and particles. Possibly it does not apply to distant ether-elements directly, but these, perhaps, influence each other's motion only indirectly by directly influencing the motion of their immediate neighbours. In this case the "action at a distance" generally asserted of corpuscles of gross "matter," may very probably be conceived as due to the action between adjacent ether-elements. We should then have to state the first law as follows :—

Every corpuscle, whether of ether or gross " matter," influences the motion of the adjacent ether corpuscles, and through them of every other corpuscle, however distant; the influence thus spread is nevertheless very insignificant at great as compared with small distances.

§ 4.—*The Second Law of Motion, or the Principle of Inertia.*

Now, in constructing the universe conceptually from our corpuscles, it is impossible to take into account the influence of all the corpuscles upon each other at one and the same time. Accordingly we neglect at once influences which even in the aggregate are beyond our powers of measurement. Further, we purposely exclude from consideration slight, if measurable, variations of motion due to more distant groups. We isolate a particular group of corpuscles, and this group which we deal with conceptually apart from the rest we term, for the purposes of some particular discussion, the *field.*

The most limited field that we can conceive is that of a single corpuscle. If we could isolate such a corpuscle from the rest of the conceptual universe, how would it move ? At first sight the question is absurd, because in Chapter VI. (p. 247) we saw that motion is mean-ingless if it be not relative to something. The moment, however, we introduce a second corpuscle into the field in order to measure the motion of the first, they begin to pay regard to each other's presence, and we are no longer dealing with the motion of an isolated corpuscle. But we have seen that the greater the distance between the corpuscles, the less this influ-ence must be conceived to be ; hence we may take the conceptual limit by supposing that the corpuscles are so far off that their mutual influence is negligible, while their mutual presence will still suffice to mark a relative motion.[1] Now in order that the laws which govern the motion of corpuscles shall lead to the con-struction of complex motions, fully describing the phases of our perceptual experience, we are compelled to suppose that the more and more completely we separate one corpuscle from the influence of a second corpuscle, the more and more nearly does its motion relative to the second corpuscle cease to vary. The first corpuscle either remains at rest relatively to the second or continues to move with the same speed —the same number of miles per minute—in the same direction. But this is what we term uniform motion, or motion without acceleration (pp. 276–7), and we are thus endowing our corpuscles with a very important property, namely, we assert that they will not dance, that is, alter their motion, unless they have partners to dance with. This characteristic of cor-

[1] The reader must remember that relative position is conceptualized by a *directed* step and that it is a series of directed steps which form the path of the relative motion (p. 250). Each directed step is to be con-ceived as " fixed " in direction, *i.e.*, its points are to be considered as having no accelerations relative to each other. See Appendix, *Note I.*

puscles, that they do not alter their uniform motion except in the presence of other corpuscles, is scientifically termed their *inertia*.

With regard to this law of inertia it must probably be conceived as holding from the prime-atom to the particle, but a difficulty comes in when we consider ether-elements. If the prime-atom be a particular type of ether-motion, for example an ether vortex-ring or ether-squirt, then the very existence of the corpuscles of gross " matter " depends upon the presence of the ether-elements, not only in their own constitution but in their immediate neighbourhood. It becomes, therefore, hopelessly absurd to consider what a corpuscle of gross " matter " would do if it were isolated from the influence of ether-elements. The law of inertia for gross " matter " must then flow from the peculiar structure of gross " matter." The mutual presence of ether-elements and of an isolated prime-atom will then be seen to involve the inertia of the latter, but the ether-elements themselves will, while the prime-atom moves uniformly, be varying their motion with due regard to the presence of the prime-atom.[1] What the law of inertia is to be considered as meaning when applied to isolated ether-elements, it is again difficult to say. Possibly it is idle to inquire so long, at any rate, as the conceptual ether remains as little defined as at present. Our notions of the ether are so essentially bound up with the conception of its *continuity*, while our notions of gross

---

[1] For example, it may be shown that an *isolated* vortex-ring in an infinite fluid moves without sensible change of size with uniform velocity perpendicular to its plane ; on the other hand, the ether-elements vary their velocity according to their position relative to the ring (See A. B. Basset, *A Treatise on Hydrodynamics*, vol. ii. pp. 59–62).

" matter " are, on the other hand, so closely associated with the idea of the discontinuity of matter, that we are inclined to treat as fundamental for ether-elements the method in which they act in each other's presence, and for gross " matter " corpuscles the method in which they act when isolated. On this account the law of inertia, as we postulate it for gross " matter " corpuscles, may be considered as a feature of mechanism very probably flowing from the structure of the prime-atom itself.

§ 5.—*The Third Law of Motion. Acceleration is determined by Position.*

Let us now proceed a stage further and postulate the next simplest field ; let us suppose two corpuscles taken and their motions determined relatively (p. 250) to a third corpuscle which, however, like that on p. 343 we will consider to be at such a distance as to be quite isolated from their influence. What must we conceive as happening? In the first place, because two corpuscles are in the same field must we consider them as having a certain definite position relative to each other? Certainly not. We find ourselves compelled to consider them as capable of taking up a great variety of positions with regard to each other. Does, then, the fact that they are in the same field, or in a certain relative position in that field, determine with what velocities we are to consider them as moving? Again we must answer : No—at any rate for particles. In order to construct motions which will effectively describe our sequences of sense-impression we are forced to suppose that particles may move through the same relative position with every variety of velocity. What, then, must we consider as determined when we know the relative position of two corpuscles? It is their accelerations, the rates at which they are changing their relative position. *Two corpuscles may be moving through the same position with any velocities, but they will spurt and shunt each other's motions in a perfectly definite manner, depending on their relative position.*

If A and B represent two corpuscles moving (relative to the isolated third corpuscle) in the directions AT and BT' with the velocities V and V' given by the steps OQ and O'Q' of their respective hodographs (p. 263), then the spurt and shunt of V and V', or, as we have seen (p. 265), the velocities of Q and Q' along their hodograph paths will be deter-

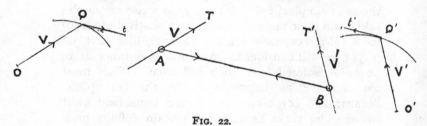

FIG. 22.

mined at each instant by the relative position of A and B. Let these velocities of Q and Q', or the accelerations of A and B be represented by the steps Qt and Q't' taken along the tangents at Q and Q' (pp. 259 and 265). Then the question naturally arises, How are we to consider the spurts and shunts given by Qt and Q't' (p. 268) to depend on the relative position of A and B? In the first place we conceive Qt and Q't' to be *parallel, but in opposite senses* (p. 248). We find it needful to suppose universally that the mutual accelerations of corpuscles have the same direction but opposite senses.[1] In the next place it is usually

[1] That is, if A spurts B in the direction from B toward A, then B will spurt A in the direction from A to B and *vice versâ*.

assumed that this direction is that of the line joining the points which represent the corpuscles A and B. Now this assumption is possibly correct enough [1] when we are dealing with particles of gross " matter," at any rate when we are discussing the motion of non-adjacent particles, or those for which we are not compelled to consider the distance AB vanishingly small like the dimensions of the particles themselves.[2] On the other hand there appear to be many physical and even chemical phenomena which cannot be described by replacing the motion of a prime-atom, chemical atom, or molecule by the motion of a point. In this case the line joining the two corpuscles becomes a meaningless term, and we have really to deal with the relative motion of groups of elements, constructed very probably from the motion of simple ether-elements.

When, however, we ask of ether-elements whether we are to consider them as mutually accelerating each other in the line joining them, we are at once stopped by the difficulty that we have reason for supposing non-adjacent ether-elements do not influence each other's motion at all (p. 342). But if we turn to adjacent ether-elements, the line joining them vanishes with the dimensions of the elements when we try to conceive the ether as absolutely continuous (pp. 213, 324 and 344). Discontinuity of the ether may carry us over

[1] See Appendix, *Note II.*

[2] It will be noticed in this case that if we take the motion of A relative to B, the ray and tangent to the path or orbit of A are respectively parallel to the tangent and ray to the hodograph or path of Q. This is expressed in technical language by saying that the orbit of such a motion is a link-polygon (funicular polygon) for the hodograph as a vector-polygon (force-polygon), and it forms the basis of a powerful graphical method of dealing with central accelerations.

this difficulty and allow us to consider ether-elements as mutually accelerating each other's motion in the direction of the line joining them, but such discontinuity reintroduces one of the problems which the conception of the ether was invented to solve (pp. 213 and 328). We may be quite safe in postulating that when an ideal geometrical surface is supposed drawn and fixed in the ether its *points* will have a motion relative to each other upon its form being changed ; the points of the surface will tend to return to their original positions with accelerations depending on their change of relative position.   But when we assert that this is due to ether-elements mutually accelerating each other's motion in the line joining them, we may, after all, be postulating a phase of mechanism for the ether which is only true for gross " matter," and which may indeed flow from the particular type of ether-motion which constitutes gross " matter."   If the prime-atom be a vortex-ring it would be impossible to describe in general the action between two prime-atoms as a " mutual acceleration in the line joining them."   On the other hand, if the prime-atom be an ether-squirt, this phrase would effectively describe the action between two prime-atoms.   In both cases the statement that particles mutually accelerate each other's motion in the line joining them would flow either as an absolute or an approximate law from the particular structure of gross " matter," and would not be a mechanical truth for all corpuscles from ether-element up to particle.

There are still several points to be noticed with regard to the nature of the manner in which corpuscles spurt and shunt each other's motion.   We have said that this depends on the relative position of the corpuscles—but is the mutual acceleration never influenced by the velocities of the corpuscles? Do two of our conceptual dancers influence each other solely by their relative position and never by the speed and direction with which they pass through that position? It has been supposed that the introduction of the relative velocity as a factor determining the mutual acceleration of two particles would be contrary to a well-established physical principle termed the conservation of energy. It is indeed a fact that many writers, from Helmholtz downwards, have given a mathematical *proof* of the conservation of energy which depends on mutual acceleration being a function of relative position and not of relative velocity. But if two moving bodies be placed in a fluid they will apparently accelerate each other with accelerations depending upon their velocities as well as on their relative position. The conservation of energy still holds in this case for the entire system of fluid and moving bodies, and yet to the observer unconscious of the fluid the mutual accelerations of the bodies would certainly appear to be determined by their velocities as well as by their position.[1] Something of this kind may well occur when we regard the action between corpuscles of gross " matter " without regard to the ether in which we conceive them floating. We cannot assume that the mutual accelerations of prime-atoms, chemical atoms, and molecules depends solely on their relative positions ; it may depend also on their velocities relative to each other or relative to the ether in which we suppose them to be moving.

[1] The ether being neglected, its unregarded kinetic energy appears as potential energy of the moving bodies, and is generally expressible in terms of the velocities of those bodies. Hence those bodies appear to have a mutual acceleration depending not only on their relative position but on their velocities.

This remark is of special importance when we try to describe electric and magnetic phenomena by the mutual accelerations of particles at a distance.

It is usually assumed by physicists, however, that the action between particles at a distance is to be considered as taking place in the line joining them and as depending only on relative position. There have not indeed been wanting scientific writers who have asserted that the whole universe could be described mechanically by aid of a system of particles or points, the mutual accelerations of which depended solely on their mutual distances. But simple as such an hypothesis would be, its propounders have hitherto failed to demonstrate its sufficiency.[1] Nevertheless it has played a great part in physical research, and its influence may still be seen in much that is written at the present time about the laws of motion and the conservation of energy.

The above discussion puts us in a better position for appreciating the statements that we may legitimately make with regard to the dance not only of two but of any number of corpuscles. In general we may assert that whether we are dealing with the continuous ether or with discontinuous atoms and molecules, then if we fix our attention on a geometrical point which symbolizes an element of ether, atom, or

[1] The impulse to this mode of describing the physical universe certainly arose from the Newtonian law of gravitation. It was perhaps pushed as far as it could possibly be of service in the writings of Poisson, Cauchy, and the great French analysts at the beginning of the century. Traces of its persistency may be still found in modern writers; for example we may cite Clausius—one of the most distinguished of modern German physicists—who considered that all the phenomena of nature can probably be reduced to points mutually accelerating each other in the lines joining them with accelerations which are functions only of their mutual distances (*Die mechanische Wärmetheorie*, Bd. i. S. 17).

molecule, the acceleration (*not* the velocity) of this point will depend on the position of this point or element relative to other points or elements (and possibly in certain cases on its velocities relative to those points or elements). For particles of gross " matter," on the other hand, we find it as a general (if not invariable) rule sufficient to assert that the mode in which their velocity is being spurted and shunted depends solely on their position relative to other particles. In particular, if two particles be alone in the field, their mutual accelerations will depend on their relative position and may be conceived as taking place in the line joining them, but in opposite senses.

§ 6.—*Velocity as an Epitome of Past History.   Mechanism and Materialism.*

There are one or two points in these statements which deserve special notice. If we avoid the metaphysical idea of force, and consider causation as pure antecedence in phenomena (pp. 155-6), then the cause of change of motion or acceleration must in our conceptual model of the phenomenal world be associated with *relative position.* The given velocities of a system at any time may be looked upon as the sum of the past changes of motion ; or the causes of a given motion can only be conceived as lying in the totality of all past relative positions of the system. Thus force, as the conceptual idea of moving cause, could only be defined as the history of the relative positions of a system.   This history determines the actual velocities of the parts of the system, while actual position determines how the velocities are actually changing.   The "actual position," however, is the conceptual equivalent of the mode in which we perceptually distinguish coexisting sense-impressions, while "past history" is the conceptual equivalent of the perceptual sequence in sense-impressions. "Actual position" and "past history" taken in conjunction thus symbolize what we have termed the routine of perceptions (p. 122). We conclude, therefore, that if with Professor Tait and other metaphysical physicists we even project our conceptions into the perceptual sphere, we still shall not find in "force," as either the cause of motion, or the cause of change in motion, anything more than that routine of perceptions which we have already seen is the basis of the scientific definition of causation (p. 153).

The idea that the past history of a corpuscle is resumed in its present velocity is an important one. If we knew the actual velocities of all existing corpuscles and how their accelerations depend on relative position (or it may be also on relative velocity), then *theoretically*, by aid of the process indicated on our p. 278, or by an extension of this process, we should be able to trace out the whole of the past, or, on the other hand, the whole of the future history of our conceptual model of the universe. The data would be sufficient to theoretically solve these problems, although our brains would be quite insufficient to manipulate the necessary analysis. Portions of it they do, however, manage. From the present velocities of earth and moon and their known accelerations relative to the sun and to each other, we calculate the eclipses of two or three thousand years ago, and rectify our chronology by determining the dates of eclipses which are recorded in the history of past human experience. Or, again, from thermal or tidal data we describe the condition of the universe as we conceive it to have been millions of years back, or as we conceive it will be millions of years hence. In all such cases we consider that because our conceptual model describes very accurately our limited perceptual experience of past and present, it will continue to do so if we apply it to describe sequences which cannot be verified as immediate sense-impressions. In this case we are clearly making inferences, but inferences which are logically justifiable (pp. 72 and 420) ; we assume that because our conceptual model describes very accurately our immediate perceptual experience, it would also describe the antecedents and consequents of that experience, did they exist perceptually ; it is logical to infer when we see the panorama of a river, one portion of which accurately depicts all we know of the River Thames, that the rest of the panorama depicts parts of the *same* river, with which we are unacquainted. In the necessarily limited verifiable correspondence of our perceptual experience with our conceptual model lies the basis of our mechanical description of the universe. As a shorthand *résumé* of our perceptual experience, and as a co-ordination of that experience with stored sense-impresses, the only objective element of this mechanical theory is seen to lie in the similar perceptive and reasoning faculties of two human minds. Thus the sole support of that materialism which, " proceeding from the fixed relation between matter and force as an indestructible basis," finds " mechanical laws inherent in the things themselves," collapses under the slightest pressure of logical criticism.[1]

[1] The chief German representatives of this materialism are J. Moleschott and L. Büchner, and it has found its warmest supporters in Eng

§ 7.—*The Fourth Law of Motion.*

It is needful, however, that we should return to our discussion on the laws of motion, and, assuming for the present that relative position is the principal factor in the determination of mutual accelerations, we must ask what more exact laws may be postulated with regard to these accelerations. We have in the first place to investigate how far the *individuality* of the dancers is to be conceived as influencing the manner in which they spurt each other's motion. Do *any* two dancers, whatever their race and family, and under whatever surroundings they may meet, always dance in the same fashion whenever they come to the same position ? Or must we consider it necessary to classify our corpuscles by some scale which may itself indeed change with a change in the field ? Again, are two dancers to be conceived as dancing in the same manner whatever aspect (p. 240) they bear to each other, whether they come to the same position face to face, or back to back, as it were ? Lastly, if we know how A and B influence each other's motions when they are alone in the field, and how A and C dance when alone together, shall we be able to tell how A will act in the presence of *both* B and C? Here are a number of ideas which we must try and express in scientific language with the view of determining what answers are to be given to the problems they suggest.

In the first place we ask the question:—

land among the followers of the late Mr. Bradlaugh. It is perhaps needless to add that the gifted lady, who speaks of secularists as holding the "creed of Clifford and Charles Bradlaugh," has failed to see the irreconcilable divergence between the inventor of " mind-stuff " and the follower of Büchner.

Is there any relation between the mutual accelerations of two corpuscles A and B, which is independent (1) of their relative position, and (2) of their possible companions in the field ? Is there any relation, in fact, which depends on the *individualities* of the corpuscles A and B ?

This problem may be termed that of the *Kinetic Scale*.[1] Let us see how we might solve this problem ideally. We might take two corpuscles and put them at different distances in a field in which they alone exerted influence, and we might measure their mutual accelerations. Then we might repeat this process with other corpuscles in the field,[2] and vary the field itself in every possible manner. We should thus obtain two series of numbers, the one series representing the acceleration of A due to B,[3] and the other the acceleration of B due to A. In the sphere of conception we should then be applying the scientific method of classifying facts and trying by careful examination of these facts to discover a law or formula by aid of which they might be described. And we should very soon find a fundamental relation between these mutual accelerations of A and B. Returning to our Fig. 22, we should discover that the number of units of length in Qt (if this represents

---

[1] *Kinetic* is an adjective formed from Greek κίνησις, a *dance*, a movement ; the kinetic scale signifies a scale of movement.

[2] The manner in which the part of A's acceleration due to B might be separated from that due to the other corpuscles in the same field cannot be fully discussed in this work. In many cases it could be discriminated by aid of the parallelogram of accelerations (p. 282).

[3] By the expression "acceleration of A *due* to B," frequently used in this chapter, the reader is not to understand that B *enforces* A's change in motion. The term is solely used as shorthand for the conceptual idea that A and B, when in each other's presence, are to be considered as changing their relative motions in a certain manner.

the acceleration of A due to B) was always in a constant ratio to the number of units of length in $Q't'$ (or the acceleration of B due to A). If Qt were 7 units and $Q't'$ 3 units, then whatever other corpuscles were brought into the field, or however the relative position of A and B might be altered, still Qt and $Q't'$, be they both large or both small, would always have the ratio of 7 to 3. Now here is the beginning of the answer to our first question, and we may state our immediate conclusion in the following words :—

The ratio of the acceleration of A due to B to the acceleration of B due to A must always be considered to be the same whatever be the position of A and B, and whatever be the surrounding field.

The ratio of mutual accelerations is thus seen to depend on the individual pair of dancers, and not on their relative position, or the presence and character of their neighbours.

But the reader may ask: How can science possibly have drawn such a wide-reaching conclusion as this, since even the most metaphysical of physicists has never caught one corpuscle, let alone two, and could not therefore have experimented upon them in every possible field. The answer is of the same character as that to the problem of the gravitating particles (p. 336). Physicists have experimented on perceptual bodies in all sorts of fields ; they have electrified, magnetized, warmed, or mechanically united by strings or rods bodies of finite dimensions ; but, whatever the nature of the field, they have found that the smaller the bodies—the more nearly they approached the conceptual limit of particle,—the more nearly they have been able to describe the sequence of their sense-impressions by aid of conceptual particles obeying the above law. They then postulated the above law as true for particles, and, inverting the process, proceeded by aid of this law to describe the motion of those aggregates of particles which are our symbols for perceptual bodies. The validity of the law was then demonstrated by the power it was found to give us of predicting the future routine of our sense-impressions with regard to perceptual bodies. Once established as a mechanical principle for particles, it was natural to investigate whether its application to the whole range of corpuscles would give results in agreement with our perceptual experience. In so far as it did so, it became recognized as a universal law of mechanism. This process of discovering and then justifying the conceptual law by aid of our perceptual experience applies to all our further statements with regard to the laws of motion, and I shall not think it necessary for my present purposes to refer in each individual case to the experimental discovery and justification.

§ 8.— *The Scientific Conception of Mass.*

This fourth law of motion carries us a long way in our description of the dance of corpuscles, but I have now to ask the reader to follow me in a rather more difficult investigation. This will, however, eventually repay us by the number of new ideas to which it introduces us. As the fourth law stands at present we should have to make experiments on every possible pair of corpuscles in order to form a scale of the ratios of their mutual accelerations. In order to avoid this very laborious process we conceive a standard corpuscle taken, which we will represent by the letter Q, and we suppose a record formed of the ratio of the mutual accelerations of Q and of each of the other corpuscles with which we populate conceptual space.

By the third law of motion the acceleration of Q due to A will always be in the same ratio to the acceleration of A due to Q, whatever be the field. Now we are going to give a name to this ratio; we shall call it the *mass* of A relative to the standard Q, or more simply the mass of A. Thus we have :—

$$\text{Mass of A} = \frac{\text{Acceleration of Q due to A}}{\text{Acceleration of A due to Q}} \quad . \quad (a).$$

And similarly, if B be a second corpuscle, we have :—

$$\text{Mass of B} = \frac{\text{Acceleration of Q due to B}}{\text{Acceleration of B due to Q}} \quad . \quad (\beta).$$

This definition leads us to two important points. We see, namely, that the mass of a corpuscle has relation to some standard corpuscle, or mass is always a *relative* quantity ; and, further, mass is a mere number representing a ratio of accelerations. We have here, then, a perfectly clear and intelligible definition ; we can grasp what velocity means, and we can understand how its change is measured by acceleration. Mass, accordingly, as the ratio of the numbers of units in two accelerations, is a conception which can easily be appreciated. It is in this manner that mass is invariably determined scientifically, yet nevertheless the reader will frequently find mass defined in text-books of physics as " the quantity of matter in a body." After our discussion of matter in Chapter VII. the reader will easily appreciate how idle is a definition of mass in terms of matter.[1]

[1] *Quantity* belongs essentially to the sphere of sense-impression. We cannot consider it to have any meaning when projected beyond that sphere. It seems, therefore, illogical to apply the word quantity to the metaphysical " source " of sense-impressions.

§ 9.— *The Fifth Law of Motion. The Definition of Force.*

We can now pass to the next stage in our investigation of the corpuscular dance. Having selected a standard corpuscle Q, we conceive the masses relative to it of many other corpuscles—A, B, C, &c.—measured. If we tabulated these masses and then compared them with the ratio of the mutual accelerations of A and B, B and C, C and A, &c., with a view of ascertaining whether there were any relation between the mutual accelerations of each pair and their masses, we should very soon discover a fifth important law of motion, namely, *that the ratio of the acceleration of* A *due to* B *to the acceleration of* B *due to* A *is exactly equal to the ratio of the mass of* B *to the mass of* A, or in simple algebraical notation :—

$$\frac{\text{Acceleration of A due to B}}{\text{Acceleration of B due to A}} = \frac{\text{Mass of B}}{\text{Mass of A}} \quad . \quad . \quad (\gamma).$$

This is expressed briefly by the statement that mutual accelerations are *inversely* as masses. The validity of this statement is demonstrated in precisely the same manner as the fourth law of motion. We note that if unity be taken as representing the mass of the standard corpuscle,[1] Q, the definition of mass on p. 358 may be replaced by the formula :—

$$\frac{\text{Acceleration of Q due to A}}{\text{Acceleration of A due to Q}} = \frac{\text{Mass of A}}{\text{Mass of Q}} \quad . \quad . \quad (\delta),$$

a result in perfect accordance with the law just stated.

Now this law may be put into a slightly different form. By a well-known proposition [2] the product of

---

[1] That is, the ratio of the mutual accelerations of Q and an absolutely identical corpuscle. These accelerations must by symmetry be exactly equal, and hence their ratio, the mass of Q, must be taken as unity.

[2] *Euclid*, vi. 16, interpreted arithmetically.

the means in any proportion is equal to that of the
extremes.   Hence it follows that :—

Mass of A × Acceleration of A due to B is equal to
Mass of B × Acceleration of B due to A.

We will, then, give a name to this product of mass
into acceleration ; we will term the product of the
mass of A into the acceleration of A due to the
presence of B, the *force of* B *on* A.   This force will
be considered to have the direction and sense of the
acceleration of A due to B, while its magnitude will
be obtained by multiplying the number of units in the
acceleration of A due to B by the number of units in
the mass of A.   Thus the proper measure of a force
will be its number of units of mass-acceleration.
Remembering that the accelerations of A and B are
of opposite sense, we can now restate our fifth law in
new language, thus :—

The force of B *on* A *is equal and opposite to the force
of* A *on* B ;

Or, as it was originally stated by Newton himself :—

"Action and Reaction are equal and opposite"[1] . . (ε).

Now it is clear that with our definition force is
a certain measure of *how* a corpuscle is dancing
relative to a second corpuscle, this measure depending partly on the individual character of the first
corpuscle (its mass) and partly on the attention it is
paying to the presence of a second corpuscle (its
acceleration due to the second corpuscle).   That
this measure is scientifically a convenient one is
proven by its general use, and may be almost fore-
seen by comparing the simplicity of the statement

---

[1] *" Actioni contrariam semper et æqualem esse reactionem."*

(ε) with the complexity of (γ). The definition of force we have reached is a perfectly intelligible one ; it is completely freed from any notion of matter as "the moving thing," or from any notion of a metaphysical "cause of motion." We have only to take the step which represents the acceleration of A due to B's presence and to stretch or magnify its length in the ratio of A's mass to the mass of the standard body Q, and we have a new step which represents B's force on A. Force is accordingly an arbitrary conceptual measure of motion without any perceptual equivalent.

The distinction between the definition of force thus given and that to be found in the ordinary text-books [1] may at first sight seem slight to the reader, but the writer ventures to think that the distinction makes all the difference between an intelligible and an unintelligible theory of life, between sound physical science and crude metaphysical materialism. Causation, as we have had occasion more than once to point out, is only intelligible in the perceptual sphere as antecedence in a routine of sense-impressions. In the conceptual sphere, on the other hand, the cause of change in the motion of our corpuscles lies solely in our desire to form an accurate mechanical model of the world of phenomena. For every definite configuration of the corpuscles we postulate certain mutual accelerations as a mode of bringing our mechanism into tune with our sense-impressions of

---

[1] " Force is any cause which tends to alter a body's natural (*sic !*) state of rest, or of uniform motion in a straight line " (Tait's *Dynamics of a Particle*, art. 53). It is perhaps unnecessary to remark that we cannot conceive any body to be *naturally* at rest or moving in a straight line, unless the word *natural* be re-defined in some artificial sense.

change. Force as an arbitrary measure of these conceptual changes in motion is intelligible. On the other hand, to project the cause of motion into something behind sense-impression is to dogmatically assert causation where we cannot know, to illogically infer from the like to the unlike (pp. 72, 186). The only alternative is to consider force as an antecedent group of sense-impressions; this, however, is not only to project our purely conceptual notions of motion into the perceptual field, but it throws upon us the duty of defining the particular group of sense-impressions to which force corresponds. We have already spoken of the "muscular sensation of force" (p. 327), which, if we project conceptions into the perceptual field, is more accurately to be described as a sense-impression of mutual acceleration indissolubly linked to the fact of consciousness. It throws absolutely no light on the cause of motion in such "automata without consciousness," as we must conceive "phenomenal corpuscles" to be. Hence, whichever way we turn, the current definitions of both mass and force lead us only into metaphysical obscurity. Mass as the quantity of matter in a body, matter as that which perceptually moves, force as that which changes its motion, are solely and purely names which serve to cloak human ignorance. This ignorance is at bottom the ignorance of *why* there is routine in our sense-impressions, and with this question of routine we have already fully dealt (pp. 122-8). But science answers no *why*—it simply provides a shorthand description of the *how* of our sense-impressions ; and it therefore follows that if mass and force are to be used as scientific terms they must be symbols by aid of which we describe this *how*. It is thus that

I have dealt with them; we have seen that to briefly describe the corpuscular dance, which forms our conceptual model of the universe, the notions of mass and force as based on mutual accelerations arise naturally and with intelligible definitions.

§ 10.—*Equality of Masses Tested by Weighing.*

Although it is impossible for us to review the whole field of mechanics, it is still necessary to indicate to the reader that our definitions of mass and force would ultimately lead us to the same conclusions as he will find in current physical text-books. In the first place we will investigate an elementary problem which will lead us to a mode of testing the *equality of masses.* Suppose we had two particles A and B of masses m_a and m_b in the same field, and we will suppose them placed in a horizontal line, A to the left and B to the right. Now, owing to the presence of some system to the left of A, which we need not definitely describe, we will suppose A to have an acceleration represented by g units horizontally to the *left.* Similarly B, owing to some other system, shall have a horizontal acceleration of g units to the *right.* Further, A and B will mutually accelerate each other, and we will represent B's acceleration of A from left to right by the symbol f_{ba} and A's of B by f_{ab}, which will be in the opposite sense. We are going to choose a particular " physical field " for the acceleration of A and B ; they shall be linked together so that their distance cannot change, but the link itself shall be conceived as producing no accelerations in either A or B. We might conceptualize this link by aid of a limit to actual perception, namely, by a fine weightless and inextensible string. Such a

string would not in itself produce sensible accelerations in A or B.   Since the string is inextensible, the
whole system must move in the *same* direction, say
from right to left.   Then clearly the velocity of A
must be at all times equal to the velocity of B, or the
string would be stretched.   But if the velocities of A
and B are always equal, their accelerations must also

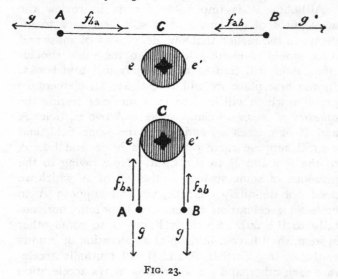

FIG. 23.

be equal, or their velocities, being differently spurted,
would begin to differ.   Hence we conclude that the
total acceleration of A towards the left must be equal
to the total acceleration of B in the same direction, or
in symbols :—

$$g - f_{ba} = f_{ab} - g \quad \ldots \ldots \ldots \quad \text{(i.).}$$

But by the fifth law of motion (*i.e.* (γ), p. 359)

$$\frac{f_{ba}}{f_{ab}} = \frac{m_b}{m_a} \quad \ldots \ldots \ldots \quad \text{(ii.).}$$

Thus (i.) and (ii.) are two simple relations to find f_{ba} and f_{ab}. By elementary algebra we have :—

$$f_{ab} = 2 \frac{m_a}{m_a + m_b} g, \text{ and } f_{ba} = 2 \frac{m_b}{m_a + m_b} g.$$

Hence we deduce :—

Acceleration of A or B to the left $= g - f_{ba} = \dfrac{m_a - m_b}{m_a + m_b} g$ . . (iii.)

Further :—

Force of B on A = mass of A × acceleration of A due to B.

$$= m_a \times f_{ba},$$
$$= 2 \frac{m_a \, m_b}{m_a + m_b} g,$$
$$= m_b \times f_{ab}, \text{ or Force of A on B.}$$

Now this force of B on A is what we usually term the *tension in the string*. Hence we have :—

Tension in the string $= 2 \dfrac{m_a \, m_b}{m_a + m_b} g$ . . . . . (iv.).

A further important point has now to be noticed. In order that A and B should be at rest relative to the field which produces the acceleration g, it will be necessary that their velocities should always be zero, and this involves that the changes in their velocities, or their accelerations, should always be zero. But the only way in which these accelerations can be zero is seen at once from (iii.) to arise from m_a and m_b or the masses of A and B, being equal, for then the difference, $m_a - m_b$ is zero. Thus rest will depend on the *equality of the masses* of A and B.

A further conceptual notion can now be introduced, namely, that the terminal physical effects—consequent sense-impressions—are not altered in magnitude, only in direction, by carrying a weightless inextensible string round any "perfectly smooth" body. This again is a purely conceptual limit to a very real perceptual experience. Now we will suppose our string placed round a perfectly smooth horizontal cylinder or peg inserted under it at its mid-point C, so that the portions eA, e'B of the string hang vertically downwards. We can further suppose that the particular systems, which produce the acceleration g in both A and B, are now replaced by the single system of the earth, for Galilei has demonstrated that all particles at the same place on the surface of the earth are to be conceived as having the same vertical acceleration (g) towards the surface. We conclude, therefore, that if two particles be connected by a weightless inextensible string placed over a perfectly smooth cylinder, the acceleration of one downwards and the other upwards is given by the relation (iii.) and the tension in the string by (iv.). Hence, if the particles are to be at rest, or to "balance each other," their masses must be equal. In this case, since $m_a = m_b$, the tension in the string equals $m_a \times g$, or equals the product of the mass of A into the acceleration of A due to the earth; that is, equals the *force of the earth on* A. This force is termed the *weight* of A, and since $m_a = m_b$, it follows that the weight of A is equal to the weight of B.

In this investigation, therefore, we have reached the simplest conceptual notion of a weighing-machine— an inextensible string, with the particles suspended from its extremities, placed over a smooth cylinder. If the weights of the particles are equal, their masses will also be equal, and they will balance. Thus equality of masses may be tested by *weighing*. Another important result also flows from this discussion. If a particle suspended by a string be at rest relative to the earth, then its weight will be equal to the tension in the string. Hence, if the earth-acceleration g at any place be known, we have a means of measuring mass in terms of tension. A further development of this principle forms the basis of important methods of determining the equality of masses by the equality of strains (p. 242) due to equal tensions.

§ 11.—*How far does the Mechanism of the Fourth and Fifth Laws of Motion extend?*

Before we conclude this discussion of mass, there are still several points with regard to it which must be elucidated even in an elementary work like the present. We have first to ask whether our fourth and fifth laws of motion, with the definitions of mass and force involved in them, must be conceived as holding for the whole range of corpuscles from ether-element to particle. The same difficulty, of course, arises with regard to force as arose with regard to acceleration, if we conceive prime-atoms as possibly, and chemical atoms and molecules as almost certainly, extended bodies. There cease to be definite points between which the mutual accelerations, and accordingly the forces, act. We are thrown back on the conception that if these laws are to be applied to atoms and molecules, it must be to the action and reaction between the elementary parts of those corpuscles and to the masses of the elementary parts that our laws refer. From the action of these elementary parts on each other we must, then, deduce by aid of the above laws the total action between two atoms or two molecules. This will not necessarily be measurable by a single force acting between two definite points.

Further difficulties, however, arise with regard to our conception of mass. Is the mass of an ether-element of the same character as the mass of an atom, or a molecule, or a particle ? This seems very doubtful indeed. If the ratios of the mutual accelerations of two ether-elements, of two atoms and of two particles be each in themselves constant and capable of leading us to a clear definition of mass for each type, it is still by no means certain whether the ratio of the mutual accelerations of an ether-element and a particle are inversely as the ratio of the ether-element mass to the particle mass. *Possibly we cannot conceive these masses measurable by the same standard.*

If the prime-atom consist of ether in motion, then its mass would certainly vanish with this motion ; but the ether-elements which formed the prime-atom would still retain their ether-mass. Hence it seems likely that the possibility of a velocity entering into the mass of gross " matter " may hinder us from asserting that the ratio of the mutual accelerations of ether-element and particle is "inversely as their masses." Thus the idea of mechanical action and reaction between ether and gross " matter " becomes very obscure. Of the validity of postulating these laws for particles there can be small doubt; they may possibly suffice to describe the relation of ether-elements to each other, but they cannot be dogmatically asserted of the action between ether and gross " matter." I have purposely led the reader to these difficult and still unsettled points, because physicists finding that certain laws of motion applied to par-

ticles will suffice to describe our perceptual experience
of physical bodies, are, I venture to think, too apt to
assert that these same laws hold throughout the
whole of the conceptual model by which they
describe the universe. They would admit that special
modes of acceleration like gravitation, magnetization, &c., &c., probably flow from the manner in
which the prime-atom and the particle are to be
conceived as constituted. But there may be more
than this to be admitted—the greater part of the laws
of motion as we state them for particles may also
flow from the peculiar structure of the particle.
They may largely result from the nature we postulate for the ether and from the particular types of
ether-motion by aid of which we construct the various
phases of gross " matter."

It is not, therefore, questioning the well-established
results of modern physics when we ask whether to
conceive the ether as a pure mechanism [1] is, after all,
scientific. The object of science is to describe in
the fewest words the widest range of phenomena, and
it is quite possible that a conception of the ether
may one day be formed in which the mechanism of
gross " matter " itself may, to a great extent, be resumed. Indeed, it is on these points of the constitution of the ether and the structure of the prime-atom
that physical theory is at present chiefly at fault.
There is plenty of opportunity for careful experiments to define more narrowly the perceptual facts
we want to describe scientifically ; but there is still
more need for a brilliant use of the scientific imagina-

---

[1] By a *pure* mechanism the writer means the reader to understand a
system which is conceived to obey *all* the fundamental laws of motion
as stated in mechanical treatises.

tion (p. 36). There are greater conceptions yet to be formed than the law of gravitation or the evolution of species by natural selection. It is not problems that are wanting, but the inspiration to solve them ; and those who shall unravel them will stand the compeers of Newton and Darwin.

§ 12.—*Density as the Basis of the Kinetic Scale.*

. If our mechanism as it is formulated in the above laws of motion can only be definitely asserted as true for particles, we have still to ask how the geometrical forms by which we symbolize perceptual bodies are to be conceived as constructed from particles, and how many different families of particles we are to postulate. Now in order to appreciate the answer to this question, we must define what we mean by *sameness of substance.* Suppose we take two portions of different bodies, or of the same body, and suppose we find these portions, however we test them, present to us the same groupings of physical and chemical sense-impressions, then we shall term these portions of the *same substance.* Further, if portions of a body, taken from any part of it whatever, always appear of the same substance, so that, if we could postulate exactly the same perceptions of shape, any one portion might be mistaken for any other, then we shall say that the body is *homogeneous.* Now although we cannot realize a particle in perception, still we conceive that if particles were to be formed by taking smaller and smaller elements from every part of such a homogeneous substance, all these particles would be of *equal mass.* We thus come to look upon our conceptual symbol for a homogeneous body as a uniform distribution of particles of equal mass throughout a geometrical surface. Applying our laws as to the motion of particles to such a uniform distribution of particles, we construct a motion for the geometrical form which closely describes our routine of sense-impressions in the case of those perceptual bodies which approximate to the conceptual ideal of homogeneity. We then define the sum of the masses of the particles contained in any portion of our geometrical form as the mass of this portion. From this it follows at once that : *the masses of any two portions of the same homogeneous substance are proportional to their volumes.*

This result is not a truism [1] ; it flows only from the uniform distribution of particles which we postulate for a homogeneous substance, and this distribution is a conception only justified, like the law of gravitation, by the results which it describes being in accordance with our perceptual experience. If we take two small and equal volumes of a homogeneous substance, then the smaller they are the more nearly we can describe our perceptual experience of them by the conceptual symbols, "particles of *equal* mass." If we take two small and equal volumes of two *different* homogeneous substances, then, the smaller they are, the more nearly we can describe our perceptual experience of them by the conceptual symbols of "particles of *different* mass." Thus in conception each independent substance must be looked upon as individualized for the purposes of our mechanical model of the universe by a special mass for its fundamental particle. If we take any homogeneous substance as a standard substance, then if we take

---

[1] It might well be described as the *sixth* fundamental law of motion.

small and equal volumes of any given homogeneous substance and of the standard substance, the ratio of the masses of the particles by which we represent conceptually these volumes as they become smaller and smaller is termed the *density* of the given homogeneous substance.[1] It follows, from the above statement as to the masses of two portions of the same homogeneous substance being proportional to their volumes, that : *the density of a given homogeneous substance is the ratio of the masses of equal volumes of it and of the standard substance.*

If a body be not such that its portions, anywhere taken, present to us the same groupings of physical and chemical sense-impressions, then the body is said to be *heterogeneous.* If we take small and equal volumes of this body from different parts, then the smaller we take them the more nearly we find that our perceptual experience of them can be described by particles of *different* masses. If we take small and equal volumes " from a given point " of a heterogeneous body and from the standard homogeneous substance, then the smaller we take them the more nearly our perceptual experience can be described by the mutual action of two particles. The ratio of the mass of this particle of the heterogeneous substance to that of the particle of the standard substance is termed the *density* of the heterogeneous substance *at the given point.* The density of such a substance is therefore not, as in the case of a homogeneous substance, the ratio of the masses of finite volumes of the given and of the

---

[1] The name adopted in the text-books is " specific gravity," but I think this term unfortunately chosen and I prefer to use the word *density* in this sense.

standard substances, it is a quantity which varies *from point to point* of the heterogeneous body.

Clearly the notion of density thus discussed affords a key to the manner in which we are to conceive the symbols for physical bodies constructed from aggregates of particles. By means of density we individualize substances and kinetically classify the particles which are the conceptual elements of bodies. Density forms the *kinetic scale* we have been in search of (p. 355) ; it is the fundamental means by which we measure the relative magnitude of the accelerations which we conceive the ideal elements of bodies to experience in each other's presence. It throws life into the geometrical forms by means of which we conceptualize the phenomenal universe.

The reader must, however, be careful to note that the whole of this discussion of density abounds in purely ideal notions. I have defined homogeneity ; but homogeneity thus defined is a limit drawn purely in conception to a process of comparison which can be begun but not completed perceptually. No perceptual substance is accurately homogeneous. Further, I have spoken about taking " equal volumes," a process which is a geometrical conception, and never exactly realizable in perception, where continuous boundaries cannot be postulated (p. 205). Then, again, I have spoken of taking a " volume at a point," and of the " density of a heterogeneous body at a point," conceptual limits again having no exact perceptual equivalents. Lastly, I have spoken of density as equal to the ratio of the masses of " certain volumes," and of aggregates of particles as filling " geometrical forms." These indications will be sufficient to show the reader that density, like mass, is a conceptual notion, an ideal means of classifying the symbols of our conceptual model of the universe. We do, indeed, choose these densities so that our model shall describe as accurately as possible our perceptual experience, but the density itself belongs to the conceptual sphere, and is defined with regard to the geometrical forms by which we symbolize physical bodies. It is a conceptual link between those geometrical forms and the accelerations with which we endow them. The importance of this point must be insisted upon, for it is this relation between geometrical volume and mass in the case of homogeneous substances which led physicists to the definition of mass as the "quantity of matter in a body" (p. 358). The geometrical form was first projected into the phenomenal world, and then this form filled with the metaphysical source of sense-impressions—matter. Mass as proportional to volume thus became mass as a measure of matter, and the sluice-gate was opened for that flood of metaphysics which has threatened to undermine the solid basis of physical science.

§ 13.—*The Influence of Aspect on the Corpuscular Dance.*

Hitherto I have only been dealing with the value of the *ratio* of the mutual accelerations of two corpuscles. The discussion of the absolute values of these mutual accelerations for each individual field would carry us through the whole range of modern physics ; we should have to deal with those special laws of motion which describe the phenomena we class under the heads of cohesion, gravitation, capillarity, electrification, magnetization, &c., &c. To discuss these does not fall within the scope of our present work, but there are one or two general points I must notice here. I proceed, in the first place, to state in accurate terms the second problem suggested on p. 354. I ask : *Are the absolute magnitudes of the mutual accelerations of two corpuscles influenced by the aspect they present to each other ?*

Now no very decisive answer can yet be given to this very important question of aspect influence. If we discriminate between the various types of corpuscles, there seem no facts of our perceptual experience that would lead us to suppose that aspect plays any part in the mutual action of ether-elements With regard to the prime-atom, we can only leave the matter unsettled ; if this atom were a vortex-ring aspect would be of importance, but if it were an ether-squirt it would not. On the other hand, in both cases, and probably in most other conceivable mechanisms, aspect would play a great *rôle* in the mutual actions between chemical atoms and between molecules. These groups, built up of comparatively few prime-atoms, can hardly accelerate each other's motion in the same manner however they turn towards each other. It is to this change of mutual acceleration with change of aspect that we have probably to look for aid in our conceptual attempts to describe such phenomena as crystallization and magnetization. As to the particle, aspect has probably little influence when we are dealing with particles at distances great compared with their vanishingly small size ; but it is still conceivable that if all the molecules in a particle had a similar aspect, aspect might be important in determining the action of this particle on an *adjacent* particle. In the phenomenon of gravitation aspect does not, however, play any part that we can perceptually appreciate. On the whole we conclude that aspect must be considered as a significant factor in determining the absolute magnitudes of mutual accelerations, but the exact influence which the "posture" of our dancers has upon the mode in which they dance remains still one of the obscure points of physics (see pp. 369, 386).

§ 14.—*The Hypothesis of Modified Action and the Synthesis of Motion.*

The next problem that we have to consider is one that is of extreme importance when we are dealing with the synthesis of motion, or the construction of the motion of complex from simple groups of corpuscles (p. 283). It is the problem of *modified action.* I may state it thus :—

If we have found the acceleration of A *in the presence of* B, *will the magnitude*[1] *of this acceleration be altered when* C *is introduced into the presence of* A *and* B ? This problem may be put a little differently, thus : Suppose we find when A and B are alone in the field that the acceleration of A due to B is represented by the step *b*, and that when A and C are alone in the field the acceleration of A due to C is represented by the step *c*, then when both B and C are in the field will these accelerations remain the same, and consequently will the total accelerating effect of B and C be represented, owing to the law we have stated for combining accelerations (p. 282), by the diagonal step *d* of the

---

[1] We have already seen that the *ratio* of the mutual accelerations, or of the masses of A and B, is not to be conceived as altered by the presence of other corpuscles in the field ; but this leaves the question of absolute magnitudes unsettled

parallelogram, whose sides are b and c? Or, on the
other hand, are we to conceive that when B and C are
both in the field the former acceleration b due to B
is altered to b' and the acceleration c due to C to c',
so that the total acceleration of A is now the diagonal
d'? Clearly if the latter statement be correct the
synthesis of motion becomes much more complex.
It will still be true that the acceleration of A is com-
pounded of the accelerations due to B and C, but
these accelerations will depend not on the respective
positions of B and C relative to A, but on the configuration of the entire system A, B, C. It will thus

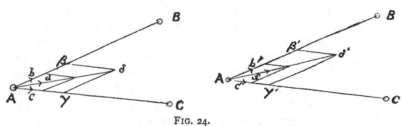

FIG. 24.

be impossible to form complex motions from the
combination of simple ones, until we have determined
how the actions b and c of B and C alone are modified
into b' and c' by being superposed. Now this question may also be looked at from the standpoint of
force. If m be the mass of A, then $m \times b$ and $m \times c$
will be the forces of B and C on A, and will be represented by steps m times the steps b and c in length
(p. 360) If B and C do not modify each other's influence, then their combined action, given by the
acceleration d corresponds to a force which, measured
by the product of mass and acceleration, or by $m \times d$,
is m times the step d. This force is termed the

resultant force; and we see that, since the resultant
and component forces are respectively m times the
diagonal and the sides of the acceleration-parallelogram, these forces must themselves form the diagonal
and sides of a parallelogram $A \, \beta \, \delta \, \gamma$ which is a magnified picture of the acceleration-parallelogram. This
is the famous *parallelogram of forces*, and we notice
that it follows at once from the parallelogram of
accelerations when we assume that B and C do not
modify each other's action.[1]

If they do modify each other's action there will
still be a parallelogram $(A \, \beta' \, \delta' \, \gamma')$ of forces, namely,
the resultant force $m \times d'$ will be the diagonal of the
parallelogram on the sides $m \times b'$ and $m \times c'$. But if
we mean, as physicists generally do, by the *force of*
B *on* A, the force when A and B are alone in the
field, and similarly by the force of C on A the force
when A and C are alone in the field, then we must
assert that on the hypothesis of modified action : *the
parallelogram of forces is not a synthesis by which we
can truly combine forces.*

This conclusion may appear to the reader so entirely
opposed to all that he has read of mechanics, that he
may be led at once to reject the hypothesis of modified action. One of Newton's laws of motion distinctly excludes indeed this hypothesis, and a great
simplification in our process of constructing complex
from simple mechanical systems undoubtedly arises
when we exclude it ; we have not to deal with every
new field afresh, and to re-measure accelerations for
each variation of its constituent elements : we simply

---

[1] This, for the purposes of the physics of the *particle*, might be
spoken of as the *seventh* law of motion.

analyze it, break it up into simple fields the individual motions of which have been previously discussed. Yet it is not scientific to assert that the simplest hypothesis is necessarily correct (Appendix, *Note III.*) ; we must ask, when we proceed to extend it beyond the range where it has been found to describe experience, whether it still suffices to simplify our conceptions, or leaves undescribed certain recognized phases of perception. Newton's law appears perfectly sufficient, and may therefore be said to be verified, when we are dealing with particles of gross " matter." The mutual accelerations, for example, of two gravitating particles seem to be uninfluenced by the presence of a third particle ; there is nothing, to take a still more concrete example, yet observed which would compel us to conceive that the mutual accelerations, by which we describe the mutual dance of sun and earth, are in the least influenced by the presence of the moon. Yet when we come to extend this law of Newton's, invaluable as it is for dealing with particles of gross " matter," to the mutual action of molecules, atoms, and ether-elements, there appears to be considerable reason for doubting its accuracy.

We can conceive atomic structures—for example, the ether-squirt—for which modified action is essentially true. There are phenomena of cohesion which can hardly be described without supposing the action of two molecules A and B to be modified by the presence of a third molecule C.[1] There are chemical facts which suggest that the introduction of a third

---

[1] A fuller discussion of "aspect" and "modified action" by the author will be found in Todhunter's *History of Elasticity*, vol. i. arts. 921–31, 1527, and vol. ii. arts. 276, 304–6. See also the *American Journal of Mathematics*, vol. xiii. pp. 321–2, 345, 353, 361.

atom C may even reverse the sense of the mutual accelerations of two atoms A and B. Nay, those who, in order to describe the radiation of light, treat the ether as an elastic jelly (p. 315), will find that it is very difficult to conceptualize its elastic structure, without asserting that the hypothesis of modified action is true of the ether-elements. The parallelogram of forces, then, as a synthesis of motion must be considered as applying in the first place to particles of gross " matter " ; its extension to other corpuscles can only be made cautiously and with continual reservation. Like so many other features of mechanism it cannot be dogmatically asserted to hold for all corpuscles, but it may in itself flow from the constitution we postulate for the ether and the structures we assume for the various types of gross " matter."

§ 15.—*Criticism of the Newtonian Laws of Motion.*

Before we close our discussion of the laws of motion it is only just to the reader to state that the method adopted differs widely from the customary physical treatment ; and in deference to the authority on which that treatment is based some comparison and criticism seems called for. We have already dealt with the current definitions of force, matter, and mass, and shown reasons for rejecting them as involving metaphysical obscurity. When, therefore, we come across these terms in the statement of the laws of motion we must endeavour to interpret them in our own sense. To the reader on first examination the Newtonian statement of the laws of motion may seem simpler than that of the present chapter. They are stated generally of *bodies*, and appear to describe the mechanism under which all bodies move, and therefore presumably describe the motion of the whole range of corpuscles from ether-element to particle. Now this loses sight of what the present writer thinks a very important possibility, namely, that not only special modes of motion, but much of the mechanism which describes the action of sensible bodies, will be found ultimately to be involved in some wide-reaching conception of ether and atom. It is not logically satisfactory to describe one mechanism by another of equal complexity; and we must hope to ultimately conceptualize an ether from the simple structure of which several of the laws of motion postulated for particles of gross " matter " may directly flow. Remembering these points we now turn to the version of the Newtonian laws given by Thomson and Tait.[1]

Law I.—*Every body continues in its state of rest or of uniform motion in a straight line, except in so far as it may be compelled by force to change that state.*

Now the reader who is acquainted with treatises on dynamics will remember that one of the most difficult chapters is frequently entitled, *Motion of a Body under the Action of no Forces.* The motion described is of an extremely complex kind. For example, the body may not only be spinning about an axis, but may be, and as a general rule is, conceived as continually changing the axis about which it spins.

---

[1] *A Treatise on Natural Philosophy*, part ii. pp. 241-7. The writer will not admit that he is second to any one in his admiration for the genius of Newton, or in his respect for the authors of the above classical *Treatise.* Yet he cannot believe that the two centuries which have elapsed since Newton stated his *Leges Motûs* " have not shown a necessity for any addition or modification " ! Old words grow as men are compelled to express new ideas in terms of them, and few definitions have a virile life of even a score years.

The "state of rest or of uniform motion in a straight line" is thus *not* that which the physicist postulates to describe the motion of a body under the action of no forces. It is quite true that we conceive a certain point termed the *centre of mass* of such a body to be either at rest or moving uniformly in a straight line; this, however, is not a conception which is itself axiomatic, but arises from an application of the principle of the equality of action and reaction to the *particles* by which we conceptually construct the body. In the first place, therefore, the use of the word *body* does not really give generality to the law, but introduces obscurity ; we ought at least to replace it by the word *particle*. In the next place the law is very wanting in explicitness as to what we are to understand by state of rest or of uniform motion in a straight line. All motion must be *relative* to something, but Newton does not indicate with regard to what, for example, the relative path is a straight line. Force is also a relative term (p. 360), but Newton nowhere tells us what the force on the body is related to. Thus, until a second body (or other particles) be introduced (p. 343), the law remains meaningless. In the last place, what are we to understand by the words, "compelled by force to change that state"? We take force to be a certain measure of motion, namely, the product of mass into acceleration ; then to assert the absence of force is to assert the absence of acceleration, or the law would merely contain the platitude that without change of motion a particle moves uniformly. But Newton certainly meant something more than this, for he was thinking of force in the sense of mediæval metaphysics as "a cause of change in motion." Now the nearest approach we can get to his idea is that position relative to surrounding particles determines a given particle's acceleration, and thus the first law is seen, liberally interpreted, to amount to the statement that surrounding circumstances determine acceleration— that without the presence of other particles there is no acceleration. This is the important principle of inertia to which we have already referred (p. 342), but it certainly appears to be stated with great obscurity in Newton's first law of motion. Further, even in this law, as I have restated it, no hint is given as to what application the principle may have to other corpuscles than particles of gross " matter " (p. 344).

Law II.—*Change of motion is proportional to force applied, and takes place in the direction of the straight line in which force acts.*

This is a veritable metaphysical somersault. How the imperceptible cause of change in motion can be applied in a straight line surpasses comprehension ; the only straight line that can be conceived, or, as some physicists would have it, *perceived*, is the direction of change of motion. We may assert that the imperceptible has this direction, but to postulate that the imperceptible will determine this direction for us seems to be pure metaphysics. We come down on our feet again, however, when we interpret this law as simply indicating that physically force is going to be taken as a measure for some change in motion (p. 360). As to the exact meaning of change of motion taking place in a straight line, all the real difficulties as to what thing we are to suppose changing its motion, and what is the presence associated with this change of motion, *i.e.*, the difficulties about the line joining two corpuscles (p. 367), are concealed by talking vaguely about force as an entity "acting in a straight line." Furthermore, if the "change of motion" is to be that of a body, not a particle, then we naturally ask which point of the body will have its motion changed in the direction of a straight line. We are thus again brought face to face with the fact that the motion of "bodies" is far more complex than is in the least indicated by this law.

Sir William Thomson and Professor Tait have restated the *Second Law* in the following form :—

When any forces whatever act on a body, then, whether the body be originally at rest or moving with any velocity and in any direction, each force produces in the body the exact change of motion which it would have produced had it acted singly on the body originally at rest.

These conclusions they consider really involved in Newton's *Second Law*. The same difficulty repeats itself here with regard to the interpretation of the term "body." Further, the law thus expressed denies the possibility of "modified action" (pp. 376–80), and the likelihood that in certain cases the velocity of corpuscles may help to determine their mutual accelerations (p. 349). It thus asserts the absolute validity of that synthesis, which we have termed the parallelogram of forces, and which we have ventured to suggest cannot be dogmatically asserted of corpuscles of all types.

Law III.—*To every action there is always an equal and contrary reaction, or the mutual actions of any two bodies are always equal and oppositely directed.*

If we replace "bodies" by "particles"—for the mutual action of two bodies is more complex than a reader just starting his study of mechanism would imagine, if he naturally interpreted mutual action as corresponding to mutual acceleration in some one line—the above law is identical with our *Fifth Law* (p. 359), and therefore we need not repeat the qualifying discussion of our §11. See Appendix, *Note II.*

The Newtonian laws of motion form the starting-point of most modern treatises on dynamics, and it seems to me that physical science, thus started, resembles the mighty genius of an Arabian tale emerging amid metaphysical exhalations from the bottle in which for long centuries it has been corked down. When the mists have quite cleared off we shall see more clearly its proportions, and there is special need for a strong breeze to clear away our confused notions as to matter, mass, and force. The writer is far from imagining that he can accomplish this clearance, but he is convinced that a firm basis for physics will only be found when scientists recognize that mechanism is no reality of the phenomenal world—that it is solely the mode by which we conceptually mimic the routine of our perceptions. The semblance is, indeed, so striking that we are able with astonishing accuracy to predict in vast ranges of phenomena what will be the exact sequence of our future sense-impressions. If, however, the scientist projects the whole of his conceptual machinery into the perceptual world he throws himself open to the charge of being as dogmatic as either theologian or metaphysician. On the other hand, when he simply postulates the conceptual value of his symbols as a mode of describing past and predicting future perceptual experience, then his position is unassailable, for he asserts nothing as to the *why* of phenomena. But as soon as he does this, matter as that which moves, and force as the cause of change in motion, disappear into the limbo of self-contradictory notions. What moves is only a geometrical ideal, and it moves only in conception. Why things move thus becomes an idle question, and *how* things are to be *conceived* as moving the true problem of physical science.[1]

In this field we know much, but our account of the laws of motion has been specially intended to emphasize how great is the room both for further investigation and for the exercise of disciplined imagination. In the vagueness of our conceptions of ether and atom lies the ill-explored continent which, by clearer definition, the Galilei and Newton of the future will annex. But before this annexation there is work for the unpretending pioneer in helping to clear away the jungle of metaphysical notions which impedes the progress of physical science.

SUMMARY.

The physicist forms a conceptual model of the universe by aid of corpuscles. These corpuscles are only symbols for the component parts of perceptual bodies and are not to be considered as resembling definite perceptual equivalents. The corpuscles with which we have to deal are ether-element, prime-atom, atom, molecule, and particle. We conceive them to move in the manner which enables us most accurately to describe the sequences of our sense-impressions. This manner of motion is summed up in the so-called laws of motion. These laws hold in the first place for particles, but they have been frequently assumed to be true for all corpuscles. It is more reasonable, however, to conceive that a great part of mechanism flows from the structure of gross " matter."

The proper measure of mass is found to be a ratio of mutual acceler-

---

[1] " Such demonstrations, however, only show how all these things may be ingeniously made out and disentangled, not how they may truly subsist in nature ; and indicate the apparent motions only, and a system of machinery arbitrarily devised and arranged to produce them—not the very causes and truth of things " (Bacon, *De Augmentis*, bk. iii. chap. iv.).

ations, and force is seen as a certain measure of motion, and not its cause. The customary definitions of mass and force, as well as the Newtonian statement of the laws of motion, are shown to abound in metaphysical obscurities. It is also questionable whether the principles involved in the current statements as to the superposition and combination of forces are scientifically correct when applied to atoms and molecules. The hope for future progress lies in clearer conceptions of the nature of ether and of the structure of gross " matter."

LITERATURE.

The views put forward in this chapter were reached when the author was studying the laws of motion for teaching purposes in 1882, and were developed for the purpose of college lectures in 1884 and subsequent years. A brief account of them was published in 1885, on pp. 267–71 of Clifford's Common Sense of the Exact Sciences, but the only published work in which the author has found any indication of similar opinions, or from the perusal of which he has received any help or encouragement, and the only work he can therefore heartily recommend to the reader is :—

MACH, E.—Die Mechanik in ihrer Entwicklung, S. 174–228. Leipzig, 1883.

The customary physical view of the Laws of Motion will be found in :—

CLERK-MAXWELL, J.—Matter and Motion, pp. 33–48.   London, 1876.

THOMSON, Sir W., and TAIT, P. G.—Treatise on Natural Philosophy, part i. pp. 219–24, 240–49. Cambridge, 1879.

CHAPTER IX.

LIFE.

§ 1.—*The Relation of Biology to Physics.*

IT does not fall within the range of the present work, still less within the power of its author, to discuss at any length the fundamental principles of biological science. The object of our *Grammar* has been to investigate the radical concepts of physics, the basis of that "dead" mechanism to which science is popularly supposed to reduce the universe. In the course of this investigation we have had occasion to call in question several of the notions commonly associated with these physical concepts ; we have seen that in speaking of matter and force much of our current language requires to be remodelled for scientific purposes Now physics is a much older branch of science than biology, and biologists have been so wont to look with something of awe and a little of envy to the presumed exactness both in language and in conclusions of mechanical science, that it may come with rather a shock to them when they hear that physics, like biology, is solely a description and not a fundamental explanation. While on the one hand, however, physicists can get on very well without biology, at any rate within a certain limited field of observation, biologists, on the other, have not only adopted many of the physicist's notions as to *matter, force,* and *eternity,* as modes of describing biological facts, but they are further, whether they wish it or not, inevitably bound to physics by the fact that life is never found apart from physical associations. Mechanism, on its side does not as a theory involve a discussion of biological phenomena, but biology without a discussion of mechanism is necessarily incomplete.[1]

" The elements of living matter are identical with those of mineral bodies ; and the fundamental laws of matter and motion apply as much to living matter as to mineral matter ; but every living body is, as it were, a complicated piece of mechanism which 'goes,' or lives only under certain conditions."

So wrote Professor Huxley in 1880.

The use of physical terms abounds in biology, often, I fear, with scarcely accurate definition. Nägeli talks of the " known forces of the organism, heredity and variability " ; Weismann speaks of the impossi-bility of the egg being " controlled by two forces of different kinds in the same manner as it would have been by one of them alone " ; he further talks of " forces residing in the organism " influencing the germ-plasm, which imperceptible entity he halves and divides as if it were a physical quantity.[2] Lan-kester speaks of " that first protoplasm which was

---

[1] From the author's standpoint, of course, conceptions as representing the products of the perceptive faculty are largely conditioned by the perceptive faculty of an individual genus, man (pp. 99-104, 211), and therefore their nature may be ultimately elucidated by biological, in particular psychological, inquiry.

[2] If Spencer can be included in the list of biologists, it will be found that' he uses force without special definition in the following senses : (i.) As cause of change in motion ; (ii.) as a biological process ; (iii.) as a name for kinetic energy ; (iv.) as a name for potential energy; (v.) as a general name for physical sense-impressions, such as light and heat, &c.

the result of a long and gradual evolution of chemical structure and the starting-point of the development of organic form." Biologists lay the greatest weight on the "chemical structure" of protoplasm and the chemical processes which are or accompany physiological functions, while free use is made of such terms as "unit-mass of living matter," "resultant of organic forces," "molecular stimuli," "continuity of organic substance," "conditions of tension and movement," "physical constitution necessary for immortality," &c., &c. Now either these terms are used figuratively, in which case we ought to find them re-defined, or else biologists have adopted them from physics and intend to use them in the sense of the latter science.

But there is small doubt that the latter alternative represents the true state of the case. The biologist considers his organic matter to be inexorably united to the "matter" of the physicist, and he uses, or considers he uses, such terms as matter, force, mechanism, &c., in the sense of the sister science. This dependence of biology on physics is so well brought out in the following passage that the reader must pardon our quoting it at this stage of our investigations :—

Experience cannot help us to decide this question ; we do not know whether spontaneous generation was the commencement of life on the earth, nor have we any direct evidence for the idea that the process of development of the living world carries the end within itself, or for the converse idea that the end can only be brought about by means of some external force. I admit that spontaneous generation, in spite of all vain efforts to demonstrate it, remains for me a logical necessity. We cannot regard organic and inorganic matter as independent of each other and both eternal, for organic matter is continually passing without residuum, into the inorganic. If the eternal and indestructible are alone without beginning, then the non-eternal and destructible must have had a beginning. But the organic world is certainly not eternal and indestructible in that absolute sense in which we apply these terms to matter itself. We can, indeed, kill all organic beings and thus render them inorganic at will. But these changes are not the same as those which we induce in a piece of chalk by pouring sulphuric acid upon it ; in this case we only change the form, and the inorganic matter remains. But when we pour sulphuric acid upon a worm, or when we burn an oak-tree, these organisms are not changed into some other animal and tree, but they disappear entirely as organized beings and are resolved into inorganic elements. But that which can be completely resolved into inorganic matter must have also arisen from it, and must owe its ultimate foundation to it. The organic might be considered eternal if we could only destroy its form, but not its nature. It therefore follows that the organic world must once have arisen, and further, that it will some time come to an end.[1]

Now this passage is extremely instructive, for we have the notion of the " eternal and indestructible " character of inorganic " matter " used to demonstrate the " logical necessity " of spontaneous generation. The reader who is in sympathy with the results of our discussion on " matter " and has recognized : (1) that " matter " as a substratum of our sense-impressions is a metaphysical dogma, not a scientific concept (p. 311) ; (2) that eternity is an idle phrase in the field of nomena (pp. 221, 227) ; and (3) that indestructibility relates to certain groupings of sense-impressions and not to an undefinable something behind them (p. 304), will be inclined to admit that the physicist is not wholly free from responsibility for the intrusion of metaphysics into biology. The physicist is therefore hardly warranted in demanding that the biologist shall accurately define his use of such terms as matter and force, for the physicist himself is not above reproach. At the same time the author is free to confess that the concepts of physics as defined, and he believes logically defined, in the present work

[1] Weismann : *Essays on Heredity*, p. 33. Oxford, 1889

scarcely lend themselves to the reasoning of the above passage. Nor can he think that, when physics has impressed upon biology that force is only a certain measure of motion, and not an explanation of anything whatever, biologists will be so ready to ascribe the phenomena of life to "forces residing in the organism." It is with the intention of suggesting how the view of mechanism, discussed in this work, can be conceived as applying to life rather than of dealing with the fundamental principles of biology, that the present chapter has been included in our volume.

§ 2.—*Mechanism and Life.*

In previous chapters we have seen how the phenomenal world is a world of groups of sense-impressions distinguished by the perceptive faculty under the two modes of space and time, or the mixed mode of *change*. This change or shifting of sense-impressions occurs in repeated sequences, or what we have characterized as *routine*. In the sense-impression itself there is nothing to suggest or enforce a routine, nor have we sufficient grounds as yet to definitely attribute this routine to the perceptive faculty. It remains for the present the fundamental mystery of perception, but it is the basis upon which all scientific knowledge is built. Science is the description in conceptual shorthand (never the explanation) of the routine of our perceptual experience. If this be true, it follows that the task of the biologist is to describe in conceptual shorthand (not to explain) the sequences of certain classes of sense-impressions. The problem of whether life is or is not a mechanism is thus not a question of whether the same things, " matter " and " force," are or are not at the back of organic and inorganic phenomena

—of what is at the back of either class of sense-impressions we know absolutely nothing—but of whether the conceptual shorthand of the physicist, his ideal world of ether, atom, and molecule, will, or will not, also suffice to describe the biologist's perceptions of life.

The mystery in the routine of sense-impressions is precisely the same whether those sense-impressions belong to the class of living or to that of lifeless groups. Life as a mechanism would be purely an economy of thought; it would provide the great advantages which flow from the use of one instead of two conceptual shorthands, but it would not "explain" life any more than the law of gravitation explains the elliptic path of a planet (p. 160). As we have—to speak paradoxically—no sense which can reach anything behind sense-impressions, no "metaphysical sense" which enables us to perceive that supposed entity "matter," so we have no special sense which enables us to perceive another supposed entity, "life." [1] Life and lifeless are merely class names for special groups of sense-impressions. When, therefore, we assert "matter" as the substratum of one group of sense-impressions and "life" as the substratum of another, and "explain" life by aid of matter and its attribute "force," we are simply, albeit often unconsciously, wallowing in the Stygian creek of metaphysic dogma. If the biologist gives us an accurate account of the development of the ovum and then remarks that the changes are *due* to "forces resident in the egg," he certainly cannot mean that the chemist and physicist are capable of *explaining* what has taken place. He

---

[1] The "sense of consciousness," if so it can be called, is hardly a special sense of life, for consciousness and life are not equivalent terms.

probably considers that the conceptual shorthand of chemistry and physics would suffice to *describe* what he has himself described in other language. If we always remember that the physicist's fundamental conception of change of motion is that the change of motion of one particle is associated with its position relative to other particles, and that force is a certain convenient measure of this change, then, I think, we shall be in a safer position to interpret clearly the numerous biological statements which involve an appeal to the conception of force. We must in each case ask what individual thing it is which is conceptualized as moving, what is the field with regard to which it is considered as moving, and how its motion is conceived to be measured. When we have completed this investigation then we shall be better able to appreciate the real substance which lies beneath the metaphysical clothing with which biological, like physical, statements are too often draped.[1]

Admitting, therefore, that our object in biology is identical with that in physics, namely, to describe the widest ranges of phenomena in the briefest possible formulæ (p. 116), we see that the biologist cannot throw back life for an explanation on physics. Whether he

[1] We are told, for example, that "force is always bound up with matter," that too small an "amount of matter" may be present to exercise a "controlling agency" over the development of the embryo, and when we seek to associate this "amount of matter" with some definite group of sense-impressions we find that no perceptual equivalent has been found for it. What the biologist is clearly striving to do is to form a conceptual model of the embryo by aid of the relative motions of the parts of a geometrical or rather kinetic structure (p. 373), but it is difficult to reach his ideas beneath the metaphysical language in which he projects matter, force, and germ-plasm into real substrata of sense-impression (see Weismann : *Essays cn Heredity*, pp. 226–7).

can hope to describe life in physical shorthand is a point to which we shall return a little later. If we look upon biology as a conceptual description of organic phenomena, then nearly all the statements we have made with regard to physics will serve as canons for determining the validity of biological ideas. In particular, any biological concept will be scientifically valid if it enables us to briefly summarize without internal contradiction any range of our perceptual experience. But the moment the biologist goes a step further, and asserts on the ground of the validity of his concept that it is a reality of the phenomenal world, although no perceptual equivalent has yet been found for it, then he at once passes from the solid ground of science to the quicksands of metaphysics. He takes his stand with the physicist who asserts the phenomenal existence of the concepts atom and molecule.

§ 3.—*Mechanism and Metaphysics in Theories of Heredity.*

I cannot bring home to the reader the difficulties with which the projection of conceptions into the phenomenal world is attended better than by briefly referring to two well-known biological theories of heredity. Of the change in those groups of sense-impressions which the biologist sets himself to describe there are two prominent features which at first sight might seem to correspond to nomic and anomic changes (p. 114, *footnote*), to routine and to breaches of routine. These features are the recurrence in our experience of the offspring of sense-impressions associated with the parental organism, and the occurrence in our experience of the offspring of sense-impressions not associated with the parental organism. These features are termed inheritance and variation. The apparent anomy, involved in variation is very probably like the anomy of the weather, a result of our not yet having formed a sufficiently wide or fundamental classification of facts. Be this as it may, inheritance and variation form the basis upon which biologists construct the evolution of life. Theories which endeavour to resume inheritance and variation under a single and simple formula are termed theories of heredity, and two of the most important of these theories are due respectively to Darwin and Weismann.

On Darwin's hypothesis of pangenesis every cell of the body throws off particles or gemmules which collect in the reproductive cells. These gemmules, or "undeveloped atoms," are transmitted by the parent to the offspring, they multiply by self-division, they may remain undeveloped during early life, or even during several generations, but when under the influence of suitable environment they do develop, they become cells like those from which they were derived. By aid of this hypothesis Darwin was able to resume a great many of the facts of heredity. Inheritance was simply the development of the parental gemmules in the offspring; variation could be described partly by a commingling of the gemmules of two parents, partly by a modification of the gemmules of the parental cells due to their use or disuse.[1] Now it is quite clear that no biologist would have propounded this hypothesis, but for the currency of corpuscular theories in physics. Indeed, Weismann actually re-states Darwin's hypothesis in terms of molecules, and speaks of unknown forces drawing these molecules

[1] *Variation of Animals and Plants under Domestication*, vol. ii. chap. xxviii.

to the reproductive cells and marshalling them there.[1]
But as no physicist ever caught an atom, so no biologist ever caught an "undeveloped atom," or gemmule.
The validity of the conception can only be tested by
the power it gives us of resuming the facts of heredity,
and it is no more disproved by the statement that
"gemmules have not been found in the blood," than
the atomic theory is disproved by the fact that no
atoms have been found in the air. If the biologist has
once grasped that the physicist is making a metaphysical statement when he asserts the phenomenal
existence of corpuscles, then he will be the more ready
to admit that the non-finding of gemmules and the
"unknown forces necessary to control them" are not
arguments against a conceptual description of heredity,
but against a metaphysical projection of its concepts
into the phenomenal world.

Weismann, who I think projects Darwin's gemmules
into the phenomenal world, and then rather oddly
states that they compel us to suspend all physical
conceptions, has, on the other hand, shown good reason
for Darwin's theory not being valid as a full description
of the phenomena of heredity, notably because the
transmission of acquired characteristics receives support from that theory, but hardly from our perceptual
experience. He has in his turn endeavoured to
formulate a theory which shall more accurately
describe the facts of heredity, especially those relating
to the non-transmission of characters acquired by
parents, owing either to use or accident during their
lives. This theory is summed up in the formula of
the "continuity of the germ-plasm." According to
this theory there exists a substance of a *definite*

[1] *Essays on Heredity*, pp. 75-8.

chemical and molecular structure termed germ-plasm, which resides somewhere in the germ-cells, from which reproduction takes place. In each reproduction a part of the germ-plasm "contained in the parent egg-cell is not used up in the construction of the body of the offspring, but is reserved unchanged for the formation of the germ-cells of the following generation." This constitutes the continuity of the germ-plasm.[1] Variation arises from the mixture of parental germ-plasms ; similarity of characteristics in parent and offspring—inheritance—from their both being developed under the control of the same germ-plasm. The " immortal " part of the organism which descends from generation to generation is the germ-plasm.[2] Now this hypothesis of Weismann as a conceptual mode of describing our perceptual experience seems to be of considerable value, but the author weakens his position throughout by projecting his conceptions into the phenomenal world, where up to the present nothing has been identified as the perceptual equivalent of germ-plasm. It is this transition from science as a conceptual description of the sequences of sense-impressions to metaphysics as a discussion of the imperceptible substrata of sense-impressions, which mars biological as well as physical literature. But the physicist is here to blame, for he has projected without perceptual evidence his molecule and atom into the phenomenal world, and the biologist only

[1] The reader must be careful to note that it is not a continuity of the germ-cells, but of a hitherto unidentified substance contained in these cells. Cells, we know, nuclei we know, with complicated networks of nucleoli; but what is *germ-plasm ?* Not to be seen and not to be caught by aniline stain or acetic acid.

[2] *The Continuity of the Germ-plasm as the Foundation of a Theory of Heredity*, 1885. *Essays on Heredity*, pp. 165–248.

follows the physicist's example when he asserts the reality of gemmule or germ-plasm. Finding the ground behind sense-impressions already occupied by molecule and atom, by matter and force, he not unnaturally gives his metaphysical products molecular or atomic structure ; he endows them with force and "explains" life by mechanism. In both the theories of Darwin and Weismann a metaphysical element seems to enter owing to a misinterpretation of the concepts of physics.[1] Only when we have fully recognized that physical science is solely a conceptual description, that matter as that which moves, and force as the why of its motion are meaningless, will this recognition begin to react on the fundamental conceptions of biology.

Our object hitherto has been to suggest that if the physicist withdraws, as we trust he may do, from the metaphysical limbo beyond sense-impression, then the biologist who has followed him there will retreat also. The problem as to whether life is or is not a mechanism will then have to be restated. We shall then have to ask whether organic and inorganic phenomena are capable of being described by the same conceptual shorthand. In order to understand more clearly the exact nature of this question we must stay for a moment to consider what we mean when we speak of organic and inorganic phenomena. What

[1] There are still stronger metaphysical aspects in Weismann's doctrine. That a substance which possesses continuity and sameness should indefinitely reproduce itself, or if it increases by absorption of foreign substances should remain the same, and this owing to a definite molecular structure, can hardly be looked upon even as a conceptual limit to any perceptual experience. We may ask, as Weismann does of Darwin's gemmule, whether it does not compel us "to suspend all known physical and physiological conceptions"?

groups of sense-impressions do we classify as living, what groups as lifeless?

§ 4.—*The Definition of Living and Lifeless.*

Now the first point to be noted is that there is no single sense-impression which can be said to be that of life. We do, indeed, seem in our own individual cases to have in consciousness a direct sense of life. But in the first place we have not at present any perception of consciousness except in our own individual case (p. 58), and in the next place we cannot even infer that consciousness is associated with all types of life (p. 69). We still find it reasonable to speak of human beings as living when they are asleep, or as living when they are completely paralyzed ; we speak of organisms as living when there is none of that hesitation between immediate sense-impression and exertion which constitutes thought and is the essential factor in human consciousness (p. 51). We cannot, indeed, say where consciousness must be taken to cease in the scale of life, but it would be ridiculous to question whether fungus spores had consciousness or not as a means of settling whether they were to be classified as living or dead substance. The less we find exertion conditioned by stored sense-impresses, the less degree of consciousness can we infer. The lowliest organisms appear to respond directly to their environment, and in this they resemble very closely the ideal corpuscle of the physicist, which dances in response to its surroundings. Seeds which have been preserved for fifty or a hundred years without losing their power of germination (see Appendix, *Note IV.*) are organic substance and contain life, at least in a dormant form, yet it is idle here to postulate consciousness as a means of classifying living and lifeless organisms.

The moment we accept without reservation the theory that all life has been evolved from some simple organism, then we are bound to recognize that consciousness has gradually become part of life, as forms of life grow more and more complex. This does not explain consciousness, but it is the only consistent description we can give of its evolution. The correlation of thought and consciousness seems to indicate that this complexity of the organism is to be sought in the inception and development of its capacity for storing sense-impressions. We can mark where this storage fails, we can mark where it exists ; but where it exactly begins we can hardly assert. This apparent continuity has led to some rather metaphysical reasoning on the part of biologists seeking for a distinguishing characteristic between living and lifeless groups. As in some types of life consciousness may be evolved, it is argued that there must be in life " something-which-is-not-yet-consciousness-but-which-may-develop-into-consciousness," and to this something Professor Lloyd Morgan has given the name of *metakinesis*.[1] This metakinesis does not appear to be more than a metaphysical name for nonconscious life, for there is no sense-impression that we have of such life that we can describe as metakinetic. Metakinesis is as intangible as the germ-plasm of the biologist or the molecule of the physicist, but less conceptually valuable as it describes no phenomenal side of life except the fact that it may or may not be associated with consciousness. Those who believe that the organic has been developed from inorganic,

[1] See in particular his letter to *Nature*, vol. xliv. p. 319.

that living has proceeded from dead "matter," may then assert that there must be in matter "something-which-is-not-yet-life-but-which-may-develop-into-life," and may fitly term this side of matter *supermateriality.* It is quite true that we have no direct series of sense-impressions to which this supermateriality corresponds, but as we mark some forms of matter associated with life (just as we mark some forms of life associated with consciousness), so we have the same reason for postulating its existence as we have in the case of meta-kinesis. How metakinesis develops from super-materiality will of course be the next stage in metaphysical investigation !

Now, I hope that Professor Lloyd Morgan will not think I am laughing at him, for this is far from being the case. I believe that no biologist is so patient with the physicist, even when the latter waxes paradoxical ; and I recognize that to look upon the mechanical and the conscious as two aspects of one and the same process may be a distinct simplification of our descrip-tion of life, and therefore scientifically valid. But I want to point out, and this very earnestly, how the physicist too often entices the biologist into a meta-physical slough by postulating mechanism as the substratum and not as the conceptual description of certain groups of sense-impressions. Had the physicist asserted that the reality of the external world lies for him in the sphere of sense-impressions, and that of the beyond of sense-impression physics knows nothing—had he said : " What I term mechanism and Professor Lloyd Morgan *kinesis* (see our p. 355) is purely a mode of describing conceptually the sequences of my sense-impressions," then the door would not have been opened for the metaphysician to parody metakinesis by supermateriality. So long as the biologist is taught to look upon mechanism as a series of imperceptible motions undertaken by imperceptible bodies under the guidance of imperceptible " molecular forces," he cannot be criticized for introducing another imperceptible element—" metakinesis "—into this process. But when the physicist ceases to postulate any of these imperceptibles and boldly states that mechanism is a conceptual process, by aid of which he is able to describe at any rate certain phases in those sequences of sense-impressions which we classify as unconscious life, then he may fairly ask what sense-impressions of unconscious life the biologist classifies by aid of metakinesis. If the biologist replies it is the potentiality of consciousness, then this is not the equivalent of the mechanism of primitive forms of life. The latter corresponds not only to the potentiality of all the complex nervous system of a conscious organism, but it actually describes some of our perceptual experience of primitive life. It thus does more than describe a potentiality, it describes a reality, and thus cannot be classed like metakinesis with supermateriality as a metaphysical "being," " essence," or " aspect."

The biologist therefore may describe for us the various stages in the evolution of consciousness, reducing them to scientific formulæ or laws, but he cannot postulate metakinesis, still less consciousness, as that which separates living from lifeless groups. All types of life do not appear capable of developing into conscious types ; and a potentiality not bearing any outward " recognition marks " will not lead us to a definition of life any more than the potentiality of becoming a bishop would lead us to a definition of man.

§ 5.—*Do the Laws of Motion apply to Life?*

If we seek for the characteristics of life apart from the possibility of consciousness, we can only seek them in some special features of those sequences of sense-impressions which we associate with living organisms. Now we have seen that groups of sense-impressions are all distinguished under the two modes of space and time, and we are thus able to conceptualize all change as a motion of ideal corpuscles. Now "currents," "vibrations of filaments," "moving masses of proto-plasm," "contraction," "change of form," "strain," &c., &c., are all terms in current biological use adopted to describe sequences or changes in sense-impressions. As to what are the symbolic bodies to which these motions are attributed and how they are to be built up from the most elementary organic corpuscles—"unit-masses of living matter" as one biologist terms them—there appears to be some diversity of opinion. But there is practical agreement among biologists that the organic corpuscles—the "physiological units" of Spencer or the "plastidules" of Haeckel—must be conceived as con-structed from the atom and molecule, the inorganic corpuscles of the physicist. Hence, if all we are to understand by mechanism, is something which we conceive as being constructed of atom and molecule and in motion, then life can only be conceived as mechanical.

How, therefore, we must ask, is it possible for us to distinguish the living from the lifeless, if we can describe both conceptually by the motion of inorganic corpuscles? The only answer that can be given to this must be that the nature of the motions by which we conceptualize organic and inorganic phenomena are very different. We mean by mechanism something more than the conceptual description of change by aid of the motion of physical corpuscles ; we mean that this motion is itself summed up in the laws of motion discussed in the preceding chapter. Herein lies the apparent kernel of the problem. Before we assert that life can be described mechanically, we must determine whether the motion by which we conceptualize organic phenomena can be resumed in the same laws as the motion by which we conceptualize inorganic phenomena.

But we soon find that we are only at the beginning of our investigation. In Chapter VIII. we have seen that the complex laws of motion which hold for particles of gross " matter " do not necessarily hold throughout the whole range of physical corpuscles ; they vary in character and probably increase in complexity from ether-element up to particle. We cannot therefore, without further consideration, determine what are the laws of motion which are to be postulated of the organic corpuscle, if life is to be dealt with as a mechanism. The laws which describe the motion of two groups of molecules are not necessarily the same as those which describe the motion of two isolated molecules, or of two atoms. If the laws by aid of which we might describe the motion of ideal organic corpuscles were found to differ from those which describe the motion of particles of heavy " matter," it would not settle the problem as to whether we could describe life mechanically or not.

The atomic system by which we conceptualize even the simplest unit of life is far too complex to allow, in the present state of mathematical analysis, of any synthesis of its motions in the presence of other systems by which we conceptualize either living or lifeless "matter." We cannot at present assert that the peculiar atomic structure of the life-germ and its *environment*, or field (p. 342), would not be sufficient to enable us on the basis of the laws of atomic motion to describe our perceptual experience of life. Such a broad generalization as that of the conservation of energy does not appear to be contradicted by our experience of the action of living organisms ; but then the conservation of energy is not the sole factor of mechanism, as some fetish-worshippers nowadays imagine it to be.

For example, there is the principle of inertia, the statement that no physical corpuscle need be conceived as changing its motion except in the presence of other corpuscles, that there is no need of attributing to it any power of self-determination (p. 343). There are probably those who think some power of self-determination must be ascribed to the elementary organic corpuscle, but this seems very doubtful. Placed in a certain field, environed with other organic or inorganic corpuscles, the life-germ moves relatively to them in a certain manner, but there seems no reason to assert (indeed there are facts pointing in the exactly opposite direction) that any change of movement need be postulated were the life-germ entirely removed from this environment. Indeed the whole notion of self-determination as an attribute of living organisms seems to have arisen from those extremely complex systems of organic corpuscles, where the environment in the form of immediate sense-impressions determines change through a chain of stored sense-impresses peculiar to the individual or self (p. 149). But if this be self-determination we can hardly consider it to have any bearing on the simplest forms of life.

We see, then, that biological change can probably be conceptually described by the change of motion of certain organic corpuscles in the presence of other corpuscles, either organic or inorganic. The structure of these organic corpuscles can further, to a great extent, be described in terms of physical corpuscles. But whether the laws of this motion can be deduced from the laws of motion of physical corpuscles remains at present, and may long remain, an unsolved problem. If the one set of laws could be deduced from the other, it would greatly simplify scientific description, but it would not lessen the mystery of life. Those who project their conceptions into the phenomenal sphere would still be puzzled to know why corpuscles dance in each other's presence, and the mystery would be no less or no greater because a dance of organic corpuscles is at bottom a dance of inorganic atoms. Those who treat all motion as conceptual (p. 329) would still have the mystery of why sense-impressions change and change with routine as insoluble as ever. Clearly those who say mechanism cannot *explain* life are perfectly correct, but then mechanism does not explain anything. Those, on the other hand, who say mechanism cannot *describe* life are going far beyond what is justifiable in the present state of our knowledge. We must content ourselves for the time being by saying that organic phenomena may be described by aid of organic corpuscles constructed out of inorganic corpuscles, and that the organic corpuscles move in certain characteristic manners, but that whether this motion follows or does not follow laws deducible from those dealt with in Chapter VIII. we have not at present the means of determining.

§ 6.—*Life Defined by Secondary Characteristics.*

The distinction, therefore, between the inorganic and the organic cannot be defined by saying that the one is mechanical and the other is not. We are ultimately obliged, in order to define life, to take secondary characteristics—to describe the structure by which we conceptualize the organic corpuscle, the motions which are peculiar to it, and the environment in which alone we perceive life to exist. Thus we note that its atomic structure is based upon complex compounds (p. 333) of carbon, hydrogen, nitrogen, and oxygen, a substance termed protein peculiar to organic bodies, together with water. The combination is termed *protoplasm,* but although its chemical constitution has in some measure been investigated, it has not been, and there at present appears no probability of its being, obtained except from organic substances. Turning to the characteristic movements of life, we note that organic substance is conceived as growing differently from inorganic substance. When crystals increase in size we conceive them to set molecule to molecule, building up from the outside. Organisms, on the other hand, we suppose to grow by an inner growth or the addition of new organic corpuscles in between and not on the surface of the old ones. Life further undergoes cyclical changes or movements in which some process of reproduction or division renews the individual. Lastly, a peculiar environment, certain conditions of moisture and temperature are necessary to maintain life. All these characteristics suffice to mark off the organic from the inorganic, and the distinction thus drawn appears to be absolutely rigid.[1]

---

[1] These are the distinctions of biology (see, for example, the article *Biology* in the *Encyclopædia Britannica*). Of course a physical state-

There is at the present time, so far as we know, *no generation of living from lifeless substance.* Thus our endeavour to define life has led, through some perhaps not unprofitable byways, to the consideration that the distinction between organic and inorganic is not so marked that we can separate the one from the other by anything but a lengthy statement of secondary characteristics.

The axiom *omne vivum e vivo* is one which deserves the reader's special attention, for it is closely associated with many important problems on the borderland of biology and physics. In the language of this *Grammar,* living and lifeless are class names for certain groups of sense-impressions, fundamentally distinguished from each other by requiring for their conceptual description different atomic structures and different types of motion. So far as our present experience goes there is no routine of sense-impressions which, starting from the lifeless class, concludes with the living class. On the other hand the converse transition from the living to the lifeless is an everyday routine.[1] We have seen (p. 390) that the latter fact has been used by Weismann as an argument in favour of the spontaneous generation of life—"that which can be completely resolved into inorganic matter must also have arisen from it and must owe its ultimate foundation to it," he writes. This passage seems to be rather too dogmatic and to

ment as to the laws under which organic corpuscles are to be conceived as moving in each other's presence and in that of inorganic corpuscles, might, could it be found, resume many of these characteristics in a simple formula.

[1] For example, in the boiling of impure water or in the pouring of acid on vegetable matter, but hardly in the ordinary "death" of a complex animal organism.

suggest a metaphysical subtratum to sense-impression which is "completely resolved." The argument would only be a valid one if we could assert that *all sequences of sense-impressions are reversible*, but this is too wide a statement to be laid down unrestrictedly in the present state of scientific knowledge. Physicists will recall processes like the *degradation of energy*, of which they are unable to at present conceive any reversion. It may be that their perceptual experience is not wide enough, and that their geometrical and mechanical laws are only applicable to a certain portion of the universe, or it may be, after all, that sequences are irreversible. Hence the spontaneous generation of life does not follow as a "logical necessity" from the transition of living into lifeless substance, at least as long as we cannot reasonably infer the reversibility of all sequences of sense-impressions.

§ 7.—*The Origin of Life.*

Those who accept the evolution of all forms of life from some simple unit, a protoplasmic drop or grain— and this scientific formula is so powerful as a means of classification and description that no rational mind is likely to discard it—will hardly feel satisfied to stop at this stage. They will demand some still more wide-embracing formula, which will bring under one statement their perceptual experience of both the living and the lifeless. Here the physicist comes in with some very definite conclusions. He tells us that in order to classify his perceptions with regard to the earth he is compelled to postulate a period, distant, it is true, many millions of years back, in which, owing to conditions of fluidity and temperature, no life, *such as we now know life*, not even the protoplasmic grain, could have existed on the earth. This period has been termed the *azoic* or lifeless period, but we must be careful to note that we mean by lifeless only " without life as we now know it." Bearing these facts in mind there are three hypotheses by which we can conceptually describe and classify our present experiences of the living and the lifeless. They are as follows :—

(*a*) Life may be conceived as based upon an organic corpuscle which is immortal—that is to say, it will, with suitable environment, continue to exist for ever. This hypothesis may be termed the *perpetuity of life*.

(*b*) Life may be conceived as generated from a special union of inorganic corpuscles, which union may take place under favourable environment. This hypothesis is termed the *spontaneous generation of life.*[1]

(*c*) Life may have arisen from the " operation in time of some ultra-scientific cause." This is the hypothesis of a *special creation of life*.

We will briefly consider these hypotheses in succession.

§ 8.—*The Perpetuity of Life, or Biogenesis.*

The perpetuity of life at first sight appears to contradict what physicists tell us of the azoic condition of the earth. A reconciliation of the two hypotheses has, however, been found by Von Helmholtz and Sir William Thomson, who suggest that a meteorite like an ethereal gondola might have brought in a crevice the protoplasmic drop to our earth when the azoic stage was passed. But our experience of meteorites —especially the intense cold they are subjected to in

---

[1] In more technical language the hypotheses (*a*) and (*b*), are spoken of as *biogenesis* and *abiogenesis* respectively. In using the popular term " spontaneous generation " I must not be supposed to suggest that life (any more than consciousness) can be *suddenly* generated.

space and the intense heat they undergo in passing through our atmosphere, together with the probability that they are fragments of azoic rather than zoic bodies—does not allow of much significance being attributed to this pleasant conceit. The perpetuity of life seems to involve the conception of forms of life anterior to the protoplasmic grain and capable of withstanding an environment totally unlike what protoplasm as we know it can endure. Now it is highly probable that protoplasm itself must be conceived as having had a long development anterior to any stage in which we now find it. These stages may have been eliminated in the struggle for existence, or they may have been peculiar to conditions of moisture and temperature which have long passed away on our earth. We might indeed be forced to conceive them as imperceptible like the atom, or, indeed, as indistinguishable from inorganic substance, which would lead us remarkably close to the second hypothesis of spontaneous generation.

This theory of the perpetuity of life, we must remember, is stated in purely conceptual language. As " eternity " is a meaningless term in the perceptual universe of physical phenomena, so it must be in the perceptual universe of biological phenomena. Time is a mode of distinguishing our sense-impressions, and it extends only so far as we have sense-impressions to distinguish (p. 221). The perpetuity of some primitive life unit is therefore a pure conception which, like that of the indestructibility of the atom (p. 304), helps us to classify and describe our perceptual experience, but for which it is meaningless to assert any phenomenal reality.

The perpetuity of life, however, involves some

rather extensive inferences—in particular, that life in
its earliest protoplasmic forms (which we must conceive to have resembled in many respects existing protoplasm), was yet capable of subsisting under a totally
unlike environment,[1] an environment in which only
what we term inorganic substances have hitherto been
perceived to exist. Such an hypothesis must accordingly be less adequate than any other which without
greater inference, brings under a single formula our
perceptual experience of both the living and the
lifeless.

§ 9.—*The Spontaneous Generation of Life, or Abiogenesis.*

Such a formula is that of the spontaneous generation of life. In the first place, this formula involves
the conception of forms of protoplasm anterior to
those with which we are at present acquainted, but it
does not suppose these like forms to have existed in
unlike conditions. It postulates that if we were to go
backwards the organic would have disappeared into
the inorganic before we reached the azoic age. After
the azoic age the physical conditions must be conceived as such that the various chemical compounds
were evolved which ultimately culminated in the first
protoplasmic unit.[2] But if this be so, it may be asked :

[1] Compare the Second Canon of Logical Inference (p. 72).

[2] Lankester (Article "Protozoa"), remarking on the steps which brought
the earliest type of protoplasm into existence, writes :—" A conceivable
state of things is that a vast amount of albuminoids and other such
compounds had been brought into existence by those processes which
culminated in the development of the first protoplasm, and it seems
therefore likely enough that the first protoplasm fed upon these antecedent steps in its own evolution just as animals feed on organic
compounds at the present day, more especially as the large creeping
plasmodia of some Mycetozoa feed on vegetable refuse." These words
suffice to indicate the long stages of development that probably lie
behind protoplasm as we know it.

Why cannot we find this sequence of sense-impressions in our present experience, why cannot we repeat the spontaneous generation of life in our laboratories? The reply probably lies in the statement that we seek to reverse a process which is irreversible (p. 410). In five or ten minutes we convert living into lifeless substance, but there is no reason for asserting that the reverse process can be gone through even in the lifetime of a man. On the contrary, it probably took millions of years, with complex and varying conditions of temperature, to pass from the chemical *substance* of life to that complex *structure* which may have been the first stage of organic being. Let us for a moment consider that there is possibly as long an evolution from the chemical *substance* to the protoplasm we now know, as from protoplasm to conscious animal life. Let us suppose that all the existing links between protoplasmic life, and that of the highest mammals had disappeared, and then let us set the biologist to demonstrate in his laboratory the spontaneous generation of consciousness by experiments on protoplasm! We cannot assert where consciousness begins or ends, but we can trace back in continuous series the conscious to the unconscious, and it is no argument against the truth of the hypothesis that consciousness is spontaneously generated to say that we cannot repeat the process at our will. In precisely the same manner spontaneous generation of life could only be perceptually demonstrated by filling in the long terms of a series between the complex forms of inorganic and the simplest forms of organic substance. Were this done, it is quite possible that we should be unable to say (especially considering the vagueness of our definitions of life) where life began or ended. The failure to reproduce the spontaneous generation of life in a laboratory has thrown some discredit on the hypothesis ; but we ought to wonder that any one should have hoped for an experimental demonstration of such an hypothesis rather than be surprised at its absence. At the very best, physicists will have to give us far more definite information than we have at present, both with regard to the physical changes at the close of the azoic period, and with regard not only to the chemical constitution but the physical structure of protoplasm, before it would be advisable even to think of further experiments on the spontaneous generation of life.

Even in the face of laboratory failure this second hypothesis seems far more satisfactory than that of the perpetuity of life. For in the latter case we carry back life through a continuous evolution to a stage where change seems to cease and we are left with a primordial life-germ and no antecedent state. Yet our whole perception of the phenomenal universe is continuous change. It cannot be said that this primordial germ is comparable with the physicist's prime-atom. The latter is a pure concept by aid of which the physicist constructs his symbols for phenomenal bodies, but he does not assert that these bodies have been evolved from prime-atoms. Bodies, he considers, may at any time be formed by aggregates of atoms, or again dissolved, but he does not postulate that the whole physical universe was ever in such a condition that it would have to be conceived of as resolved into simple disaggregated prime atoms. Indeed it is clear if he did so, that the primordial life-germ, if anything akin to protoplasm, would be non-extant, and the perpetuity of life be contrary to physical theory. In order to compare at all the primordial germ with the atom, we ought to take the former as the basis of the most complex extant organisms and suppose that on their dissolution they were resolved again into germs. But this would practically involve the indestructibility of the unit·of life—an hypothesis which appears to be at once confuted by our perceptual experience. The physical history of the universe does not lead us back to an evolution from a prime-atom and then stop at that point. The hypothesis of the perpetuity of life does lead us back to a primordial germ and then stop there. What is more, this germ appears placed in surroundings where it is destructible, while no environment, so far as our experience goes, need be conceived to have this effect on the atom. The two hypotheses, of the perpetuity of life and of the indestructibility of the atom, are therefore, if superficially alike, in reality far from comparable. It is an inference from the like to the unlike when we assert an evolution up to the primordial germ, and then a cessation of that evolution. On the other hand, it is no argument against spontaneous generation to assert that it, in its turn, leads us back to the prime-atom, at which we must again stop. For this is not the fact. It only leads us back to bodies conceptually constituted of prime-atoms, but which in physical evolution may be *continually* passing from one condition of aggregation to another. On the hypothesis of spontaneous generation we must conceive life as reappearing and again disappearing when and wherever the physical conditions are suitable. The hypothesis does not in the least *explain* the appearance of life ; it merely formulates its appearance as a routine on the occurrence of certain phenomena. Whenever a planet passing through the azoic stage begins to consolidate and cool, then begins the chemical evolution which ends in the first stage of life ; but *why* this succession of stages takes place is no more a subject of knowledge than *why* the sun rises daily. As we *describe* the latter so we could *describe* the former, were we capable of closely watching for millions of years the physical history of a planet.

§ 10.—*The Origin of Life in an "ultra-scientific" Cause.*

As to the hypothesis of a "special creation," science could not accept it as a contribution to *knowledge* had it even been able to cross-examine the only witness to the proceeding. The object of science is to classify and resume in brief formulæ the phases of our perceptual experience. It has to knit together all our sense-impressions by conceptual links, and thus to enable us to take a wide survey of the universe with the least possible expenditure of thought. Since time is a mode under which we perceive things, we cannot accurately assert of the earth that such and such changes occurred "between one and two hundred million years ago." What we really mean is this : that in order to resume and classify our perceptual experience of the earth, we form a conceptual model of it, and such a model we conceive to have passed through certain changes one or two hundred million years ago in *absolute time* (p. 226). Such a statement is ultimately involved in the formulæ by which we resume our immediate sense-impressions, and its scientific validity does not depend upon its describing something which took place beyond the

28

sphere of our perceptions, but upon its flowing from laws which accurately describe the whole of our present perceptual experience in the same field. Now the hypothesis of a " special creation " cannot be accepted as part of a conceptual model of the universe ; it cannot serve—like the formula of evolution for example—as a means of linking together phases of our perceptual experience : it would not bring unity into the phenomena of life nor enable us to economize thought.   Had the universe been created, just as it is, yesterday, the scientific mind would describe and classify its immediate sense-impressions and its stored sense-impresses far better by aid of the theory of evolution than by aid of a " special creation," and in this sense science cannot accept the hypothesis of a special creation as any contribution to knowledge at all.   Knowledge is the description in conceptual shorthand of the various phases of our perceptual experience, and the very statement of the hypothesis —as " the operation in time of some ultra-scientific cause "[1]—shows us that we have gone beyond knowledge, and are metaphysically separating time from perception and projecting causation beyond the sphere of sense-impression (p. 186).

The history of human thought shows us that at whatever stage men's power of describing the sequence of phenomena fails, that is, wherever their knowledge ends and their ignorance begins, there, to fill the place of the unknown antecedent, they call in a " special creation " or an " ultra-scientific cause."   To the untrained minds of earlier ages this cloak to

[1] This form of the statement is due to Sir G. G. Stokes : *On the Beneficial Effects of Light*, p. 85.   (Third Course of Burnett Lectures.) London, 1887.

ignorance seemed natural enough, but in a scientific age it is only an excuse for intellectual inertia; it shows that we have given up trying to know, where to strive to know is the first duty of science. For many centuries a seven days' creation of the world sufficed to screen our ignorance of the physical history of the earth, and of organic evolution, or the origin of species. On these points science is now perfectly definite, but it has had a hard struggle to get rid of the obstacles across the path of knowledge. The slight plantation by which mythology sought to screen human ignorance had become a forest, the special preserve of a caste, which it was sacrilege to hew down. Whether the battle will be now transferred to a "special creation" of the ultimate element of life remains to be seen, but in saying that science is at present ignorant as to the ultimate origin of life, we must be careful to allow no metaphysical hypothesis of an "ultra-scientific cause" to take root. We trust that light will come to science here, as it has come in equally difficult problems in the past; and not impossibly this light will come in the direction of the spontaneous generation of life. It is not before or behind in the sequence of cause and effect that we must insert the supernatural full stop. There is no need to cloak ignorance at distant stages with mystery; the mystery lies at hand in every change of sense-impression, in the fact that knowledge is at all times a description, but never an explanation of that change. The spontaneous generations of life and of consciousness are not conceptions which reduce the mystery of being; they but knit more closely together the veil of sense-impressions which bounds the field of knowledge and enshrouds the fundamental

mysteries of why we perceive at all and why we
perceive by routine.

§ 11.—*On the Relation of the Conceptual Description to the Phenomenal World.*

The reader will have noticed that the standpoint
which the author of this volume has reached through
an analysis of physical conceptions is largely confirmed when we turn to biological science.
Hypotheses of heredity, of the generation of life, and
of the origin of consciousness, are clearly formulæ
which attempt to describe the routine of our percep-

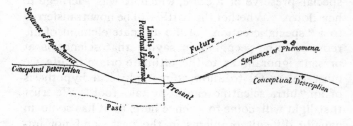

FIG. 25.

tual experience; and they do this by aid of a
conceptual model which not only resumes our present
perceptions, but enables us to carry back into the
past, or forward into the future, the sequence of scientific
causation (p. 153). That the conceptual model and our
perceptual experience agree at all points where we
can compare them, forms the sole basis of our assertion
that the model can be used to describe the non-
perceptible past and future.   If two curves were to be
in contact along the whole of that portion of the arc
which we were capable of examining, it would be
valid to replace one curve by the other; and to calcu-

late the probability that the curves would continue to touch, would be to measure the belief we ought to put in our scientific predictions as to the future (p. 177). The capacity of the conceptual curve for representing the phenomenal curve within the sphere of our perceptions would not be in the least invalidated, if the phenomenal curve came to a full stop beyond the sphere of perception.[1]

It is only when the symbols of our conceptual description are treated as the substrata of perception, or converted into what may truly be described as " ultra-scientific causes " of the routine of phenomena, it is only when the scientist becomes metaphysical, that difficulty arises. In biology this projection seems invariably to occur through the channel of physics ; the biologist looks to force, chemical constitution, molecular structure, for an *explanation*, where at best they can merely provide conceptual shorthand for descriptive purposes. It seems all the more necessary to emphasize and repeat this important distinction, because the failure to grasp it has been made the ground for what is really a metaphysical attack on the Darwinian theory of evolution. As I interpret that theory it is truly scientific, for the very reason that it does not attempt to explain anything. It takes the facts of life as we perceive them, and attempts to describe them in a brief formula involving such conceptions as " variation," " inheritance," " natural selection," and " sexual selection." But no more than

[1] The analogy to the laws of science may be still better brought home, at least to the mathematician, by supposing the *equation* to the conceptual curve known, but not that to the fragment of a curve AB (Fig. 25). The points A and B would not lend themselves to scientific description, they would fall outside the field of knowledge.

the law of gravitation explains our routine of perceptions with regard to the sun, does Darwin's theory
of the origin of species explain our perceptions
of change in living forms. Perhaps some of the
modern critics of Darwin will be less ready to consider adaptations as "not explicable" by natural
selection, but due to the " precise chemical nature of
protoplasmic metabolism," or to " an internal fate,
expressible in terms of dominant chemical constitution," if they once grasp that physics and chemistry
in their turn render nothing " explicable," but merely,
like natural selection itself, are shorthand descriptions
of changes in our sense-impressions.

§ 12.—*Natural Selection in the Inorganic World.*

There is a problem, however, with regard to natural
selection which deserves special attention from both
physicist and biologist, namely : Within what limits
is the Darwinian formula a valid description ? Assuming the spontaneous generation of life as a
plausible, if yet unproven, hypothesis, where are we to
consider that selection as a result of the struggle for
existence began ? Again, for what, if any, forms of
life are we to consider it as ceasing to be an essential
factor in descriptive history ? We may not be able
to answer these questions definitely, but some few
words at least must be said with regard to their
purport.

In the first place we notice that as soon as we conceive a perfectly gradual and continuous change from
inorganic to organic substance, then we must either
call upon the physicist to admit that natural selection
applies to inorganic substances, or else we must seek
from the biologist a description of how it came to be a

factor in organic evolution. Now there are two elements in natural selection—environment,which may be either organic or inorganic, and death, as a process of eliminating those less fitted to this environment. In the case of purely inorganic substances we can conceive that, under the physical conditions which follow the azoic period of a planet, all sorts of chemical products with varying physical structures might appear. Scientifically we might describe these products as the complex dances of corpuscular groups. In the meeting of group and group some groups would retain their individuality, others would lose it or be dissolved and possibly re-combined in new forms. Any group which retained its individuality would be spoken of physically as a *stable* product; and in the early history of a planet, although we are far from being able to describe accurately what might actually take place, it is not unreasonable to suppose that a physical selection of stable and destruction of unstable products might go on. We do not know why one element is more stable than a second, why it is better suited to its environment (we might describe the stability by aid of atomic accelerations, but this would not *explain*, only *resume* it) ; we can only suggest a selection of certain compounds which, because they are selected, we describe as more stable. Now this selection of stable compounds is a very possible feature of physical evolution,[1] but it must be noted that it is not precisely the same as natural selection. The environment is in this case purely

[1] It has been applied with remarkable power by Crookes (*British Association Address*, Section B, 1886), to give a suggestive sketch of how even the chemical elements might be conceived as evolved from *protyle* or prime-atoms.

inorganic, and "death" corresponds to the dissolution and ultimate re-absorption into more stable compounds. The competing substances form, indeed, their own environment ; and it is the special structure, not the corpuscle, which is conceived to disappear in the struggle. This physical selection is possibly what led up to the complex chemical substances endowed with special molecular structure, the hypothetical albuminoids in which some biologists suggest that life originated.

We are, then, face to face with the problem of how far this physical selection continued to act on the evolution of the earliest organic substances. How far was it the chief factor in the processes which we conceive as modelling both the chemical constitution and the physical structure of the earliest life-germs ? The first organic corpuscles must have been so close to the inorganic, and must have had an environment so essentially inorganic and not organic, that the test of relative physical stability must surely have been more important than the competition of superabundant organisms of *varying types* with each other. To those who have accustomed themselves to look upon organic substance as essentially differing from inorganic only by complexity of chemical and physical structure, the notions of organic and inorganic environment, of the elimination of the unfit and the destruction of less stable compounds—in short, the notions of biological and physical selection —shade insensibly one into the other. Selection will be physical when the environment is more inorganic than organic, and biological or natural in the converse case. But those naturalists who postulate a special organic corpuscle are certainly called upon to decide how and when the formula of natural selection begins to govern its evolution, and what part, if any, physical selection has played in the determination of its chemical and physical constitution.

§ 13.—*Natural Selection and the History of Man.*

Passing to the superior limit we have next to ask, How far are the principles of natural selection to be applied to the historical evolution of man? To judge by the author's experience of historical literature, we should have to say that up till very recent times historians have assumed that the historical development of man cannot be briefly resumed in wide-reaching formulæ; that history is all facts and no factors. That natural history, the evolution of organic nature, is at the basis of human history is the unwavering belief of the present writer. History can never become science, can never be aught but a catalogue of facts rehearsed in more or less pleasing language, until these facts are seen to fall into sequences which can be briefly resumed in scientific formulæ. These formulæ can hardly be other than those which so effectually describe the relations of organic to organic and of organic to inorganic phenomena in the earlier phases of their development. The growth of national and social life can give us the most wonderful insight into natural selection, and the elimination of the unstable on the widest and most impressive scale.[1] Only when history is interpreted in this sense of *natural* history does it pass from the sphere of nar-

[1] This view is far from being held by the majority of sociologists and historians. One example typical of many may be cited here: "Every phase of the history of the development of organisms, which Darwin brings forward as an hypothesis, remains, in any case, quite unsuited for

rative and become science. But, on the other hand, in this sense of a description of facts resumed in brief formulæ, all science is history. It may take a long training in scientific modes of thought before the literary historian is converted, but his conversion must come sooner or later in an age when the reading public is becoming more and more imbued with the scientific spirit.[1]

comparison with the constantly and uniformly progressive and never-resting history of the human race."—Dr. Georg Mayr : *Die Gesetz-mässigkeit im Gesellschaftsleben.*

[1] The present confusion of thought on this subject cannot be illustrated better than by referring to a recent work and the remarks upon it of a well-known critic. Dr. E. Westermarck has lately published a book entitled : *The History of Human Marriage* (Macmillan, 1891). The introduction to this work states in clear and fairly accurate language the scientific method of historical investigation, but when we come to the material of the book we find a singular absence of scientific method. There is a great collection of facts under different headings from every quarter of the globe, but it does not seem to have struck the writer that to find sequences of facts—a growth or evolution expressible by a scientific law—we must follow the changes of one tribe or people at a time. We cannot trace the successive stages of social life except by the minute investigation of facts relating to one social unit, which may, and indeed must, be afterwards compared with like investigations for other units. We have, then, in Dr. Westermarck an excellent example of good theory and bad practice.

In his critic, Professor Robertson Smith (*Nature*, vol. xliv. p. 270), we have a writer who has done unsurpassed work in the natural history of religions and of marriage. Yet this critic is so unconscious of the character of his own work that he considers Dr. Westermarck confuses "history" and "natural history"! "The history of an institution," he writes, "which is controlled by public opinion and regulated by law is not natural history. The true history of marriage begins where the natural history of pairing ends." And again : "To treat these topics [polyandry, kinship through female only, infanticide, exogamy] as essentially a part of the natural history of pairing involves a tacit assumption that the laws of society are at bottom mere formulated instincts ; and this assumption really underlies all our author's theories. His fundamental position compels him, if he will be consistent with

It is peculiarly in "prehistoric history" that we are for the time being best able to apply the scientific method. That the earliest history of each individual people follows general laws of human development which are capable of accurate scientific statement is a view which is being daily confirmed by the discoveries of comparative anthropology, folklore, and mythology. It is true that the application of these laws varies to a certain extent with the physical environment, with the climate and geographical surroundings. Nevertheless, in broad outline the development of man, whether in Europe, Africa, or Australasia, has followed the same course. The divergencies from this uniformity of development appear indeed to be less the farther we penetrate into the nascent history of the human race. This uniformity is to some degree of course only apparent and must be attributed to the obscurity in which all early history is involved. Yet it is for the greater part real, and due to the fact that in the early stages of civilization the physical environment and the more animal instincts of mankind are the dominating factors of evolution.

Primitive history is not a history of individual men, nor of individual nations in the modern sense ; it is a description of the growth of a typical social group of human beings under the influences of a definite

himself, to hold that every institution connected with marriage that has universal validity, or forms an integral part of the main line of development, is rooted in instinct, and that institutions which are not based on instinct are necessarily exceptional and unimportant for scientific history." When a really scientific historian can in a scientific journal reject an unscientifically executed investigation because it starts from an unexceptional scientific theory, we are truly in topsy-turvydom. Science has yet to do a pioneer's work in the field of historical method.

physical environment, and of characteristic physiological instincts. Food, sex, geographical position,
are the facts with which the scientific historian has to
deal. These influences are just as strongly at work
in more fully civilized societies, but their action is
more difficult to trace, and is frequently obscured by
the temporary action of individual men and individual
groups. The obscurity only disappears when we deal
with average results, long periods, and large areas.
The savage who fights with his neighbour in order to
kill and eat him, is an obvious example of the struggle
for existence. The contest of modern nations for
markets in Africa and Asia, their strife for the possession of trade routes, their attempts to cheapen their
manufactures, and to better educate their artizans,
may in reality be described by the same laws of
evolution, but the manifestation of these laws is far
more complex and difficult to analyze. This rivalry
is at bottom the struggle for existence, which is
still moulding the growth of nations ; but history, as
it is now written, conceals, under the formal cloak of
dynasties, wars, and foreign policies, those physical
and physiological principles by which science will
ultimately resume human growth.

§ 14.—*Primitive History describable in terms of the Principles
of Evolution.*

The economical condition of any nation during a
given period is closely associated with its rate of
reproductivity and with its indirect struggle against
its neighbours for land and food. Not less important
for the stability of any nation is the nature of the
prevailing forms of ownership, marriage, and family
life. But the continual variations in these forms are

in modern history usually hidden under problems of
trade and exchange, under civil laws as to ownership,
inheritance, marriage and divorce, or under statistics
of pauperism, emigration, and sexual morality.   The
old factors of evolution are there, but they are dis-
guised.   It is only when we turn to a less complex
stage of social growth that we fully grasp the direct
bearing which the struggle for food and for the gratification of the sexual instincts has had in moulding
human development.   It is this struggle which is the
fundamental formula for the description of all existing
systems of ownership and of marriage in its widest
sense.   In ownership and marriage are further rooted
the laws and institutions of competing modern states.
Sexual instinct and the struggle for food have both
separated and combined individual men ; in them we
find the basis of both the egoistic and the altruistic
instincts, of both individualism and socialism in the
most fundamental sense of these terms.

Systems of ownership and marriage have indeed
been modified by climate and geographical surroundings, but, speaking generally, they have passed
through much the same development, it may be at
very different periods, in all quarters of the world.
Fragments of the primitive history of one society can
often be linked together by our knowledge of another
society still existing in a backward stage of civilization.   The like sequences in the stages of social growth
exhibited by most primitive societies undoubtedly
arise from similarity in their general physical environment and from the sameness of the characteristic
physiological instincts in man, which everywhere
centre in the satisfaction of hunger and in the
gratification of the sexual appetite.   Diverse as at

first sight ownership and marriage may seem, they will yet be found on nearer investigation to be closely associated.   Broadly speaking, each particular mode of ownership has been accompanied by a particular form of marriage.   These two social institutions have acted and reacted upon each other and their changes have been nearly simultaneous.   Ownership, inheritance, common rights, are essentially connected with the structure of the family, and therefore with the nature of the sexual tie.   Thus it comes about that primitive history must be based upon a scientific investigation into the growth and correlation of the early forms of ownership and marriage.   It is only by such an investigation that we are able to show that the two great factors of evolution, the struggle for food and the instinct of sex, will suffice to resume the stages of social development. When we have learned to describe the sequences of primitive history in terms of physical and biological formulæ, then we shall hesitate less to dig deep down into our modern civilization and find its roots in the same appetites and instincts (see Appendix, *Note VI.*). We shall then be less unwilling to admit that historical science, like any other branch of science, cannot only describe the past but is capable of predicting the future course of development.   Here, in predicting from the economic and social history of the past the probable tendencies of the immediate future, seems to be the true function of those somewhat errant sciences, political economy and sociology.

§ 15.—*Morality and Natural Selection.*

Although the reader may be prepared to admit that the " survival of the fittest " is a formula describing the development of mankind even at the present, he may still question how it can possibly be a source of altruistic conduct in life.[1] If perpetual struggle for existence between all forms of life be the keynote to progress—if the individual, stronger in body or mind, does invariably push aside his weaker fellows, render them subservient to his aims, or crush them out of existence, how can we look upon life from any but the egoistic and pessimistic standpoint? Poverty and disease must then be regarded as valuable aids in the destruction of less fit human beings, wealth and luxury as the meet reward of individual fitness. Starting with this view of life as solely a war of individuals, we inevitably reach that conception of government which may be summed up in the sentences : A maximum of good must arise from a minimum of social organization ; for government to interfere between individuals is an irrational attempt to upset the principle of the survival of the fittest.

The reader must not think that I am exaggerating the pessimism of some of our modern biologists. Here, in a few words, are the views of Haeckel :—

"Darwinism is anything but socialistic. If a definite political tendency be attributed to this English theory—which is, indeed, possible —this tendency can only be aristocratic, certainly not democratic, and least of all socialistic. The theory of selection teaches us that in human life, exactly as in animal and plant life, at each place and time only a small privileged minority can continue to exist and flourish ; the great mass must starve and more or less prematurely perish in misery. Innumerable are the germs of every form of animal and plant life, and the young individuals, which spring from these germs. The number of fortunate individuals, on the other hand, who develop to their full age and actually attain their goal in life is out of all proportion small. The cruel and relentless struggle for existence which rages throughout all

---

[1] The substance of the remainder of this chapter is taken from a lecture delivered in 1888, and afterwards published as a pamphlet.

living Nature, and in accordance with Nature must rage, this ceaseless
and pitiless competition of all living things, is an undeniable fact ; only the
select minority of the privileged fit is in a position to successfully survive
this competition, the great majority of competitors must meanwhile of
necessity perish miserably !   We may deeply mourn this tragic fact, but
we cannot deny or alter it.   ' Many are called but few are chosen ! '
This selection, this picking out of the chosen is necessarily combined
with the languishing and perishing of the remaining majority.   Another
English investigator even denotes the kernel of Darwinism as ' the
survival of the fittest,' the ' triumph of the best.'   Obviously the principle
of selection is anything but democratic, it is aristocratic in the precise
sense of the word." [1]

Spencer and Huxley have taught much the same
gospel.   Yet, if the creed of science be based on this
law of evolution, how can it inculcate aught but
pessimism for the weak, how can it ever be the faith
of any but the privileged few ?   I venture to think
that the view of the survival of the fittest propounded
by Haeckel is in reality a very insufficient analysis,
and that it requires much qualifying statement.

The struggle for existence involves not only the
struggle of individual man against individual man,
but also the struggle of individual society against
individual society, as well as the struggle of the
totality of humanity with its organic and inorganic
environment.   To include these omitted factors, might
at first sight appear only to enlarge the battle-field, to
extend the chaos of opposing interests.   But in reality
it alters the whole aspect of life.   The interest the
individual has in developing to the utmost his own
powers is a very important factor of change—let us call
it *Individualism*.   But the interest individual societies
have in developing their resources, in organizing themselves owing to the intense struggle which is ever

[1] *Freie Wissenschaft und freie Lehre*, S. 73.

waging between society and society, this is an equally important factor of evolution and one too often forgotten when the doctrines of Darwin are applied to human history. Individual societies have the strongest interest in educating, training, and organizing the powers of all their individual members, for these are the sole conditions under which a society can survive in the battle for life. This tendency to social organization, always prominent in progressive communities, may be termed, in the best and widest sense of the word, *Socialism*. The socialistic as much as the individualistic tendency is a direct outcome of the fundamental principle of evolution. Finally, there is a third factor of evolution, namely, the profit that arises to humanity at large from common organization against organic and inorganic foes. The interdependence of mankind throughout the world is becoming a more and more clearly recognized fact. The failure of human beings in one part of the world to master their physical environment may lead to a famine at their antipodes ; the triumph of the scientists of one nation over a minute bacillus, is a victory for all humanity. The development of human control over man's physical and biological environment in all parts of the world is thus of real importance to each individual group. This solidarity of humanity in the struggle with its environment is no less a feature than Individualism or Socialism of the law of evolution. We may perhaps term it *Humanism*.

If our analysis has been a correct one it has led us from the simple law of the survival of the fittest to three great factors—Individualism, Socialism, and Humanism—tending to modify human life. Our strong inherited instincts to Individualism, to Socialism, and in a less extent, to Humanism,[1] guide us to those principles of conduct, duty to self, duty to society, and duty to humanity, which our forefathers were taught to think of as the outcome of supersensuous decrees or of divine dispensations, and which some even of their children still regard as due to mysterious tendencies to righteousness, or to some moral purpose in the universe at large.

§ 16.—*Individualism, Socialism, and Humanism.*

We may fitly conclude this chapter on *Life* by a few remarks on the extent to which Individualism, Socialism, and Humanism respectively describe the features of human development. The great part played in life by the self-asserting instinct of the individual does not need much emphasizing at the present time. It has been for long the over-shrill keynote of much of English thought. All forms of progress, some of our writers have asserted, could be expressed in terms of the individualistic tendency. The one-sided emphasis which our moralists and publicists placed upon individualism at a time when the revolution of industry relieved us from the stress of foreign competition, may indeed have gone some way towards relaxing that strict training by which a hard-pressed society supplements the inherited social instinct. This emphasis of individualism has undoubtedly led to great advances in knowledge and even in the standards of comfort. Self-help, thrift,

[1] A good deal of the humanistic instinct as developed in modern times is practically a product of socialism. As the tribal recognition-marks grew feebler and localization less definite, the social sympathies were extended to the stranger whose habits and modes of thought were not too widely divergent from those of the society in which he found himself.

personal physique, ingenuity, intellect, and even cunning have been first extolled and then endowed with the most splendid rewards of wealth, influence, and popular admiration. The chief motor of modern life with all its really great achievements has been sought—and perhaps not unreasonably sought—in the individualistic instinct. The success of individual effort in the fields of knowledge and invention has led some of our foremost biologists to see in individualism the sole factor of evolution, and they have accordingly propounded a social policy which would place us in the position of the farmer who spends all his energies in producing prize specimens of fat cattle, forgetting that his object should be to improve his stock all round.[1]

I fancy science will ultimately balance the individualistic and socialistic tendencies in evolution better than Haeckel and Spencer seem to have done. The power of the individualistic formula to describe human growth has been overrated, and the evolutionary origin of the socialistic instinct has been too frequently overlooked.[2] In the face of the severe struggle, physical and commercial, the fight for land, for food, and for mineral wealth between existing nations, we have every need to strengthen by training the partially dormant socialistic spirit, if we as a nation are to be among the surviving fit. The importance of organizing society, of making the individual subservient to the whole, grows with the intensity of the struggle. We

[1] R. H. Newton : *Social Studies*, p. 365.

[2] It may be rash to prophesy, but the socialistic and individualistic tendencies seem the only clear and reasonable lines upon which parliamentary parties will be able in the future to differentiate themselves. The due balance of these tendencies seems the essential condition for healthy social development.

shall need all our clearness of vision, all our reasoned insight into human growth and social efficiency in order to discipline the powers of labour, to train and educate the powers of mind. This organization and this education must largely proceed from the state, for it is in the battle of society with society, rather than of individual with individual, that these weapons are of service. Here it is that science relentlessly proclaims : A nation needs not only a few prize individuals ; it needs a finely regulated social system—of which the members as a whole respond to each external stress by organized reaction—if it is to survive in the struggle for existence.

If the individual asks : Why should I act socially ? there is, indeed, no argument by which it can be shown that it is always to his own profit or pleasure to do so. Whether an individual takes pleasure in social action or not will depend upon his character (pp. 57, 150)—that product of inherited instincts and past experience—and the extent to which the "tribal conscience" has been developed by early training. If the struggle for existence has not led to the dominant portion of a given community having strong social instincts, then that community, if not already in a decadent condition, is wanting in the chief element of permanent stability. Where this element exists, there society will itself repress those whose conduct is anti-social and develop by training the social instincts of its younger members. Herein lies the only method in which a strong and efficient society, capable of holding its own in the struggle for life, can be built up. It is the prevalence of social instinct in the dominant portion of a given community which is the sole and yet perfectly efficient sanction to the observance of social, that is moral, lines of conduct.

Besides the individualistic and socialistic factors of evolution there remains what we have termed the humanistic factor. Like the socialistic it has been occasionally overlooked, but at the same time occasionally overrated, as for example, in the formal statements of Positivism. We have always to remember that, hidden beneath diplomacy, trade, adventure, there is a struggle raging between modern nations, which is none the less real if it does not take the form of open warfare. The individualistic instinct may be as strong or stronger than the socialistic, but the latter is always far stronger than any feeling towards humanity as a whole. Indeed the "solidarity of humanity," so far as it is real, is felt to exist rather between civilized men of European race in the presence of nature and of human barbarism, than between all men on all occasions.[1]

"The whole earth is mine, and no one shall rob me of any corner of it," is the cry of civilized man. No nation can go its own way and deprive the rest of mankind of its soil and its mineral wealth, its labour-power, and its culture—no nation can refuse to develop its mental or physical resources—without detriment to civilization at large in its struggle with organic and inorganic nature. It is not a matter of indifference to other nations that the intellect of any people should lie fallow, or that any folk should not take its part in the labour of research. It cannot be indifferent to mankind as a whole whether the

[1] The feeling of European to Red Indian is hardly the same as that of European to European. The philosopher may tell us it "ought" to be, but the fact that it is not is the important element in history.

occupants of a country leave its fields untilled, and its natural resources undeveloped. It is a false view of human solidarity, a weak humanitarianism, not a true humanism, which regrets that a capable and stalwart race of white men should replace a dark-skinned tribe which can neither utilize its land for the full benefit of mankind, nor contribute its quota to the common stock of human knowledge.[1] The struggle of civilized man against uncivilized man and against nature produces a certain partial " solidarity of humanity " which involves a prohibition against any indivi- dual community wasting the resources of man- kind.

The development of the individual, a product of the struggle of man against man, is seen to be con- trolled by the organization of the social unit, a product of the struggle of society against society. The development of the individual society is again influenced, if to a less extent, by the instinct of a human solidarity in civilized mankind, a product of the struggle of civilization against barbarism and against inorganic and organic nature. The principle of the survival of the fittest, describing by aid of the three factors of individualism, socialism, and humanism the continual struggle of individuals, of societies of civilization and barbarism, is from the standpoint of science the sole account we can give of the origin of those purely human faculties of healthy activity, of

[1] This sentence must not be taken to justify a brutalizing destruction of human life. The anti-social effects of such a mode of accelerating the survival of the fittest may go far to destroy the preponderating fitness of the survivor. At the same time, there is cause for human satisfaction in the replacement of the aborigines throughout America and Australia by white races of far higher civilization.

sympathy, of love, and of social action which men value as their chief heritage.

SUMMARY.

1. Owing to the metaphysical character of the language of much of modern physics, metaphysics has found a foothold in biology. Peculiarly in the idea of life as a mechanism do we find confusion reigning. The problem ought to be expressed in words to the following effect : Can we describe the changes in organic phenomena by the same conceptual shorthand of motion as suffices to describe inorganic phenomena ? There is difficulty in answering this question because we are unable to assert what are the exact laws of motion which would apply to the complex physical structure by which we conceptualize the simplest organic germ.

2. The distinction between living and lifeless is not capable of brief definition, consciousness and self-determination give us no assistance, and we are thrown back on special characteristics of structure and motion.

3. Of the three hypotheses which have been invented to describe the origin of life—its perpetuity, spontaneous generation, and origin from an "ultra-scientific cause"—the second seems the most valuable. Like the "spontaneous generation of consciousness," it is only a conceptual description, and not an explanation of the sequence of phenomena.

4. Biologists are called upon to define the limits within which they suppose the formula of natural selection to be a valid description : in particular, how it is related to that physical selection of more stable inorganic compounds which we may conceive to have taken place during and after the azoic period. At the other end of the scale we have again to ask how far the survival of the fittest describes the sequences of human history. While it seems probable that human history may be resumed in the brief formulæ of biology and physics, still several leading biologists who have examined human progress from this standpoint do not appear to have paid sufficient regard to the socialistic instinct, which, as much as the individualistic instinct, is a factor of the principle of evolution.

LITERATURE.

CLAUS and SEDGWICK.—Elementary Text-Book of Zoology (General Part, Chapter I.). London, 1884.

HAECKEL, E.—Natürliche Schöpfungs-Geschichte (Zwölfter Vortrag, S. 250–310), 4th ed. Berlin, 1873. (History of Creation, revised by Ray Lankester, London, 1883.)—On the Development of Life. Particles and the Perigenesis of the Plastidule, 1875 (pp. 211–57 of The Pedigree of Man and Other Essays, London, 1883).—Freie Wissenschaft und freie Lehre. Stuttgart, 1878. (Freedom of Science and Teaching, with Note by Huxley. London, 1879.)

HUXLEY, T. H.—On our Knowledge of the Causes of the Phenomena of Organic Nature. London, 1863.—Lay Sermons, Addresses, and Reviews. (On the Physical Basis of Life, pp. 132–61). London, 1870.—Nineteenth Century, vol. xxiii. (The Struggle for Existence).

SANDERSON, J. S. BURDON. — Opening Address to the British Association, 1889, Nature, vol. xl. p. 521.

SPENCER, H.—Principles of Biology, vol. i. (to be read with caution). London, 1864. The Man versus the State, London, 1890.

See also the articles in The Encyclopædia Britannica on "Biology," "Physiology" (Part i., General View), and "Protoplasm," and those in the new edition of Chambers's Encyclopædia on "Abiogenesis," "Biogenesis," and "Development."

CHAPTER X.

The Classification of the Sciences.

§ 1.—*Summary as to the Material of Science.*

In the first chapter of this *Grammar* we saw that science claims for its heritage the whole domain to which the word knowledge can be legitimately applied; that it refuses to admit any co-heirs to its possessions, and asserts that its own slow and laborious processes of research are the sole profitable modes of cultivation, the only tillage from which we can reach a harvest of truth unchoked by dogmatic tares. In the further course of our volume we have seen that knowledge is essentially a description and not an explanation—that the object of science is to describe in conceptual shorthand the routine of our past experience, with a view of predicting the future. The work of science viewed from the psychological standpoint is thus essentially that of association, and from the physical standpoint the development of the various excitatory connections between the several portions of the cortex or the centres of brain activity. We have immediate sense-impressions; these are in part retained as stored sense-impresses, and are capable of being revived by kindred immediate sense-impressions. From the stored sense-impresses we form by association conceptions, which may or may not be real limits to perceptual processes. These conceptions are in the latter case only ideal symbols, conceptual shorthand by aid of which we index or classify immediate sense-impressions, stored sense-impresses, or other conceptions themselves.   This is the process of scientific thought, which probably has for its physical aspect the development or establishment of what the physiologist would term "commissural" links between the physical centres of thought.[1]

To recognize that the contents of the mind thus ultimately take their origin in sense-impressions, and in our modes of perceiving sense-impressions, may indeed limit the material which we have to classify, by removing, for example, natural theology and metaphysics from the field of knowledge ; but it still does not render the task of classifying the various departments of science an easy one.   Indeed, so soon as we approach any definite range of perceptual experience, we feel at once the need of a specialist to tell us "the lie of the land "—to describe to us how it is related to surrounding districts and what are the exact bearings of the corresponding branch of science on other problems of life and mind.   The development of the embryo before birth may be a reproduction in miniature of the evolution of the species ; the changes of minute microscopic organisms may be crucial for theories of heredity or of disease which involve momentous results for sociology ; the mathematician carried along on his flood of symbols, dealing apparently with purely formal truths, may still reach results of endless importance for our description of the

---

[1] The extent to which the *localization* of the centres of thought, or of the different elements of consciousness has already proceeded would be brought home to the reader by even a cursory inspection of II. C. Bastian : *The Brain as an Organ of Mind* (pp. 477–700) ; or J. Ross : *On Aphasia* (especially pp. 87–127).

physical universe. Such possibilities suffice to show how incapable any individual scientist must nowadays be of truly measuring the importance of each separate branch of science and of seeing its relation to the whole of human knowledge. An adequate classification could only be reached by a group of scientists having a wide appreciation of each other's fields, and a thorough knowledge of their own branches of learning. They must further be endowed with sympathy and patience enough to work out a scheme in combination. Their labours would, indeed, in course of time, come to have only historical value, but their scheme would have very great interest as a map of the field already covered by science and as a suggestion to the lay reader of the innumerable highways and byways by which we are gradually but surely reaching truth.

§ 2.—*Bacon's " Intellectual Globe."*

Failing such combined action on the part of our scientific leaders, we are compelled to turn to what individual thinkers have done by way of classifying the sciences, and in the first place we ought at least to refer to three well-known philosophers who have dealt with this subject at length. I mean to Francis Bacon, Auguste Comte, and Herbert Spencer.

Bacon has given us a classification of the sciences in his *Of the Dignity and Advancement of Learning*, and in his *Description of the Intellectual Globe*, which were originally intended as parts of that *Instauratio Magna* by which human knowledge was to be revolutionized. But Bacon, like many another reformer, was the product of the very system he denounced. While he saw the evils of mediæval scholasticism, he could never quite free himself from their modes of thought and expression. His classification, however historically interesting, is thus wanting from the standpoint of modern science, and we shall only briefly summarize it here with a view of gaining insight from its defects.

Human learning, according to Bacon, takes its origin in the three faculties of the understanding— Memory, Imagination, and Reason ; and upon this basis Bacon starts his analysis of knowledge. The accompanying scheme, in which I have modernized some of the terminology and omitted some of the details, represents Bacon's classification. The reader will observe at once that there are no clear distinctions drawn between the material of knowledge and knowledge itself, between the real and the ideal, or between the phenomenal world and the unreal products of metaphysical thought. Man is not classed under nature, and a mysterious *Philosophia Prima* or *Sapience* is postulated which deals with the " highest stages of things," divine and human. The axioms which Bacon gives as specimens of this *Sapience* are not very suggestive of what this hitherto wanting branch of science would be like; they are either logical axioms or fanciful analogies between natural theology, physics, and morals. The scheme as it stands is a curious product of a transition period of thought. With its " errors of nature,"—the anomies in which nature is driven out of its course by " the perverseness, insolence, and frowardness of matter "— and with its " purified magic," we recognize its author as on the fringe of the Middle Ages, but when we turn more closely to his analysis of History and Sociology, we feel that Bacon's classification has hardly been

without influence on the scheme of the modern
Spencer. Indeed, one essentially Baconian idea has
been adopted by Spencer. This idea will be found

HUMAN LEARNING.

MEMORY. *History.*				IMAGINATION. *Poesy.*			REASON. *Philosophy, or the Sciences.*				
Natural.		Civil.		Parabolic (Fables).	Dramatic.	Narrative, or Heroical.	Divinity.	Natural Philosophy.			
Freedom (Nomic Law). / Erors (Anomies). / Bonds (Control by Man). / Generations. / Pretergenerations (Monsters). / Astronomical Physics. Physical Geography. Physics of Matter. Organic Species.	Ecclesiastical. / Arts. / Arts. Mechanical. Experimental.	Literary. / Learning.	Political (Civil History proper). / Memorials. Antiquities. Perfect History.				Revelation.	Natural Theology. Nature of Angels and Spirits.	God. / Nature. / Operative. / Purified Magic. / Mechanics. / Metaphysics (Form and Final Causes). / Speculative. / Physics (Material and Secondary Causes). / Abstract. Concrete. Abstract. Concrete. / Mathematics.	Man. / Philosophy of Humanity (Anthropology). / Soul. Body. / Ethics. Logic. Medicine, Athletics, &c. / Civil Philosophy (Standards of Right in :) / Government. Business. Intercourse.	*Philosophia Prima, or Sapience.*

in the *Advancement of Learning*, bk. iii. chap. i.
" The divisions of knowledge," Bacon writes, " are
not like several lines that meet in one angle, but
are rather like branches of a tree that meet in one
stem." This idea, common to Bacon and Spencer,

that the sciences spring from one root, is opposed
to the view of Comte, who arranges the sciences in
a series or staircase.

§ 3.—*Comte's " Hierarchy."*

Now in some respects science owes a debt of
gratitude to Comte, not indeed for his scientific work,
nor for his classification of the sciences, but because
he taught that the basis of all knowledge is experience and succeeded in impressing this truth on a
certain number of people not yet imbued with the
scientific spirit, and possibly otherwise inaccessible to
it. The truth was not a new one — Bacon had
recalled it to men's minds with greater power than
Comte ever did ; it had been essentially the creed of
the scientists who preceded and followed Comte, and
of whom the majority never probably opened his
writings. Yet because Comte repudiated all metaphysical hypotheses as no contributions to knowledge,
and taught that the sole road to truth was through
science, he was in so far working for the cause of
human progress, and his services are not necessarily
cancelled by the peculiar religious doctrines which he
propounded at a later period of his life.

According to Comte there are six fundamental
sciences : Mathematics, Astronomy, Physics, Chemistry, Biology, Sociology, culminating in the seventh or
final science of Morals. In the supreme science of
morals lies the " synthetical terminus of the whole
scientific construction." The hierarchy of the sciences
thus postulated suffices in a very obscurely stated
manner to guide the Positivist in the subdivision of
each special science. For the *scala intellectûs*, as
propounded by Comte, I have been able to find in his

" System " no more valid argument than is contained in the following passage :—

" The conception of the hierarchy of the sciences from this point of view implies, at the outset, the admission that the systematic study of man is logically and scientifically subordinate to that of Humanity, the latter alone unveiling to us the real laws of the intelligence and activity. Paramount as the theory of our emotional nature, studied in itself, must ultimately be, without this preliminary step it would have no consistence. Morals thus objectively made dependent on Sociology, the next step is easy and similar ; objectively Sociology becomes dependent on Biology, as our cerebral existence evidently rests on our purely bodily life. These two steps carry us on to the conception of Chemistry as the normal basis of Biology, since we allow that vitality depends on the general laws of the combination of matter. Chemistry again in its turn is objectively subordinate to Physics, by virtue of the influence which the universal properties of matter must always exercise on the specific qualities of the different substances. Similarly Physics become subordinate to Astronomy when we recognize the fact that the existence of our terrestrial environment is carried on in perpetual subjection to the conditions of our planet as one of the heavenly bodies. Lastly, Astronomy is subordinated to Mathematics by virtue of the evident dependence of the geometrical and mechanical phenomena of the heavens on the universal laws of number, extension, and motion."

According to Comte nothing can ever supersede the need for the individual " to acquire successively, as the race has acquired, the knowledge of each of the seven phases which meet him in the relative conception of the order of the world." It perhaps requires little critical power to demolish a scheme so fanciful that mathematics are related to physics through astronomy, and physics to biology through chemistry ! [1] What remains, indeed, to be said of a philosopher who gravely asserts that the study of each science is to be limited by the requirements of

[1] How much, too, of the real understanding of mathematical truths is based on psychology, on a right appreciation of those modes of perception which have geometrical conceptions for ideal limits ! (p. 214-7).

the one next above it, in order that we may reach as soon as possible the supreme science of morals, for, "if carried further, the cultivation of the intellect inevitably becomes a mere idle amusement"? It is clear that we have in Comte's staircase of the intellect a purely fanciful scheme, which, like the rest of his *System of Positive Polity*, is worthless from the standpoint of modern science.[1]

§ 4.—*Spencer's Classification.*

Historically, however, Comte is an interesting link between Bacon and Spencer. For Comte deduces his hierarchy from fifteen axiomatic statements which he asserts realize the noble aspiration of Bacon for a *Philosophia Prima* (p. 444), and which were clearly not only suggested by Bacon's axioms, but surpassed them in want of scientific definition. On the other hand, it is difficult not to admit that the writings of Comte have at the very least acted as a stimulus—if of the irritant kind—to Spencer's thought.[2] Much more importance must, however, be attached to Spencer's than to Comte's scheme for classifying the sciences, in particular because he returns to Bacon's notion of the sciences as the branches of a tree spreading out from a common root, and rejects the staircase arrangement of the Positivist hierarchy. The root of this tree is to be sought in phenomena, and its trunk

[1] The reader who wishes to verify this conclusion may be referred to Chapter III., "Definitive Systematization of the Positive Doctrine," in vol. iv. of the *System of Positive Polity*, translated by Congreve (London, 1877). See for the hierarchy of the sciences, p. 160 *et seq.* Compare Huxley : "The Scientific Aspects of Positivism," *Lay Sermons, Addresses, and Reviews* (London, 1870), pp. 162–91.

[2] See his "Reasons for Dissenting from the Philosophy of M. Comte," *Essays*, vol. iii. p. 58.

at once divides into two main branches, the one corresponding to the sciences which deal solely with the forms under which phenomena are known to us, and the other to the sciences which deal with the subject-matter of phenomena. These divisions are respectively those of the *Abstract* and the *Concrete Sciences*. The former embraces *Logic* and *Mathematics*, or the sciences which deal with the modes under which we perceive things ; the latter deals with the groups of sense-impressions and the stored sense-impresses we perceive under these modes. From the standpoint taken in this *Grammar*, namely, that all science is a conceptual description, the Abstract Sciences must not be considered as dealing with the space and time of perception, but rather with the conceptual space (p. 203) and absolute time (p. 227) of the scientific description. This distinction is of importance, for Bain has called in question Spencer's language about the Abstract Sciences by asking how Time and Space can be thought of without any concrete embodiment whatever, *i.e.*, as empty forms. This objection holds with regard to the perceptual modes, space and time, but hardly with regard to the conceptual notions of geometrical space and absolute time by which the physicist represents these modes. Spencer's opening paragraph on this point may be quoted :—

" Whether as some hold, Space and Time are forms of Thought ; or whether as I hold myself, they are forms of Things, that have become forms of Thought through organized and inherited experience of Things ; it is equally true that Space and Time are contrasted absolutely with the existences disclosed to us in Space and Time ; and that the Sciences which deal exclusively with Space and Time, are separated by the profoundest of all distinctions from the Sciences which deal with the existences that Space and Time contain. Space is the abstract of all relations of coexistence. Time is the abstract of all relations of sequence. And dealing as they do entirely with relations of coexistence and sequence in their general or special forms, Logic and Mathematics form a class of the Sciences more widely unlike the rest, than any of the rest can be from one another." [1]

Now it cannot be said that this passage brings out very clearly the distinctions between the phenomenal reality of space and time, their perceptual modality and their conceptual equivalents. But what it does bring out is this, that according to Spencer the latter or conceptual values form the basis of scientific classification. And this is in complete agreement with the views expressed in this *Grammar*. That Spencer himself, admitting space and time to be forms of perception, yet considers them to be forms of things, appears to be merely an instance of that unnecessary duplication, which is met by the canon that we ought not to multiply existences beyond what are necessary to account for phenomena.[2]

Turning to the *Concrete Sciences*, or those which deal with phenomena themselves, Spencer makes a new division into *Abstract-Concrete* and *Concrete Sciences*; the former, he tells us, treat of phenomena "in their elements," and the latter of phenomena "in their totalities." This leads him to associate *Astronomy* with *Biology* and *Sociology* rather than with Mechanics and Physics. Such a classification may fit some verbal distinction of formal logic, but it is certainly not one that a student of these subjects would find helpful in directing his reading, or which would ever have been suggested by a specialist in either physics or astronomy. But this

[1] " The Classification of the Sciences," *Essays*, vol. iii. p. 10.
[2] *Entia non sunt multiplicanda praeter necessitatem.* See Appendix, *Note III.*

peculiarity of Spencer's system which separates *Astronomy* from its nearest cognates *Mechanics* and *Physics* is not its only disadvantage. His third group of *Concrete Sciences* is again subdivided on what he terms the principle of the "redistribution of force." This he states in the following words :—

"A decreasing quantity of motion, sensible or insensible, always has for its concomitant an increasing aggregation of matter, and conversely an increasing quantity of motion, sensible or insensible, has for its concomitant a decreasing aggregation of matter."[1]

Now I have cited this vague principle of the "redistribution of force" with the view of showing how dangerous it is for any individual to attempt to classify the sciences even if he possesses Spencer's ability. For this principle has, so far as I am aware, no real foundation in physics and therefore cannot form a satisfactory starting-point for classifying the *Concrete Sciences*. According to Spencer, where there is increase of motion there is decreasing aggregation of "matter." Yet we have only to drop a weight to see increase of motion accompanying increased aggregation of "matter," namely, earth and weight approaching each other. The principle of "redistribution of force" seems, so far as I can grasp it at all, to flatly contradict the modern principle of the conservation of energy. Indeed Spencer's whole discussion of the physical sciences is one which no physical specialist would be able, were he indeed willing, to accept. So I fancy it must always be, when any one individual attempts to classify the whole field of human knowledge. At

[1] *Essays*, vol. iii. p. 27.

best the result will be suggestive, but as a complete and consistent system it must be more or less of a failure. But there is a good deal to be learnt from Spencer's classification, for it combines the " tree " system of Bacon with Comte's exclusion of theology and metaphysics from the field of knowledge. Especially in the primary division into *Abstract* and *Concrete* Sciences,[1] it provides us with an excellent starting-point.

§ 5.—*Precise and Synoptic Sciences.*

The scheme I propose to lay before the reader pretends to no logical exactness, but is merely a rough outline which attempts to show how the various branches of science are related to those fundamental scientific concepts, conceptual space, absolute time, motion, molecule, atom, ether, variation, inheritance, natural selection, social evolution, which have formed the chief topics of earlier chapters. The writer is content to call it an enumeration, if the logician refuses it the title of classification ; for he readily admits that he is not likely to be successful where Bacon, Comte, and Spencer have failed.

In proceeding to discuss a scheme, we have to bear in mind the following points : Science is not a mere catalogue of facts, but is the conceptual model by which we briefly resume our experience of those facts Hence we find that many branches of science, which call for admission into a practical classification, are in reality only sciences in the making, and correspond to the *catalogue raisonné* rather than to the complete conceptual model. Their ultimate position, there-

---

[1] The germ of this division appears also to be due to Bacon : see his scheme, p. 445.

fore, cannot be absolutely fixed. The distinction between those physical sciences which have been reduced to a more or less complete conceptual model and those which remain in the *catalogue raisonné* state has been expressed by terming the former *Exact* and the latter *Descriptive*. But since in the present work we have learnt to look upon all science as a *description*, the distinction rather lies in the extent to which the *synoptic* classification has been replaced by those brief conceptual *résumés* that we term scientific formulæ or laws. Thus, while descriptive must be interpreted in the sense of synoptic, exact must be taken as equivalent to concise or precise, in the sense of the French *précis*. The distinction is now seen to be quantitative rather than qualitative; and, as a matter of fact, considerable portions of the *Descriptive* or *Synoptic Physical Sciences* already belong, or are rapidly being transferred to, the *Exact* or *Precise Physical Sciences*. Thus we shall find that, whenever we begin to subdivide the main branches of science, the boundaries are only practical and not logical. The topics classified in the subdivisions cross and recross these boundaries; and although in the tables below most sciences have been entered in one place only, they frequently belong to two or more divisions at once. Hence in the correlation of the sciences and their continual growth lies the fact of the empirical and tentative character of all schemes of classification. In so far as every branch of science passes, at one or more points, not only into the domain of adjacent, but even of distant, branches, we see a certain justification for Comte's assertion that the study of one science involves a previous study of other branches; but this justification in itself is no argument for the truth of his fantastic "hierarchy" of sciences (p. 446).

§ 6.—*Abstract and Concrete Sciences.—Abstract Science.*

Like Spencer, we may begin by distinguishing in science two groups—the *Abstract* and the *Concrete*. The former group deals with the conceptual equivalents of the modes under which the perceptive faculty discriminates objects, the latter with the concepts by aid of which we describe the contents of perception. We have then, to start with, the following division :—

Perceptions (Sense-Impressions and Stored Impresses).

Modes of Perception.	*Contents of Perception.*
Abstract Science.	Concrete Science.

Now the two modes in which we perceive things apart, or discriminate groups of sense-impressions, are time and space. Hence *Abstract Science* may deal with the general relations of discrimination, applying to both time and space without specializing the mode of perception ; or it may refer in particular to space or to time or to their mixed mode motion. The general relations of discrimination may be either *qualitative* or *quantitative*. The former branch is termed *Logic*, and discusses the general laws by which we identify and discriminate things, or what are frequently termed the laws of thought. A fundamental part of logic is the study of the right use of language, the clear definition and, if needful, invention of terms,—*Orthology*. The object of the present *Grammar* has been chiefly to show how a want of clear definition has led to the metaphysical obscurities of modern science.

Both Time and Space lead us at once to the conception of *quantity* or number, and we thus have a large and important branch of *Abstract Science* which deals with the laws of quantity. Now quantity may be either *discrete* and definite, like the numbers of arithmetic 8, 100, $^1/_{13}$, $^{17}/_4$, &c., the number of inhabitants of a town, the number of cubic feet in a room ; or it may be *continuous* and changing with other quantities—for example, like the height of the barometer with the hour of the day, the marriage or birthrate with the price of bread, the position or speed of a body with the time. We thus have a distinction between discrete quantity and quantity capable of gradual variation or change. Among the sciences which deal especially (if not entirely) with discrete quantity, the best known are probably *Arithmetic* and *Algebra ;* but there are a number of others we ought to briefly note. We want to know how to measure quantity and what errors are likely to arise in its measurement. Closely allied to this is the discussion of probable and average quantities, dealing with cases where we cannot measure individual quantity, but only approximate and average results. Hence arise the *Theory of Measurement, Theory of Errors, Theory of Probability, Theory of Statistics,* &c., &c.

Passing to change in quantity, we remark that if one quantity varies with another it is said to be a *function* of the second. Thus temperature is a function of time and of position, brightness of distance, and speed of time. To understand the mutual relationship of quantities which are functions of each other is the scope of sciences like the *Theory of Functions,* which teaches us how functions can be represented and handled. Examples of this representation will be found in our chapter on the Geometry of Motion, Figs. 10 and 13. Special branches are the *Differential Calculus* or *Calculus of Fluxions*, which deals with the rates of change, and of which we have had examples in determining speed and curvature (pp. 257 and 270) ; and the *Integral Calculus*, or *Calculus of Sums*, which passes from the relation between the rates back to the relation between the changing quantities, and of which we have had an example in the process of summation by which we passed from acceleration as a function of position to the map of the path of a moving body (p. 277).

We next turn to the special relations of Space, and we note that conceptual space may be considered from two standpoints. We may deal solely with the relative position of points and lines and surfaces without taking any quantitative measurements of distances, areas, or volumes. This forms a very important and valuable sub-division of *Geometry*, which has been much developed of recent years and has been largely used by theoretical writers on various branches of engineering practice. It is termed *Descriptive Geometry*, or the *Geometry of Position*, and a branch of it, probably familiar to the reader, is *Perspective Geometry*. On the quantitative or measuring side of the special space division of Abstract Science, we deal with size, and find such subdivisions as *Metrical Geometry*—of which a large part of Euclid's Elements is constituted,—*Trigonometry* and *Mensuration*.

The second branch of special relations ought to deal with Time, but as in reality all our spacial discrimination is associated with time, so all our temporal discrimination is associated with space ; we do in actual perception separate all things in both time and space concurrently, for the immediate groups of sense-impressions are not really simultaneous, and most things perceived in space are " constructs" involving stored sense-impresses (pp. 50, 219). When, therefore, we speak of the special relations of Time, we are referring to that discrimination by sequence which we term change, and of which the fundamental element is really the time-mode of perception—conceptually we are referring to change as measured in Absolute Time (pp. 227, 288). When changes are not measured quantitatively, but only described qualitatively, we require a theory by aid of which we may accurately observe and describe such changes. We want not only a scientific theory of measurement, but a scientific theory of observation and description. For example, in the case of organic phenomena of all sorts it requires a scientific training not only to know what it is essential to observe, but how what has been observed should be described. Some discussion of the *Theories of Observation and Description* are given in treatises on Logic, but they seem capable of much more complete treatment than they have at present received.[1]

The last branch of Abstract Science to which we must refer is the quantitative side of change. Thus we may consider change in position and develop a theory of the motion of conceptual bodies without reference to the special structures and special types of motion by which we conceptualize change in phenomena. This branch of science is termed *Kinematics*, or the *Geometry of Motion*, and, on account

[1] One of the best practical trainings in *Observation and Description* is that gained by a clinical clerk in a hospital ward.

of its fundamental importance, has been fully discussed in our Chapter VI. It has made very great advances in recent years, and not only from the theoretical standpoint; in cases of constrained motion it has become an invaluable auxiliary in the practical construction of machines.[1] Closely allied to *Kinematics*, if not more properly a branch of them, we have a science which deals with change in size and shape. This is the *Theory of Strains*, and it has a wide application in the conceptual description of many portions of physics (p. 242).

A.—*ABSTRACT SCIENCE.*	*Modes of Discrimination.*					
GENERAL RELATIONS OF DISCRIMINATION.			RELATIONS PECULIAR TO SPACE AND TIME.			
Qualitative.	Quantitative.		Space, Discrimination by Localization.		Time, Discrimination by Sequence.	
Logic. Orthology, Methodology.	Discrete Quantity.	Change in Quantity.	Qualitative (Position).	Quantitative (Size).	Qualitative.	Quantitative.
	Arithmetic, Algebra, Theorie of Measurement, Errors, Probability, Statistics, &c., &c.	*Theory of Functions, Calculus of Rates or Fluxions, Calculus of Sums, &c., &c.*	*Descriptive Geometry.*	*Metrical Geometry, Trigonometry, Mensuration, &c., &c.*	*Theories of Observation and Description (allied to Logic).*	*Theory of Strains (Change in Size and Shape). Kinematics (Change in Position).*

[1] See especially L. Burmester : *Lehrbuch der Kinematik*, Bd. i. Leipzig, 1888—a classical treatise.

With this we complete our review of *Abstract Science*. We see that it embraces all that is usually grouped as *Logic* and *Pure Mathematics*. In these branches we deal with conceptual modes of discrimination ; and since the concepts formed are in general narrowly defined and free from the infinite complexity of the contents of perception, we are able to reason with great preciseness, so that the results of these sciences are absolutely valid for all that falls under their definitions and axioms. On this account the branches of *Abstract Science* are frequently spoken of as the *Exact Sciences*. I have summarized our classification in the scheme on the opposite page.

§ 7.—*Concrete Science.—Inorganic Phenomena.*

Passing from *Abstract* to *Concrete Science*, or to the contents of perception, we recall the distinction which has been made in our Chapter IX. between the living and the lifeless, or between *Organic* and *Inorganic Phenomena*. So long as we have no perceptual experience of the genesis of the living from the lifeless we obtain a clear partition of *Concrete Science* by dividing it into branches dealing respectively with *Inorganic* and *Organic Phenomena*. The sciences which deal with inorganic phenomena are termed, as a whole, the *Physical Sciences*.

The first subdivision of these sciences may be referred to the distinction we have already drawn between the *Exact Physical Sciences* and the *Descriptive Physical Sciences*, or as we will term them the *Precise* and the *Synoptic Physical Sciences* (p. 452). Thus we find that astronomers are able to predict the precise time on a given day of a given year at which Venus will appear to an observer at a given position on the Earth's surface to begin its transit over the Sun's disc. On the other hand, we discover by everyday experience that the predictions as to the weather due to the Meteorological Office and published in the daily newspapers frequently turn out incorrect, or are only approximately verified. This distinction between *Astronomy* and *Meteorology* is just the distinction between the *Precise* and the *Synoptic Sciences*. In the one case we have not only a rational classification of facts, but we have been able to conceive a brief formula, the law of gravitation, which accurately resumes these facts. We have succeeded in constructing, by aid of ideal particles, a conceptual mechanism which describes astronomical changes. In the other case we may or may not have reached a perfect classification of facts, but we certainly have not been able to formulate our perceptual experience in a mechanism, or conceptual motion, which would enable us to precisely predict the future. The *Precise* and the *Synoptic Physical Sciences*, respectively, correspond very closely to the phenomena, of which we have constructed a conceptual model by aid of elementary corpuscles having ideal motions, and to the phenomena which have not yet been reduced to such a conceptual description. The process of analyzing inorganic phenomena by aid of ideal elementary motions forms the topic of *Applied Mathematics*.[1] This science is therefore a link between the theory of pure motion as discussed in Abstract Science and the motions of those ideal corpuscles which most closely conceptualize the sequences of

[1] " And as for the mixed Mathematics, I may only make this prediction, that there cannot fail to be more kinds of them as nature grows further disclosed "—a prophecy of Bacon's which has been fully justified.

inorganic phenomena as discussed in the precise branch of *Concrete Science*.

Where we have not yet succeeded in analyzing complex changes into ideal motions, or have only done so in part—describing without quantitative calculation the general results which might be expected to flow from such motions—there we are dealing with the *Synoptic Physical Sciences*. Thus *Synoptic Physical Science* is rather *Precise Physical Science* in the making than qualitatively distinct from it. It embraces large classifications of facts which we are continually striving to resume in simple formulæ or laws, and, as usual, these laws are laws of motion. Thus considerable portions of the *Synoptic Physical Sciences* are already precise, or in process of becoming precise. This is notably the case with *Chemistry*, *Geology*, and *Mineralogy*. So much, indeed, is this the case with *Chemistry* that the reader will find that I have included *Theoretical Chemistry* and *Spectrum Analysis* under the head of *Precise Physical Science*.

Turning to the system of corpuscles, with which we have dealt in Chapter VIII., we find in them an excellent basis for classifying the *Precise Physical Sciences*. In the first place we have the particle and groups of particles forming bodies. The division of Physics dealing with the motion of particles or bodies, or of molecules in bulk, is termed *Molar Physics* from the Latin word *moles*, a mass or bulk. In *Molar Physics* we deal with the motion which conceptualizes the changes of position in bodies at the surface of the Earth, *Mechanics ;* with the motion which conceptualizes the changes in the planetary system, *Planetary Theory ;* and with the motion by which we describe changes in the configuration of a planet and its satellites, *Lunar Theory*.

After the particle we deal with the molecule, and under *Molecular Physics* treat especially of those phenomena which can be conceptualized by the relative motion of molecules. Here we have to consider the *Elasticity*, *Plasticity* (or Viscosity), and *Cohesion* of gaseous, fluid, and solid bodies. By aid of the motion of molecules we treat of the phenomena of *Sound*, the formation of crystals or *Crystallography*, the *Figure of the Earth*, the relative motion of the parts of liquids and gases, *Hydromechanics*, *Aeromechanics*, and the *Theory of the Tides*, the theory of the temperature and pressure in gases, or the *Kinetic Theory of Gases*, &c., &c.

Passing to a still simpler corpuscle, the atom, we reach *Atomic Physics*. The motions we attribute to the concept atom form the basis of *Theoretical Chemistry*, and of those wonderful lines which appear in the light, transmitted or excited by any chemical substance. The *Theory of Spectrum Analysis*, based on the elementary motions of the atom, is the source of our knowledge of the chemical constitution of the sun and stars, or of all those descriptions of perceptual experience resumed in *Solar* and *Sidereal Physics*.

The last branch of the *Precise Physical Sciences* is termed the *Physics of the Ether* and deals with the relative motions of ether-elements, or the changes of shape we attribute to the ether (p. 313). If we consider the ether, apart from the molecules we suppose it to contain, merely as a medium transmitting various kinds of motions, we have the *Theory of Radiation*, which describes how light, heat, and electro-magnetic effects are conceived to be propagated from molecule to molecule. If we deal with the mutual action between ether and molecule (pp. 333,

368), and describe how molecules disperse, absorb, transmit, or conduct optical, thermal, or electro-magnetic effects, we have the remaining portions of the fundamental physical sciences of *Light*, *Heat*, *Electricity*, and *Magnetism*.

From the *Synoptic Physical Sciences* we demand a rational classification of those physical phenomena which have not at present been conceptualized by simple formulæ of motion.   Such phenomena we should naturally expect to find where in ordinary parlance there are "a great number of forces con-temporaneously at work," or where, in more accurate language, the number of elementary bodies by which we should have to conceptualize the phenomena is so great that we are at present unable by synthesis (p. 283) to form the complex motion, which would describe the changes of the whole system.   This is particularly the case in the sciences which deal with the evolution and structure of great and intricate bodies like a planetary system or a planet itself.   We desire to know the sequence of changes by which we can describe the evolution of a planetary system and we seek an answer in the *Nebular Theory*.   We desire to know how the inorganic structure of our Earth has developed,—*Geology* describes it.   Then we turn to the formation of the surface of the Earth, and to the continual changes going on among the gases and fluids there, and study *Physical Geography* and *Meteorology*.

Finally, we inquire into the structure of the sub-stances which form our environment and their relations to each other, thus we have *Mineralogy* and *Chemistry* completing the range of the *Synoptic Physical Sciences*.

The following table resumes our classification of the *Physical Sciences* :—

B.—*CONCRETE SCIENCE.*				*Inorganic Phenomena.*
Reduced to Ideal Motions.				Not yet Reduced to Ideal Motions.
Precise Physical Sciences.				*Synoptic Physical Sciences.*
Physics of the Ether.	Atomic Physics.	Molecular Physics.	Molar Physics.	*Nebular Theories, Evolution of Planetary System, Inorganic Evolution of the Earth.* Geology, Geography (sometimes termed Physical Geography), Meteorology, Mineralogy, Chemistry, &c.
In association with Molecule. *Light, Heat, Electricity, Magnetism (in relation to Molecular Structure), e.g., Theories of Dispersion, Absorption, Transmission, Conduction, &c., &c.* **Apart from Molecule.** *Theory of Radiation (Light, Heat, and Electro-magnetic Waves).*	*Theoretical Chemistry, Spectrum Analysis, Solar and Sidereal Physics, &c.*	*Elasticity, Plasticity, Cohesion, Sound, Crystallography, Figure of the Earth, Hydromechanics, Aeromechanics, Theory of the Tides, Kinetic Theory of Gases, &c., &c.*	*Mechanics, Planetary Theory, Lunar Theory, &c., &c.*	

§ 8.—*Concrete Science.—Organic Phenomena.*

We now turn to the third and last great field of knowledge, namely, that division of *Concrete Science* which deals with *Organic Phenomena.* Its branches are frequently summed up as the *Biological Sciences,* although the term *Biology* itself is usually applied to a subdivision. If we attempt to subdivide the Biological Sciences into *Precise* and *Synoptic* groups, we do not obtain any practically valuable division. For, with the exception of certain small portions of one or two branches, the whole of the Biological Sciences would fall under the synoptic category. It is true that certain powerful formulæ have reduced large parts of biological science from a rational classification to science in the accurate sense of the word ; but the description of organic phenomena by aid of conceptual motions (p. 330) awaits long and laborious investigation on the part of both physicist and biologist before much progress will be reported. I shall therefore return to the mode of subdivision we adopted in the case of that branch of *Abstract Science* which deals with " Special Relations." I shall subdivide *Biological Sciences* into those which deal more especially with space or the localization of life, and those which deal more especially with time or—as in the case of organic phenomena we more generally term the discrimination by sequence—with growth. In the first subdivision we shall have those branches of science which deal with the *Distribution of Living Forms* (*Chorology*) and study habits in relation to environment (*Ecology*). These form the major portion of what in the old sense was termed *Natural History*.

Turning to the second subdivision of change or

growth, we notice that these may be either *recurring* or *non-recurring*. Recurring and non-recurring changes are terms which of course have only reference to man's perceptual experience. From that standpoint we treat the evolution of complex from simple organisms as non-recurring, but in the starry universe it is a legitimate inference from the like known to the like unknown (p. 72) to conceive this evolution to be going on whenever a planetary system reaches the same stage of its development as the solar system at present has reached. Thus the evolution of life may really have recurred innumerable times, and so our division is only a practical mode of classifying our actual perceptual experience. It is not to be taken as an assertion that there is anything more inconceivable in the genesis and extinction of organic life on many planets than in the birth and death of many men.

Non-recurring growth we speak of as *History*, and recurring growth as *Biology* in the narrower sense. *Biology* falls into two main divisions : *Botany*, dealing with plant life, and *Zoology* with animal life.

Regarding the historical group of sciences, we may treat generally of all life, and we then have branches of science discussing the *Evolution* or *Origin of Species* (*Phylogeny*, *Palæontology*, &c.). More especially dealing with man we have the *Evolution* or *Descent of Man*. This evolution may be considered in different phases, although these phases cannot be kept absolutely apart and discussed quite independently. Thus we may ask how the *physique* of man has developed, and find an answer in the measurement of skulls, the comparison of skeletons and prehistoric remains of the human form—in *Crani-*

ology and *Anthropology* in its narrower sense. We may next inquire how man's mental faculties have developed, and seek knowledge in the history and structure of language, in the evolution of man's mental products, or in *Histories of Philosophy, of Science, and of Art*, &c., &c. Lastly, we may trace the evolution of social institutions, and see instinctive gregarious habits developing into customs and ultimately into laws and institutions. We may discuss the origin of human dwellings, of human societies and states. Here we seek aid from *Archæology, Folklore, Anthropology* in its wider sense, and from *Histories of Customs, of Marriage, of Ownership, of Religions, and of Laws*, &c., &c.

Next examining the recurring phases of growth or *Biology*, we seek to describe the *form* and *structure* of the various types of life and thus reach the subject-matter of those important branches of biology, termed *Morphology, Histology, Anatomy*, &c., &c. Or we may deal more especially with the *growth* and *reproduction* of living forms. We want to describe the origin of the distinction between the sexes, and the purposes we conceive this distinction serves in the economy of living forms ; then we wish to describe how the parent hands down his characteristics to the child, and how the new life itself takes its origin and develops stage by stage. These topics are dealt with in the *Evolution of Sex*, the *Theory of Heredity* and *Embryology*.

The third and last great division of *Biology* is concerned with the *functions* and *actions* of living forms. If we deal with these functions and actions from the physical side, and investigate the process of life as related to inorganic forms, we have a wide branch of science termed *Physiology*. The mental side of the functions and actions of living forms is embraced by *Psychology*. *General Psychology* treats of the develop-

C.—*CONCRETE SCIENCE.*   *Organic Phenomena.*

SPACE (Localization).	TIME (Growth or Change).							
	Non-recurring Phases. *History.*				Recurring Phases. *Biology* { Of Plants, *Botany.* Of Animals, *Zoology.*			
	General *Evolution of Species.*	Special *Evolution of Man.*			Form and Structure.	Growth and Reproduction.	Functions and Actions.	
		Physique.	Mental Faculties.	Social Institutions.			Physical.	Mental.
Geographical Distribution of Living Forms (Chorology), Habits in relation to Situation and Climate (Ecology), Natural History (in old sense).	*Genesis of Life (Phylogeny, Palaeontology, &c.). Origin of Species. Theories of Natural and Sexual Selection, &c., &c.*	*Craniology, Anthropology, &c., &c.*	*History of Language, Philology, Histories of Philosophy, Science, Literature, Art, &c., &c.*	*Archaeology, Folklore, Histories of Customs, Marriage and Ownership, of Religion, of States, of Laws, &c., &c.*	*Morphology, Histology, Anatomy, &c., &c.*	*Embryology, Theory of Sex, Theory of Heredity, &c., &c.*	*Physiology.*	*Psychology.* — Special in Man. / General. — Group. / Individual. — *Theory of Instinct, Genesis of Consciousness, &c., &c. Psychics, Physiology of Thought. Sociology. Morals, Politics, Political Economy, Jurisprudence, &c.*

ment of mental powers in life generally, of the origin of consciousness, animal intelligence, and theories of instinct. If we turn to the *Special Psychology* of man, we may either consider man as an isolated individual or as member of a group. The former branch of *Psychology* may be termed *Mental Science* or *Psychics*, and deals with the various mental phases and habits of individual man and the relation of his thinking faculty to the physical structure of his brain. The latter branch of *Psychology* dealing with men in the group is termed *Sociology*, and is concerned with man's social products and institutions—it falls into such branches as the *Science of Morals*, the *Science of Politics*, *Political Economy* and *Jurisprudence*.

With *Sociology* we conclude our enumeration of the *Biological Sciences*, which are summarized in the scheme on the opposite page.

§ 9.—*Applied Mathematics and Bio-physics as Cross-Links.*

The reader might conceive that our classification was now completed, but there still remains a branch of science to which it is necessary to refer. We have seen that we have no perceptual experience of the genesis of the living from the lifeless, although it appears to be a reasonable conceptual formula (p. 413). It might therefore seem that no definite link between the two branches of *Concrete Science*, between the *Physical* and *Biological Sciences* could at present be forthcoming. But we have to remember that life invariably occurs associated with sense-impressions similar to those of lifeless forms, organisms appear to have chemical and physical structure differing only in complexity from inorganic forms. And although we cannot definitely assert that life is a mechanism (p.404) until we know more exactly what we mean by the term mechanism as applied to organic corpuscles, there still seems little doubt that some of the generalizations of physics—notably the great principle of the conservation of energy—do describe at least part of our perceptual experience of living organisms. A branch of science is therefore needed dealing with the application of the laws of inorganic phenomena, or Physics, to the development of organic forms. This branch of science which endeavours to show that the facts of *Biology*—of *Morphology*, *Embryology* and *Physiology*—constitute particular cases of general physical laws has been termed *Ætiology*.[1] It would perhaps be better to call it *Bio-physics*. This science does not appear to have advanced very far at present, but it not improbably has an important future.

Thus just as *Applied Mathematics* link *Abstract Science* to the *Physical Sciences*, so *Bio-physics* attempt to link the *Physical* and *Biological Sciences* together.

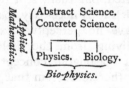

Applied Mathematics and *Bio-physics* are thus the two links between the three great divisions of science, and only when their work has been fully accomplished, shall we be able to realize von Helmholtz's prediction and conceive all scientific formulæ, all natural

[1] From the Greek *αἴτιον*, a cause. The name does not seem very aptly chosen, especially as it has a very definite meaning of older origin in medical practice.

laws, as laws of motion (p. 330). This goal we must, however, admit is at present indefinitely distant.

§ 10.—*Conclusion.*

We have passed hastily and superficially across the vast field of knowledge, omitting doubtless many things and misplacing others. But still even this survey will not have been fruitless if it has convinced the reader of the immense variety and the enormous range of facts which modern science is called·upon to classify and resume. Here before us—it may be but obscurely and as from behind a veil—we see the wide heritage of science, upon which hundreds of toilers in many countries have spent their best years and their ripest powers, for the past two centuries—and once for centuries two thousand years before these. Here we see Egyptian and Greek, American and European, alike working to a common end, alike animated by a common zeal, by the same steady enthusiasm of purpose. Here in the field of knowledge we have the one meeting-ground for all ages and for all nations ; here, indeed, age and nation cease to be ; names like those of Galilei and Keppler, Newton and Laplace, Dalton and Faraday, Linnæus and Darwin have become household words, kindling admiration, and even devotion, wherever civilized man has established his communities.

How, we may ask, has it come to pass that mankind has devoted all this time and toil in pursuit of knowledge—why should men reverence the great pioneers of science ? The answer is clear and definite. Man has mastered all other forms of life in the struggle for existence by the development of a more complex perceptive faculty and a more perfect reasoning power. In the capacity he has evolved for resuming vast ranges of phenomena in brief scientific formulæ, in his knowledge of natural law, and the foresight this knowledge gives him lie the sources of man's victory over other forms of life, from the brute power of the wild beast to the subtle power of the microscopic bacillus of some dread disease. As the bull in its horns, or the eagle in its wings, so man proudly rejoices in the strength of his mental powers, for they it is which enable him to hold his own in the struggle of life.

In this *Grammar* I have endeavoured to emphasize this side of science and scientific law ; I have striven to indicate how natural law is a product of the human reason and how the correlated growth of the reasoning and perceptive faculties in man, assisted by the survival of the fittest, may possibly have left us with a normal type of man for whom only that is perception which can be reasoned about, and for whom the reason is keen enough to appreciate and analyze what is perceived (p. 125).[1] Long and difficult must have been the evolution by which these results have been achieved ; but they ought at least to give man confidence in his own powers and assurance that with further growth will come still keener perception and still greater intellectual grasp. We have no right to assume that the development of man is completed. On the contrary we have every right to infer that the drift of evolution which we can trace from primitive

---

[1] Man certainly fails in his attempt to reason about things he does not perceive—about the "beyond" of sense-impression. We have no evidence, however, that would lead us to infer that any group of perceptions is beyond rational analysis now or after more complete classification.

man to Aristotle, and from Aristotle to the scientist of to-day, will continue the same, at least as long as man's physical environment is not materially modified.   To deny that our perception is wider and deeper, and that our analysis is more subtle than that of the great Greek philosopher is to deny the drift of man's past evolution, to deny all that gives history its deep human significance.   The growth of knowledge since the days of Aristotle ought to be sufficient to convince us that we have no reason to despair of man's ultimately mastering any problem whatever of life or mind, however obscure and difficult it may at present appear.   But we ought to remember what this mastery means ; it does not denote an explanation of the routine of perception ; it is solely the description of that routine in brief conceptual formulæ.   It is the historical *résumé*, not the transcendental exegesis of final causes.   In the latter we are not—except in honest confession of ignorance and rational definition of knowledge—one whit further advanced than Aristotle, nay, than the primitive savage.   The experience of centuries, we might hope, will at last convince the speculative that " the inquisition of Final Causes is barren, and, like a virgin consecrated to God, produces nothing." [1]

Our grandfathers stood puzzled before problems like the physical evolution of the earth, the origin of species, and the descent of man ; they were, perforce, content to cloak their ignorance with time-honoured superstition and myth.   To our fathers belongs not only the honour of solving these problems, but the credit of having borne the brunt of that long and weary battle by which science freed itself from

[1] Bacon : *De Augmentis*, bk. iii. chap. v.

the tyranny of tradition.   Their task was the difficult
one of daring to know.   We, entering upon their
heritage, no longer fear tradition, no longer find that
to know requires courage.   We too, however, stand
as our fathers did before problems which seem to us
insoluble—problems, for example, like the genesis of
living from lifeless forms, where science has as yet no
certain descriptive formula, and perhaps no hope in
the immediate future of finding one.   Here we have
a duty before us, which, if we have faith in the scientific method, is simple and obvious.   We must turn a
deaf ear to all those who would suggest that we can
enter the stronghold of truth by the burrow of superstition, or scale its walls by the ladder of metaphysics.
We must accomplish a task more difficult to many
minds than daring to know.   We must dare to be
ignorant.   *Ignoramus, laborandum est.*

SUMMARY.

An individual even with the ability of Bacon or Spencer must fail for
want of specialists' knowledge to classify the sciences satisfactorily. A
group of scientists might achieve much more, but even their system
would only have temporary value as the position of a science relative to
other changes with its development. This point is illustrated by the
Precise and Synoptic Physical Sciences.

From Bacon we learn that the best form for classification is that of a
branching tree, but from Comte that there is in reality an interdependence
in the sciences, so that a clear understanding of one may necessitate a
previous study of several others. From Spencer we may adopt the
fundamental distinction between Abstract and Concrete Science, or
those which deal respectively with the modes and the contents of perception. We then find three fundamental divisions corresponding to the
Abstract, Physical, and Biological Sciences which are united pair and
pair by Applied Mathematics and Bio-physics.

LITERATURE.

Bacon, F.—De Dignitate et Augmentis Scientiarum (London, 1623), and Descriptio Globi Intellectualis (1612). Translations will be found in J. Spedding and R. L. Ellis' edition of The Works of Francis Bacon, vols. iv. and v. London, 1858.

Comte, A.—System of Positive Polity (1854), translated by Congreve, vol. iv. chap. iii. London, 1877.

Spencer, H.—The Classification of the Sciences (Hertford, 1864), or, Essays, Scientific, Political, and Speculative, vol. iii. pp. 9–56. London, 1875.

APPENDIX.

NOTE I.

On the Principle of Inertia and "Absolute Rotation" (p. 343).

CONSIDER a very thin straight piece of material string AB, which in the conceptual limit may approach a straight line. Let C and D be two adjacent physical points of this line which in conception may approach to geometrical points. Now suppose the fact observed to be that AB remains straight and disconnected from other "matter," but that we are ignorant whether it is really in motion or not. Let us now suppose the string separated between C and D, say by

A   C D   B

a pair of scissors, without immediately altering the motion, if there be such. One of two things may now occur—either the pieces AC, DB continue to appear as parts of one unbroken piece of string AB, or else AC and DB begin to separate between C and D. Now the only thing of which we have destroyed the possibility is clearly a mechanical relation—a *tension* (p. 365) between the material points C and D. Hence, if the parts begin to separate after the application of the scissors, C and D must have had a tension between them, or have exerted mutual accelerations before the cutting in twain (p. 360). That is to say, D must initially have had an acceleration relative to C in the direction AB. Or we may assert, that in the limit two parts of a material line will tend after division to separate or not to separate according as its parts have a relative acceleration in the direction of its length. Now if we suppose the string or material line incapable of stretching, it is clear that D cannot initially have a velocity relative to C in the direction AB. Hence it follows that the acceleration of D relative to C must be of the nature of normal acceleration (p. 269), or the line AB must be spinning as a *whole* round some axis. On the other hand, if the parts AC and DB remain after being cut in twain in the same straight line, then no material particle C of AB has any acceleration relative to another particle D in the direction AB. In this case the line AB may have motion of translation as a whole, but has no spin.

A line, the points of which are conceived as having no relative accelerations in the direction of the line, is defined as having a *fixed direction* in space. Perceptually a material straight line, string or wire, removed from the influence of other matter, is to be represented on the conceptual model by a line " fixed in direction," provided that when it is cut in twain there is no tendency for its parts to separate, or they still appear as the parts of a continuous material straight line.

Given a perceptual body, which can be conceptually represented as rigid, how are we to ascertain whether it is to be conceived as spinning or not? For example, is the earth rotating about its axis, or is the whole vault of the heavens itself turning round—which will best enable us to describe our perceptual experience? The answer lies in determining whether a line drawn perpendicular to the axis of the earth is to be conceived as " fixed in direction " or not. Theoretically we might determine the problem of the earth's rotation in the following manner. Fix perpendicular to the axis of the earth a wire, the parts of which are not subjected to gravitation or to the resistance of the atmosphere, and observe on its being divided whether the parts remain the continuous parts of a material line or not. This experiment would of course be impossible, but it may bring to the reader's mind what Newton understands by *absolute rotation*. The effect, however, of the relative acceleration of the parts of the earth, if it exists, may be measured in other ways. For example, it would lead to an apparent lessening of gravitational acceleration at the equator, and, if the earth were not quite rigid, to a flattening at the poles. When, therefore, without rearranging any other portions of gross " matter " we can have a body in two states, in the one of which no mere division of the parts leads to discontinuity of the body as a whole, and in the other mere division does lead to discontinuity, then in the latter case we suppose that there will be, and in the former case that there will not be relative acceleration of the parts. When this relative acceleration of the parts manifests itself, although the elementary parts may have no relative velocity in the line joining them, we can describe it by aid of a spin about some axis. Since this spin does not seem to have reference to any external system, Newton termed it *absolute motion of rotation*. The name is an unfortunate one, as it suggests the possibility of an *absolute motion* (p. 247). What we have to deal with are perceptual facts which can only be conceptually described by supposing points at different distances from the earth's axis to have different velocities *relative* to the stellar system. The *fixity of direction* in a line which we have conceptually defined by absence of mutual acceleration between its parts, *appears* to coincide with fixity of direction relative to the stars, but it must be remembered that Galilei first stated the principle of inertia for bodies moving with regard to the earth, because the motion of the earth relative to the stars was insensible for most motions at its surface. It in no way follows that Newton's extension of the principle to the planetary system leads us to an absolute motion in an absolute space.

It has been asserted that Newton's rotating bucket of water and Foucault's pendulum [1] demonstrate an absolute rotation in an absolute space, but in the words of Professor Mach [2] :—

" The universe is not presented to us twice, with resting and again with rotating earth, but only once with its alone determinable relative motions. Accordingly we cannot say what would happen if the earth did not rotate. We can only interpret the case as it is presented to us in different ways. When we interpret it so that we are involved in a contradiction with experience, then we have interpreted it falsely. The fundamental principles of mechanics can indeed be so conceived that even for relative rotations centrifugal forces arise.

" The experiment of Newton's with the rotating bucket of water only teaches us that the rotation of the water relative to the side of the bucket gives rise to no sensible centrifugal forces, but that these forces do arise from the rotation relative to the mass of the earth and the other heavenly bodies. Nobody can

---

[1] Maxwell, *Matter and Motion*, pp. 88–92.
[2] *Die Mechanik in ihrer Entwickelung*, p. 216.

say how the experiment would turn out if the sides of the bucket became thicker and more massive till they were ultimately several miles thick. There is only the one experiment, and we have to bring the same into unison with other facts known to us and not with our arbitrary imaginings."

Allowing for the difference in terminology between Professor Mach's sentences and our *Grammar*, they show, I think, how far it is safe to go in the idea of absolute direction and absolute motion. In the conceptual model we may define lines, which are conceived as having no relative acceleration of their parts, as "fixed in direction." Take two points O and P in conceptual space; let the step OP be drawn from O, whether O be in motion or not, and let OP be supposed to remain "fixed in direction;" the tops P of such steps drawn for all instants form *the path of P relative to O*. The statement that, if O and P represent particles of gross matter sufficiently far apart from each other and from other particles, this path will be a straight line, is the principle of inertia.

The perceptual equivalent for "fixity of direction" in the conceptual step was in Galilei's day [1] represented with sufficient approximation by direction fixed with regard to the earth; since Newton we take it to sensibly coincide with direction fixed with regard to the stars. But perceptual absoluteness cannot really be asserted even in the latter case. Should the element of gross "matter," however, be ultimately conceived as a form of ether in motion, the principle of inertia will become a far more easily stated and appreciated axiom of mechanics (p. 344, *footnote*).

NOTE II.

On Newton's Third Law of Motion (pp. 347, 360, 368, and 385).

WE have seen on p. 359 that one fundamental part of Newton's third law is involved in mutual accelerations being inversely as masses. This leads at once to the equality in *magnitude* of action and reaction. In the next place we conceive mutual accelerations to be parallel and opposite in sense (p. 346).

---

[1] And even now by the writers of elementary text-books who cite bodies projected along the surface of "dry, well-swept ice" as moving in "straight lines" and illustrating Newton's first law of motion!

This does not, however, give us completely Newton's third law as it is usually interpreted, unless we suppose these mutual accelerations to be in the *same* straight line as well as parallel. In the case of particles this straight line is usually taken to be the straight line joining them.

Now it is not at all improbable that the mutual accelerations (and therefore the mutual forces) which are ascribed to corpuscles, will be ultimately found to be better described by aid of the disregarded kinetic energy of an intervening ether. For example, oscillating and pulsating bodies in a perfect fluid ether have mutual accelerations, which *may* be described by action at a distance, but are really due to the kinetic energy of the intervening ether. In the case of two small bodies moving with velocities of translation or oscillating in such an ether it by no means follows that the mutual accelerations (or the apparent action and reaction) will necessarily lie in the same straight line, and if they do, that this straight line will be the line joining the small bodies. Further, on the supposition that apparent action at a distance is due to the direct action of the ether, it does not seem likely that, if a corpuscle P be suddenly moved, the result of this motion will be immediately felt by a distant corpuscle Q, time would be required to make the change in the position of P felt at Q. The mutual actions might in this case be parallel, but it is hardly probable that they would always be in the same straight line, that is *opposite* in Newton's sense.

Thus these considerations, taken in conjunction with those referred to on p. 368 *et seq.*, suggest that greater caution is necessary than is sometimes observed in extending Newton's third law to molecules or atoms, which may really have considerable oscillatory or translatory velocities relative to the ether. For the comparatively small velocities of particles of gross "matter," the law is probably a sufficient description of our perceptual experience.

NOTE III.

William of Occam's Razor (pp. 110 and 450).

IN the course of our work we have frequently had occasion to notice the unscientific process of multiplying existences beyond what are really needful to describe phenomena. The canon of inference which forbids this is one of the most important in the whole field of logical thought. It has been very concisely expressed by William of Occam in the maxim : *Entia non sunt multiplicanda praeter necessitatem.* Sir William Hamilton in a valuable historical note (*Discussions on Philosophy,* 2nd edition, pp. 628–31, London, 1853) quotes the further scholastic axioms : *Principia non sunt cumulanda* and *Frustra sit per plura quod fieri potest per pauciora.* So far these axioms are valuable as canons of thought, they express no dogma but a fundamental principle of the economy of thought. When, however, Sir William Hamilton adds to them *Natura horret superfluum,* and says that they only embody Aristotle's dicta that God and Nature never operate superfluously and always through one rather than a plurality of causes, then it seems to me we are passing from the safe field of scientific thought to a region thickly strewn with the pitfalls of metaphysical dogma. Aristotle and Newton's opinion that—*Natura enim simplex est* is of the same character as Euler's *Mundi universi fabrica enim perfectissima est.* They either project the notions of "simple" and "perfect" beyond the sphere of sense-impression, where alone there is any meaning to the word knowledge, or else they confuse the perceptual universe with man's scientific description of it. In the latter field only is economy of principles and causes a true canon of scientific thought. On this account the "law of parsimony," as Sir William Hamilton has termed it, seems a product of scholastic thought and not due to Aristotle. As stated by Occam, it is a far more valid axiom than in Newton's version (p. 110), and I think it might well be called after the *Venerabilis Inceptor,* who first recognized that knowledge beyond the sphere of perception was only another name for unreasoning faith.

Sir William Hamilton expresses Occam's canon in the more complete and adequate form :—

Neither more, nor more onerous, causes are to be assumed, than are necessary to account for the phenomena.

NOTE IV.

On the Vitality of Seeds (p. 400).

THE determination of the maximum period during which seeds will maintain their vitality appears to be very far from settled·

In the first place, experiments lasting thirty, fifty, or one hundred years cannot be rapidly executed,[1] and secondly, well-authenticated cases of the discovery of seeds several score years or even centuries old are not very frequent. There seems, however, little doubt of seeds preserving their power of germination for periods of forty to fifty and even to one hundred and fifty years (British Association Report, 1850, p. 165; Darwin, *Origin of Species*, 4th edition, p. 430; Alph. de Candolle, *Géographie botanique raisonnée*, 1855, p. 542). With regard to still longer periods the evidence is by no means so satisfactory as might be wished. Either the finder is an archæologist and not a scientific botanist, or if the seeds have really fallen into the hands of a genuine biologist the finder may have been a questionable archæologist. In most cases the combined evidence of ancient origin and of actual germination fails to reach the point of legal testimony. The botanical evidence is doubtless complete in the case of Lindsay's raspberries, but whether the antiquarian evidence of their being found in the stomach of a man buried in Hadrian's reign is equally convincing may be doubted. In other cases the seeds may indeed have been genuine, taken by archæologists quite above suspicion, yet we find that it has been merely handed over to gardeners, "thrown out and found to grow," or even asserted by eminent botanists without trial or after an inspection with the microscope to be incapable of germinating. The question whether seeds taken from tombs (rather than from mummy wrappings) or from considerable distances below the surface of the soil might not germinate after many centuries seems still an unsettled one. The point in the text, on p. 400, is sufficiently illustrated by the known periods of fifty to a hundred years.[2]

NOTE V.

A. R. Wallace on Matter.

PERHAPS a maximum of confusion between our perceptions

---

[1] Experiments are at present being made at Kew with seeds buried in bottles.

[2] Samples of the tales and the opinions which pass for evidence will be found in J. Philipson's article: *The Vitality of Seeds found in the Wrappings of Egyptian Mummies*, Archæologia Æliana, vol. xv., 1890.

and conceptions is reached in Mr. Alfred Russel Wallace's discussion of Matter in his *Natural Selection*. It would not be needful to refer to this feeble contribution of a great naturalist to physical science, had he not recently republished it without any qualifying remarks (*Natural Selection and Tropical Nature*, pp. 207–14. London, 1891). According to Mr. Wallace matter is not a thing-in-itself, but *is* force, and all force is probably will-force. It is unnecessary here to again remark on the illegitimate inference made in this extension of the term *will* (p. 70). But as force is only evidenced in change of motion, we may well ask what it is which Mr. Wallace supposes to move. If he is talking of the perceptual sphere, he fails to distinguish between our appreciation of individual groups of sense-impressions and of change in these groups, or indeed between perceptions and the routine of perception. If he is talking of the conceptual sphere he fails to distinguish between the moving ideals (geometrical bodies, points, or Boscovich's "centres of force") and the modes of their motion. As a matter of fact he uses force for sense-impression, for sequence of sense-impressions, for moving ideal, and for mode of motion. From this confusion of the perceptual and the conceptual are drawn arguments for spiritism, exactly as Aristotle, the Stoics, and Martineau have drawn them for animism (pp. 106 and 146). The chief difference between Mr. Wallace and his predecessors lies in the fact that he has polytheistic rather than monotheistic sympathies.

NOTE VI.

On the Sufficiency of Natural Selection to Account for the History of Civilized Man (p. 430).

IT is not only literary historians, but even naturalists who deny that natural selection is a sufficiently powerful factor to describe the development of civilized man. The most noteworthy scientist who takes this view is Mr. Alfred Russel Wallace. He considers that (i.) the large brain of man, (ii.) his naked skin, (iii.) his voice, hands, and feet, (iv.) his moral sense, could never have been produced by natural selection. He holds that all these characteristics are more fully developed in the savage than are necessary for his needs. He believes, however, that they have been developed in man by selection, as man himself has developed other characteristics in the Guernsey milch cow. In other words he asserts that they are the outcome of the artificial selection of some intelligent power and not of blind natural selection. This theory of Mr. Wallace's has been well described by the phrase "man as God's domestic animal." Mr. Wallace, however, being polytheistic in conviction, has objected to the capital G in this phrase, and appears to hold that man is the domestic animal of the modern equivalents of angels and demons. According to him, therefore, "marriages are made in heaven," but by the lesser luminaries of the spirit hierarchy. No arguments in favour of the interference of this spirit hierarchy are produced except the supposed insufficiency of natural selection. The difficulties Mr. Wallace finds in natural selection do not appear of a very formidable character,[1] but surely if they were important enough to leave us in doubt as to whether we had found a sufficiently wide-embracing formula in natural selection, then the true scientific method is to remain agnostic, until it has been shown that no other sufficient perceptual formula can be found? Mr. Wallace rushes with such haste to his spirit hierarchy, that his pages read as if he had invented his difficulties in order to justify his beliefs, and not reached his "angel-made marriages" by a process of elimination, which left no other formula possible.

I have added this Note that the reader may not think that I have disregarded Mr. Wallace's views on the inapplicability of natural selection to the history of man. Such is far from being the fact, but I hold that Mr. Wallace's views as expressed in the chapter (pp. 186–214) on *The Limits of Natural Selection as applied to Man* in the recently republished "Natural Selection," and in the chapter on *Darwinism applied to Man* in the "Darwinism," will appear paralogistic enough to confute themselves if carefully studied.

[1] His whole argument, for example, with regard to the brain turns upon its *size*, whereas it appears that it is the complexity of its convolutions and the variety and efficiency of its commissures rather than its actual size, which we should psychologically expect to have grown with man's civilization.

INDEX.

ACCELERATION, 345-351 ; as a spurt and a shunt, 265-268 ; normal (shunt), 276 ; in the direction of motion (spurt), 276

Action, principle of modified, 376

Adams, J. C., 199

Aeromechanics, 462

Agnostic, 26

Alchemy, 27

Allen, Grant, " Force and Energy," 108

Anatomy, 467

Anomy, a breach in the routine of perceptions, 114, 172, 177

Anthropology, 467

Aristotle, 146-148, 161, 230, 473, 474, 482, 484

Archæology, 467

Aspect, 354, 375 ; on change of, 239-242

Astronomy, 460

Atavism, 125

Atom, 298, 299, 302-307 ; an intellectual conception, 115 ; (chemical) diagrammatically represented, 337 ; ether-squirt, 316-319 ; vortex-ring, 316-319 ; see Prime Atom

Austin, 94-98, 104, 105, 111-113 ; " Lectures on Jurisprudence," 94, 135

Bacon, Francis, 39, 444-448 ; his classification of the sciences, 445 ; " Intellectual Globe," 443, 475 ; " Instauratio Magna," 443 ;

" Novum Organum," 39, 45 ; " Of the Dignity and Advancement of Learning," 386, 443, 445, 474, 475

Bain, 449

Ball, W. P., " Are the Effects of Use and Disuse Inherited ? " 31

Ball, " Story of the Heavens," 189

Basset, A. B., " A Treatise on Hydrodynamics," 344

Bastian, H. C., " The Brain as an Organ of Mind," 442

Bearing, involves conception of direction and sense, 249, 262

Belief, use of the word changing, 71 ; as an adjunct to knowledge, 72

Beltrami, 322

Berkeley, 83, 93, 122 ; " Theory of Vision," 91 ; " Principles of Human Knowledge," 91 ; " Hylas and Philonous," 91

Biology, 450, 465 ; relation of, to physics, 388-392

Bio-physics, 469-70

Bois-Raymond, E. du, 26, 329, 330 ; " Ueber die Grenzen des Naturerkennens," 46, 330

Bois-Raymond, P. du, 26, 119, 329 ; " Ueber die Grundlagen der Erkenntniss in den exacten Wissenschaften," 26, 46

Boole, " An Investigation of the Laws of Thought," 174, 177, 180 ; on problems relating to the connexion of causes and effects. 174

Botany, 466
Boscovich, 325, 484
Bradlaugh, C., 354
Bramhall, 114
Brown-Séquard, 66
Büchner, 87, 353, 354
Burmester, "Lehrbuch der Kine-
    matik," 458

Calculus, differential, 456 ; integral,
    456
Candolle, A. de, 483
Cauchy, 350
Carnelly, Dr., negative atoms, 319
Carpenter, on will, 146
Cause, force as a, 140-143 ; meta-
    physical conception of, 146 ;
    scientific conception of, 146 ;
    scientific definition of, 155 ; will
    as a, 143-144
Causation, 136 ; is uniform ante-
    cedence, 156
Chart, time, 253-257
Chemistry, 461
Chorology, 465
Cicero, "De Legibus," 106
Clausius, "Die mechanische Wärm-
    theorie," 350
Claus and Sedgwick, "Elementary
    Text-Book of Zoology," 439
Clerk-Maxwell, 36, 100, 101, 335 ;
    article, "Atom," 318, 331, 334 ;
    article, "Ether," 331 ; "Consti-
    tution of Bodies," 331 ; "Matter
    and Motion," 284, 292, 387 ;
    "Scientific Papers," 318, 331 ;
    "Theory of Heat," 100
Clifford, W. K., 6, 45, 50, 60, 83,
    322, 323, 354 ; "Common Sense
    of the Exact Sciences," 245, 248,
    271, 284, 387 ; "Elements of
    Dynamic," 245, 253, 258, 284 ;
    "Ethics of Belief," 71 ; "Lec-
    tures and Essays," 46, 60, 91,
    123, 230, 331 ; "Seeing and
    Thinking," 91, 230
Cohesion, 462
Comte, Auguste, 443, 446, 448,
    452, 453, 475; his Hierarchy,
    446 ; "System of Positive
    Polity," 447, 448, 475

Conception, scientific validity of a,
    64-67 ; must be deducible from
    the perceptions, 65
Conscience, tribal, 6, 7
Consciousness, as the field of science,
    63 ; each individual recognizes
    his own, 58 ; evolution of, 401 ;
    measure of, 53 ; sense-impression
    does not involve, 53 ; without
    sense-impression there would be
    no, 63 ; will closely connected
    with, 70 ; see "Other-conscious-
    ness"
Construct, an external object is a,
    50-52, 77, 90
Copernicus, 25, 118
Corpuscle, 340-348
Corpuscles and their structure, 332-
    337
Cousin, Victor, 38
Craniology, 467
Crookes, Protyle, 333, 423
Crystallography, 462
Curvature, 268-276 ; radius of, 273

Darwin, 1, 2, 35, 37, 39, 40, 41, 67,
    370, 396, 397, 399, 422, 425, 433,
    472 ; "Descent of Man," 13 ;
    "Life and Letters of," 39 ;
    "Origin of Species," 13, 40, 483 ;
    "Variation of Animals and
    Plants under Domestication,"
    396
Dalton, 472
De Morgan, A., "The Theory of
    Probabilities," 180
Density, definition of, 372
Determinism, mechanical, 278
Dilatation, definition of, 243
Direction, when "fixed," 478-80
Drummond, "Natural Law in the
    Spiritual World," 108

Ecology, 465
Edgworth, 175 ; "The Philosophy
    of Chance," 175, 180
Eject, Clifford's definition of, 60,
    91 ; other-consciousness as an,
    59-61
Ejects, attitude of science towards,
    61-63
Elasticity, 462

Electricity, 463

Embryology, 467

Energy, Clerk-Maxwell's definition of, 292 ; conservation of, 349, 406 ; degradation of, 410

Erasmus, 3

Ether, 213-217, 310-316 ; -element, 328, 332, 338-348, 368, 380, 381 ; -mass, 368 ; -matter, 312 ; -motion, 338, 369 ; -sink, 319 ; -squirt, 319-322, 348, 375, 379 ; -unit, diagrammatically represented, 337

Evolution, Darwinian theory of, 137, 421 ; principles of, 428-430

Euler, 482

Faraday, 37, 38, 41, 472 ; " Experimental Researches," 13

Fischer, Kuno, " Geschichte der Philosophie," 230

Fitzgerald, G. F., " On an Electromagnetic Interpretation of Turbulent Fluid Motion," 318

Flatland : " a Romance of Many Dimensions," 321

Force, as a cause, 140-143 ; definition of, 360 ; materialists' notion of, 143 ; redistribution of, 451 ; resultant, 378 ; Thomson and Tait's definition of, 294

Forces, parallelogram of, 378, 384

Galilei, 24-26, 257, 366, 386, 472

Galloway, T., " A Treatise on Probability," 177, 180

Galvani, 35, 36

Geddes, " Evolution of Sex," 31

Gemmule, 396-399

Generation, spontaneous, 390, 409, 414, 419, 422

Geography, 463

Geology, 463

Geometry, descriptive, 456 ; metrical, 456 ; perspective, 456

Germ-plasm, 31, 389, 397, 398

Gladstone, W. E., 287 ; article in the " Nineteenth Century," 137 ; " The Impregnable Rock of Holy Scripture," 137

Gravitation, law of, 102-104, 111, 119, 120, 145, 146, 161, 336, 350 ; force of, 326

Green, " General Introduction " to Hume's works, 230

Haeckel, 330, 404, 431, 432, 435 ; " Freie Wissenschaft und freie Lehre," 46, 432, 440 ; his criticism of Darwinism, 431-432 ; " Natürliche Schöpfungs - Geschichte," 440 ; " Of the Pedigree of Man and other Essays," 440 ; " On the Development of Life Particles and the Perigenesis of the Plastidule," 440

Haldane, J. S. and R. B., " Life and Mechanism," 46

Hamilton, Sir William, 482

Harvey, " Anatomical Dissertation on the Motion of the Heart and Blood," 13

Heat, 463

Hegel, 20, 31, 34, 291 ; " Philosophy of Law," 96

Helmholtz, 322, 330, 349, 411, 471 ; " On the Relation of the Natural Sciences to the Totality of the Sciences," 46 ; " Sensations of Tone," 13

Heraclitus, 285, 286

Heredity, Weismann's theory of, 31-34, 396, 467 ; Darwin's theory of, 396

Herschel, " A Preliminary Dissertation on Natural Philosophy," 46 ; " Outlines of Astronomy ' (Force as will), 146

Hertz, 36, 214

Heterogeneity, 372

Hipparchus, 117

Histology, 467

Hodograph, the velocity diagram or, 262-265

Homogeneity, 370

Hooker, R., 107-112 ; " Ecclesiastical Polity," 107, 108

Humanism, 433

Humanity, Solidarity of, 437, 438

Humboldt, A. von, 16

Hume, 93, 192, 193, 197 ; " Dialogues concerning Natural Religion " 135 ; " Essay concerning Human Understanding," 191,

194; "Treatise on Human Nature," 197, 230

Huxley, 26, 389, 432; "Hume," 91; "Lay Sermons," 440, 448; "On our Knowledge of the Causes of the Phenomena of Organic Nature," 440

Hydromechanics, 462

Ideal, distinction between real and, 50

Imagination, disciplined use of, 41; scientific use of, 38

Individualism, 432-436

Inertia, principle of, 342-345, 383, 406, 477-80

Inference, the scientific validity of an, 67-69; limits of legitimate, 69; canons of legitimate, 71-73

Inheritance, 31, 396, 421

Jevons, Stanley, 39, 152, 166; "Elementary Lessons in Logic," 178; "Limits of Scientific Method," 41; "Principles of Science," 41, 46, 67, 71, 152, 166, 180

Judgment, æsthetic, 42-44; scientific, 43

Jurisprudence, 468

Kant, 20, 50, 83; "Kritik der reinen Vernunft," 91, 230

Keppler, 25, 118, 472

Kinematics, 232, 246, 457

Kinetic Scale, 355, 373; density as the basis of the, 370-374

Kirchhoff, 35, 139, 287; "Vorlesungen über mathematische Physik," 139

Lankester, 389, 413

Laplace, 39, 169, 170, 173-177, 472; "Théorie Analitque des Probabilités," 39, 180

Laplace's investigation, nature of, 176, 177; theory, the basis of, 171-176

Law, Austin's definition of, 94, 95; Hooker's, 107; natural, 93-108; scientific, 93-104; sequence of sense-impressions not in itself a,

102; the Stoics' conception of, 106

Leibniz, definition of space, 185

Leverrier, 199

Lewes, G. H., 287; "Aristotle," 230, 288

Life, the perpetuity of (Biogenesis), 411-413; the spontaneous generation of (Abiogenesis), 411-413; a special creation of, 411

Light, 463

Linnæus, 472

Locke, 72, 194

Logic, 449, 454, 458

Lyell, "Principles of Geology," 13

Macgregor, "Kinematics and Dynamics," 245, 284

Mach, Ernst, 77, 78, 232; "Analysis of the Sensations, Anti-metaphysical," 79, 91; "Beiträge zur Analyse der Empfindungen," 91; "Die Mechanik in ihrer Entwicklung," 387; his diagram, 78; "Sensations and the Elements of Reality," 91

Magnetism, 463

Malthus, "Essay on Population," 40

Marriage, 428-430

Martineau, on will, 146, 484

Mass, as the ratio of the number of units in two accelerations, 358; centre of, 382; the scientific conception of, 357, 358

Mathematics, 449; applied, 461; pure, 458

Matter, as non-matter in motion, 310-313; Clerk-Maxwell's definition of, 292; ether, 312; gross, 338-351, 368, 369, 379-383; "heavy," 312; Hegel's definition of, 291; J. S. Mill's definition of, 297; Tait's definition of, 296; Thomson and Tait's definition of, 293

Mayr, Georg, "Die Gesetzmässigkeit im Gesellschaftsleben," 426

Mechanics, the science of motion, 139

Mechanism, 136-140, 392-400; the limits to, 337-340

Metakinesis, 401-403

Metaphysician, definition of, 20, 21
Meteorology, 460-463
Method, philosophical, 22, 23 ; scientific, 8, 20, 28, 39-41, 45 ; methodology, 454
Mill, J. S., 39, 156; "Canons of Induction," 178 ; "System of Logic," 178, 180, 297
Mineralogy, 461
Molecule, an intellectual conception, 115 ; definition of, 210 ; diagrammatically represented, 337
Moleschott, J., 353
Morals, science of, 468
Morgan, Lloyd, 401, 402 ; "Animal Life and Intelligence," 31, 50, 91, 101
Morphology, 467
Motion, Newtonian laws of, as given by Thomson and Tait, 381-384 ; criticism of, 380-386 ; third law of, 480

Nägeli, 389
Nerve, motor, 52, 54 ; sensory, 51, 52, 54
Newton, 35, 40, 41, 102, 103, 110, 118-120, 145, 226, 258, 336, 360, 370, 378, 379, 381, 382, 386, 472, 477-81 ; "Principia," 110
Newton's Canon, 110, 482
Newton, R. H., Social Studies, 435

Object—distinction between object and eject, 60
Object, external, the stored effect of past sense-impressions, 49 ; as a construct, 50 ; actions subject to the mechanical control of the, 55
Occam's Razor, 481-2
Other-consciousness as an eject, 59-61 ; possibility of physical verification of, 60 ; the limits to, 69-71 ; recognition of, 77

Palæontology, 466
Paley, 153
Pangenesis, Darwin's hypothesis of, 396
Parallelogram Law, 253, 281 ; of velocities, 282 ; of accelerations, 282

Parsimony, law of, 482
Particle, 334 ; diagrammatically represented, 337
Pearson, K., "The Ethic of Freethought," 46, 91 ; ether-squirt, 319 ftn.
Perceptions, conceptions must be deducible from, 65 ; routine in, 152
Perceptive faculty, Nature conditioned by the, 176
Philipson, J., 483
Philosophy, Hegelian, 21
Physics, atomic, 462 ; molar, 461 ; molecular, 462 ; sidereal, 462 ; solar, 462
Phylogeny, 466
Physiology, 468
Plasticity, 462
Plastidule, Haeckel's, 404
Plato, 31, 34, 114 ; "The Laws," 114
Point-motion, 247-250
Poisson, 350
Political Economy, 468
Politics, Science of, 468
Position, 250-253
Powell, Sir J., 112
Prime-atom, as an ether vortexring, 318 ; diagrammatically represented, 337 ; the fundamental element of heavy matter, 313
Protoplasm, 389, 390, 408
Protyle, or prime-atom, 333, 423
Psychics, 468
Psychology, 468
Ptolemy, 102, 117, 118

Quetelet, 212

Radian, 272
Real, 47, 50, 77
Reality, 73, 76, 77
Reid, Thomas, 183
Riemann, 322
Rigidity, a conceptual limit, 238
Ross J., on Aphasia, 442
Rotation, motion of, 237 ; on change of aspect or, 239-242 ; Newton's absolute, 477-80
Routine, 162-166, 177-180

Sandars, T. C., "Institutes of Justinian," 106

Sanderson, J. S. Burdon, 440
Schopenhauer, 20, 83, 146, 148, 326
Schwegler, "Handbook of the History of Philosophy," 230
Science, Abstract—scheme, 459; Concrete, inorganic phenomena—scheme, 464; Concrete, organic phenomena — scheme, 469; descriptive, 140; function of, 8; goal of, 17; material of, 15, 18; prescriptive, 140; precise and synoptic, 452, 461
Seeds, vitality of, 400, 482-3
Selection, natural, 32, 39, 40, 67, 96, 421, 422-428, 430-434, 484-5
"Sense," 248
Self-determination, 406
Sense-impressions, contents of our mind based on, 62; exertion as the product of, 55; the test of identity is sameness in, 87
Sense-impresses, action conditioned by stored, 55; memories are stored, 55
Shunt, 266
Slide, or shearing strain, 245
Slope, a measure of steepness, 258; speed as a, 260
Smith, Robertson, 426
Socialism, 433
Sociology, 468
Space, 189, 229; a mode of perceiving objects, 186
Spectrum Analysis, 462
Speed, uniform, 262; variable, 262; as the slope of the tangent, 262
Spencer, Herbert, 1, 389, 404, 432, 435, 443, 445, 448-452, 454, 475; essays, 451; his classification, 448-452; his use of the term "force," 389; "Man and the State," 440; "Principles of Biology," 440; "Reasons for dissenting from the philosophy of A. Comte," 448; "The Classification of the Sciences," 450, 475
Spin, 274
Spiritualism, 28, 228
Spurt, 266
Steepness, 258

Step, 249
Stoics, the, 106-112, 131, 132, 484
Stokes, Sir G. G., 316, 318; "Burnett Lectures on Light," 109, 418; "Mathematical and Physical Papers," 316
Strain, or change of form, 242
Stretch, 243
Stuart, J., 160; "A Chapter of Science," 135, 160-162
Supermateriality, 402, 403
Supersensuousness, 114-116
Swift, 298

Tait, Professor, 309, 327, 352, 384; "Dynamics of a Particle," 361; "Properties of Matter," 295, 296, 327, 331
Tangent, definition of, 259
Tennyson, 156
Tension, 365-367
Theory, lunar, 462; nebular, 463
Theory of description, 457; of errors, 455; of functions, 455; of gases, kinetic, 462; of measurement, 455; of observation, 457; planetary, 462; of probability, 455; of radiation, 463; of statistics, 455; of strains, 458; of the tides, 462
Theosophy, 228
Thermodynamics, second law of, 99, 101
Thing-in-itself, 87
Thomson, Sir W., 309, 316, 318, 327, 384, 411; "Popular Lectures and Addresses," 124, 167, 210, 318, 331; "The Six Gateways of Knowledge," 167; vortex-atom theory, 318
Thomson and Tait, "Treatise on Natural Philosophy," 293, 295, 327, 381, 387
Thomson, J. A., "Evolution of Sex," 31
Time-chart, 255
Time, Newton's definition of conceptual, 226; perceptual, 227
Todhunter, "History of the Theory of Probability," 177, 180; "History of Elasticity," 379
Translation, motion of, 237

Trendelenburg, 230, 231; "Logische Untersuchungen," 231.
Tunzelmann, von, 214
Turgot, 40
Tycho Brahé, 118

Ueberweg, "History of Philosophy," 230
Unit (physiological), Spencer's, 404
Universe, 18, 47, 57; external, 73-76; real, 74

Variation, 396, 421
Velocity, a combination of speed and bearing, 262; an epitome of past history, 351, 352
Venn, J., "The Logic of Chance," 180
Virchow, R., "Die Freiheit der Wissenschaft im modernen Staat," 46
Vogt, J. G., "Das Wesen der Electricität und des Magnetismus," 108

Vortex-atom, Sir W. Thomson's, 316
Vortex-ring, 317, 375

Wallace, A. R., 40; "Darwinism," 232; "Contributions to Natural Selection," 484-5; on matter, 483
Weismann, 31-33, 389, 396-399, 409; "Essays on Heredity," 13, 31, 390-391, 394, 397, 398
Weight, 366
Westermarck, E., "History of Human Marriage," 157, 426
Will, 70, 83, 88, 148-152, 326; as a cause, 146; as a stage in the routine of perceptions, 148
Wordsworth, "General View of Poetry," 43
World, external, 77, 87; internal, 87; (real), is a construct, 77

Zeller, E., "Die Philosophie der Griechen," 230
Zoology, 466